Radio Wave Propagation in Ionosphere

Radio Wave Propagation in Ionosphere

Edited by **Kevin Merriman**

WILLFORD **P**RESS

New York

Published by Willford Press,
118-35 Queens Blvd., Suite 400,
Forest Hills, NY 11375, USA
www.willfordpress.com

Radio Wave Propagation in Ionosphere
Edited by Kevin Merriman

© 2016 Willford Press

International Standard Book Number: 978-1-68285-057-2 (Hardback)

Printed in the United States of America.

Contents

Preface

Every book is a source of knowledge and this one is no exception. The idea that led to the conceptualization of this book was the fact that the world is advancing rapidly; which makes it crucial to document the progress in every field. I am aware that a lot of data is already available, yet, there is a lot more to learn. Hence, I accepted the responsibility of editing this book and contributing my knowledge to the community.

The free ions and electrons of ionosphere make it an important region for the propagation of radio waves. Ionospheric measurements, cosmic magnetism, electron density, study of mesosphere, data analysis, etc. are some of the significant topics discussed in this book. It is an essential guide for both academicians and students who wish to pursue the study of ionosphere further. Coherent flow of topics, and extensive use of examples make this book an invaluable source of knowledge.

While editing this book, I had multiple visions for it. Then I finally narrowed down to make every chapter a sole standing text explaining a particular topic, so that they can be used independently. However, the umbrella subject sinews them into a common theme. This makes the book a unique platform of knowledge.

I would like to give the major credit of this book to the experts from every corner of the world, who took the time to share their expertise with us. Also, I owe the completion of this book to the never-ending support of my family, who supported me throughout the project.

Editor

Response of the Effelsberg 100 m radio telescope to signals in the near-field at 24 GHz

K. Ruf, E. Fürst, K. Grypstra, J. Neidhöfer, and M. Schumacher

Max-Planck-Institut für Radioastronomie, Auf dem Hügel 69, 53111 Bonn

Abstract. Short range radar (SRR) for cars has been proposed to operate over 5 GHz of bandwidth at the 24 GHz ISM band. To estimate the level of interference from these devices on radio telescopes, the near-field antenna pattern has to be known. We report on new measurements with the Effelsberg 100 m radio telescope. These measurements were performed with a transmitter set up at a distance of 1.7 km from the telescope. The strength of the signal picked up by the telescope sidelobes shows that the proposed SRR would interfere with sensitive radio astronomical observations.

1 Introduction

The near-field of parabolic radio antennas is usually defined by a radius R of the Fresnel zone of $R = 2D^2/\lambda$, where D is the diameter of the antenna and λ the wavelength (Balanis 1996). For the 100 m radio telescope and cm-wavelengths R is of order 1000 km. The entire terrestrial horizon is located within the near-field zone. However, all astronomical objects are located in the Fraunhofer zone (far-field). Because of this, the response of the antenna to sources in the near-field is only poorly known. At short cm-wavelengths the 100 m antenna has a very narrow beam of less than 1 arcmin. To measure the response over a larger solid angle centered on the main beam is time consuming and normally not possible.

This lack of information makes it difficult to estimate the influence of terrestrial transmitters on sensitive radio astronomical measurements. Because of the complexity of a radio astronomical antenna with focal cabins and support legs a simple calculation based of Fresnels theory in often misleading. This complexity and the different topological conditions make every radio telescope unique concerning susceptibility to artificial radio signals produced in the neighborhood.

A recent example of proposed terrestrial transmitters is connected to short range radar, SRR, for cars. Devices are proposed to operate over 5 GHz of bandwidth centered at the 24 GHz ISM band. Though the spectral power density per unit bandwidth of the individual SRR emitters considered is very low (about −90 dBm/Hz), compared to other transmitting radio devices, the passive radio services, i.e. radio astronomy and Earth exploration, also detect and analyse naturally occuring broad band noise power over large bandwidths. Many astronomical objects are observed down to the natural horizon of the particular radio telescope. At the Effelsberg 100 m radio telescope the lowest elevation of 8° is reached towards the south, the west, and the north. The location of roads in the neighborhood of the antenna is shown in Fig. 1.

The direction towards west, where the road with highest traffic density passes the telescope at a distance of about 1.7 km was choosen on 8 May 2002 to perform measurements of the antenna response to a test signal. In Sect. 2 the measurement is decribed followed by a determination of interference levels in Sect. 3.

2 The measurements

A test transmitter was set up at the above mentioned road, from where a large part of the telescope aperture is visible when pointing towards the azimuth of the transmitter. The coordinates of the test transmitter location have been determined using GPS and checked with a detailed map. The coordinates were 50° 31′ 13.1″; 6° 51′ 39.5″. The distance and the azimuth of the transmitter as seen from the radio telescope, the coordinates of which are very well known, were calculated to be 1696 m and 256.8°, respectively. Towards this azimut cosmic sources can be followed down to an elevation of 8°.

The test transmitter consisted of a sweeper HP8350B with transmitter module 83570A set to CW mode at a fixed frequency of 23.8 GHz. The output was connected to a standard gain horn with a coaxial cable. The transmission loss of the cable and the output power at the horn flange had been

Correspondence to: E. Fürst
(efuerst@mpifr-bonn.mpg.de)

Fig. 1. The location of roads in the neighborhood of the 100-m-RT and the location of the test transmitter (red circle).

Fig. 2. Set-up of the test transmitter.

measured in the laboratory before. The horn was mounted on a tripod at 1 m above ground, E-field vertical, and pointed towards the telescope. The set-up is shown in Fig. 2.

At the radio telescope the primary focus K_a band receiver was set to a receiving frequency of 23.8 GHz and the back-end was set to continuum mode. The telescope was pointed at 8.1° elevation, at which it scanned in azimuth over a range of ten degrees, centered at the calculated azimuth of the test transmitter. The received signal strength – and hence the sidelobe gain pattern – shows considerable scatter (see Fig. 3). The origin of the received signal was repeatedly verified by switching the test transmitter off and on. At the position of the maximum field strength, at 256.8° azimuth, a scan in elevation was measured, which also exhibits considerable gain variations. To the surprise of experienced observers even, a sharp peak in sidelobe gain was detected near

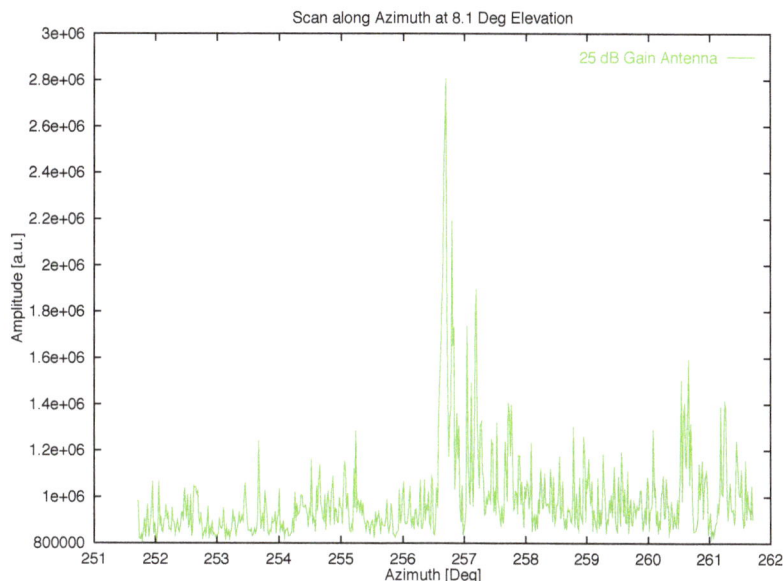

Fig. 3. Azimuth scan at 8.1° elevation.

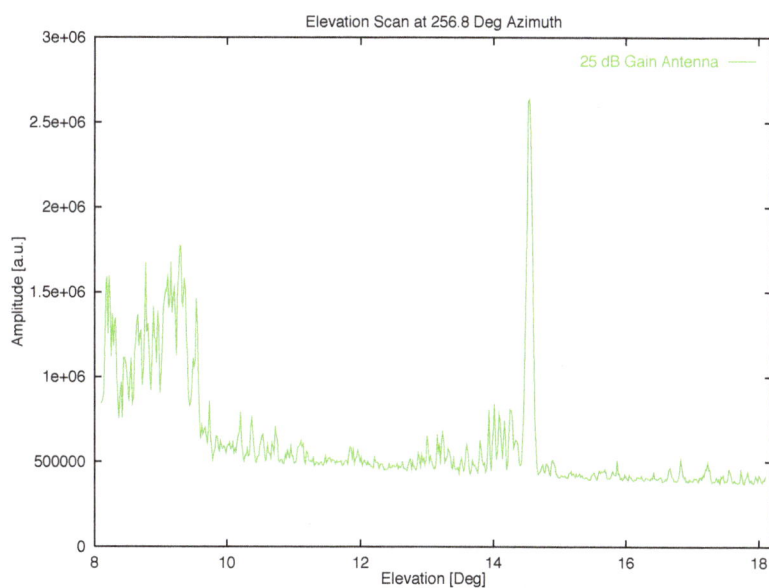

Fig. 4. Elevation scan at 256.8° azimuth.

14° elevations, which could be confirmed by serveral checks (see Fig. 4). There was, however, not sufficient time available to perform more complex side lobe structure measurements.

The autocorrelator spectrometer was then used to measure the spectrum of the test transmitter signal. The spectrum was found to be the same as the sweeper output spectrum measured with a spectrum analyser in the lab before, and is displayed in Fig. 5.

The spectrum was calibrated by comparing the output power in each spectrometer channel with the output power produced by a broadband calibration signal, which is fed into the receiver before the first amplifier stage. This is a standard procedure for calibrating astronomical measurements

and hence the signal strength is given in milli Kelvin (mK) of brightness temperature. For independent calibration and for comparing the artifical signal to an astronomical signal the telescope was then pointed towards a well-known strong source of celestial molecular line emission at 24 GHz, the Kleinmann-Low nebula in the Orion Molecular Cloud. The elevation of the cloud during the measurement was 32°. The ammonia line spectrum at 23.694 GHz is shown in Fig. 6. The signal, seen through the main beam of the 100 m antenna, is several orders of magnitude weaker than the artificial signal of the test transmitter seen through a side lobe of the radio telescope.

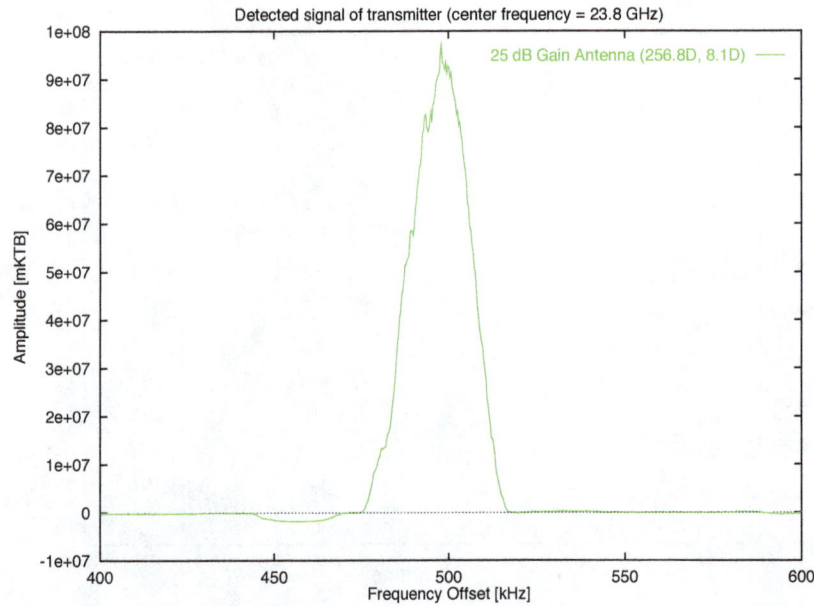

Fig. 5. Spectrum of the test transmitter measured through a sidelobe of the Effelsberg 100 m radio telescope.

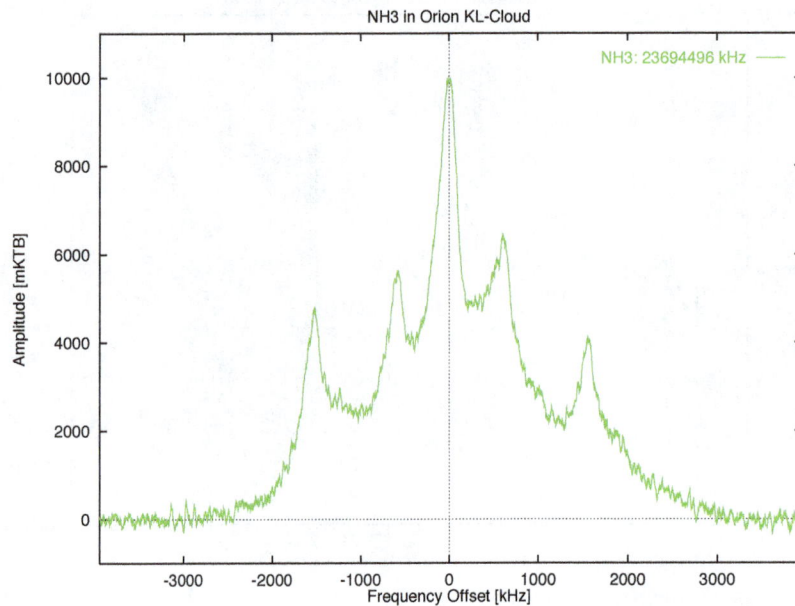

Fig. 6. NH_3 in OMC1.

3   Determination of interference levels

The output power at the transmitter horn flange had been determined in the laboratory to be -13 dBm. The horn gain is given as 25 dB, hence the eirp of the transmitted signal towards the radio telescope was 12 dBm. Taking the measured spectrum of the test transmitter, Fig. 5, the 3 dB bandwidth is $BW_{3dB} = 20\,kHz$ and $T_B = 7.5 \times 10^7\,mK$, respectively. From the NH_3 measurement a system temperature of $T_{sys} = 155\,K$ was determined, corresponding to $\approx 230\,K$ at 8° elevation. Given the air temperature of 19° and a partly cloudy sky at the time of the measurement, this

relatively high system temperature appears to be reasonable. Taking $T_{sys} = 230\,K$, the allocated bandwidth of 400 MHz, and an integration time of 2000 s as recommended for sensitive radio astronomy measurements in Recommendation ITU-R RA.769, the minimum detectable change in brightness temperature is $T_B = 0.3\,mK$.

The following calculation leads to the maximum allowable eirp for an artifical signal in order not to exceed the sensitivity limit for observations with the Effelsberg 100 m radio telescope at 24 GHz. It should be noted that the 10% criterion from Recommendation ITU-R RA.769 is not included in the calculation!

Table 1.

Output power:	−13 dBm
Antenna gain:	25 dB
3 dB bandwidth of signal (20 kHz, taken from figure 5	−43.0 dBHz
Eirp:	−31.0 dBm/Hz
Necessary attenuation ($7.5 \times 10^7 \, mK / 0.3 \, mK$)	−84.0 dB
Maximum allowable eirp:	−115 dBm/Hz

4 Conclusion

Test measurements were performed with the 100 m Effelsberg radio telescope at a frequency of 23.8 GHz under mean atmospheric conditions. A signal from a test transmitter located next to a nearby road, at a distance of 1.7 km from the telescope, picked up by a telescope sidelobe, was compared with a signal received through the main beam from a well-known astronomical source. The measurements show that the power of any devices producing artificial radio signals should be well below −115.0 dBm/Hz in order to protect routine radio astronomical measurement at the 100 m telescope.

References

Balanis C. A.: Antenna Theory and Design, John Wiley, 2nd Edition 1996.

2

Sixty years of ionospheric measurements and studies*

K. Bibl

Center for Atmospheric Research, University of Massachusetts Lowell, 600 Suffolk St., Lowell, MA 01854-3625, USA

*Dedicated to Prof. Dr. K. Rawer on the occasion of his 90th birthday.

Abstract. Professor Rawer's professional ethics allowed him to create an excellent institute with co-workers forming a successful team with creative individuals. Showing myself as an example, this paper demonstrates his guidance, his care, and his ability to create independent thinking in the areas of equipment development and the analysis of data produced by this equipment.

1 Introduction

As one of the first co-worker of Professor Karl Rawer, I want to emphasize the beginning of my relationship with him and the influence he had on me. His philosophy was "die Treue im Kleinen", which I did not understand then. Now I believe that this philosophy can be found in one of my early teacher's, Werner Heisenberg, actions, in contrast with the hate of Edward Teller. One of Rawer's actions during the war was to send a co-worker, suspected to be Jewish, to a field-station in Sicily, where he was considered safe.

Instead of promoting himself in the middle of the war, he nominated me, a young student in uniform, for a national prize, which I received for designing the first pulse receiver (Huber and Rawer, 1950). But, what would Karl Rawer do with a soldier who, after the conspiracy on Hitler's life in 1944, refused to say "Heil Hitler" in the morning, but said "Grüss Gott" instead (although the guy did not believe in God). I was that guy; in order to protect me from prosecution, Karl Rawer sent me to a farmhouse in the western part of Austria to set up a new field site. There I finished a double receiver with common oscillator for installation in an Adcock direction finding system.

After the war, although I had established a successful business, I was forced out of Austria. Karl Rawer found a place for me to stay in Bavaria and feed my family successfully by repairing radios. With his perfect mastery of the

Correspondence to: K. Bibl
(k.bibl@verizon.net)

French language, he boldly approached the Headquarter of the French Occupation Forces, persuading them to found the first French-German research institute. There he was able to reassemble many of his former co-workers; even managing to get one released from a prisoner-of-war camp.

2 Ionosphären-Institut Breisach

Karl Rawer had outstanding relations with his co-workers. Because of his multi-faceted skills in technology, mathematics, physics, and management, he liked to work with different people for joint research. But he also allowed most of them to do independent studies, leading to single-author publications. From the first publications of the institute between 1947 and 1960, 25% were under his sole authorship, 28% were jointly authored with some of his co-workers, but 47% were published with only his co-workers names. (Most were single author names.)

With me, he sometimes had a problem, because I thought I knew better in engineering. But he allowed me to prove my point in building the proto-type of the Panorama-Ionosonde (1/2 or 2 min ionograms) in the workshop after normal working hours. The Panorama-Ionosonde was based on an idea, which everybody thought was crazy. Instead of the bulky drums with bulky springs and switches, it had a single axle. On this axle three rotors of a variable capacitor, forming a 120-degree angle each, were mounted to cover three frequency bands. For independent tuning of the receiver input, the oscillator, and the synthesized transmitter, three of these sets of rotors, but also the respective inductors, were installed on the same axle. This is shown in Fig. 1. The silver coal contact with a silver ring, divided into three parts, formed also the band-switch. The oscillator rotors had different shapes for each band to allow perfect tuning for all frequencies. This Panorama-Ionosonde was built for many stations and was especially useful for measurements during solar eclipses.

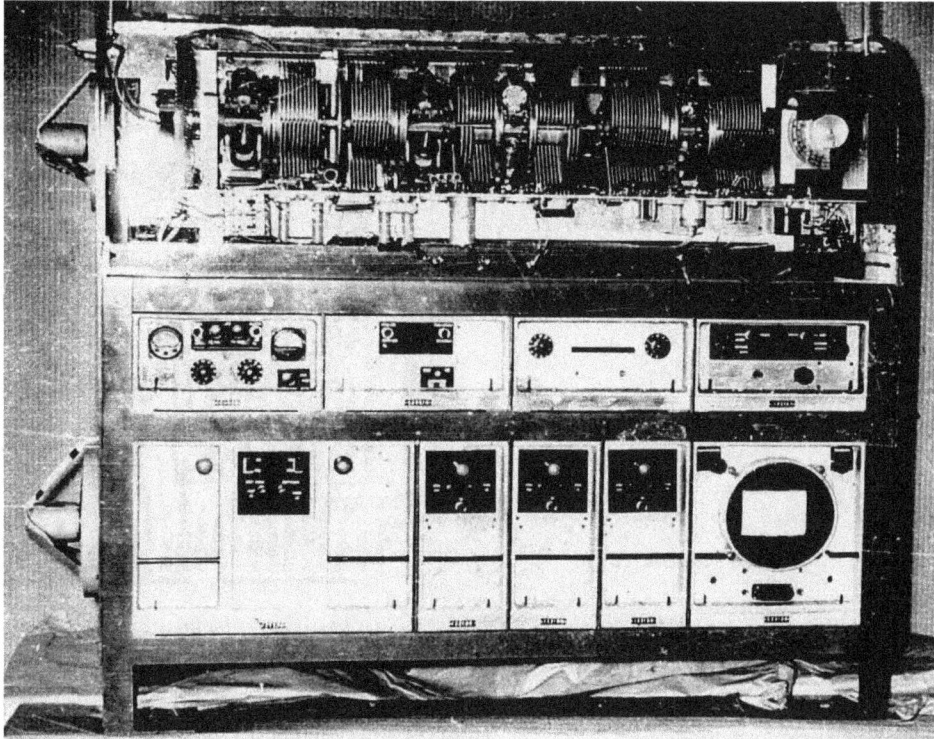

Fig. 1. Panorama-Ionosonde.

With two Ionosondes on a French frigate, sailing in May 1954 to Bergen in Norway, the Panorama-Ionosonde data, taken during a total solar eclipse under good ionospheric conditions, proved without a doubt, that the E- and F1-layer ionization follows the solar occultation without delay (Bibl and Delobeau, 1956) This finding, important for IRI, increased the recombination co-efficient in those regions by two orders of magnitude. Sir Edward Appleton remarked in writing after I sent him our report: "Very interesting."

A second solar eclipse took place over Djibouti, then in French Somalia, in 1955. This eclipse was my most frightening, but wonderful, visual experience I ever had, seeing the shadow of the eclipse raising in waves over the vast desert like the flood. Time-compressed movies from the 1/2 min ionogram recordings showed (Bibl, 1962) that the old ionization of the F2 layer stays on unchanged while the F1 ionization disappears during the eclipse and a new F2 ionization is created from the rising F1 ionization in the second half period of the annular eclipse (Fig. 2). Because the solar activity was already higher than the year before, the contribution of the corona to the total radiation from the sun was higher then the 5% experienced in the 1954 eclipse. In complementing the finding of the prior eclipse, this movie shows that the recombination coefficient in F2 layer heights, even in day time and at the equator, is much smaller than normally assumed.

The movie of the third eclipse over Genoa, Italy, was animated by using the spare camera, synchronized with the sequence of ionograms, for optical observation of the sun through a tiny hole in front of a long carton tube replacing the lens. The picture of the sequential phases of the sun's occultation was then inserted in the ionogram recordings. This eclipse confirmed the findings of the prior eclipses, although its conditions were not as good.

During the trip to Djibouti I achieved another breakthrough. One of Karl Rawer's favorite subjects (Bibl et al., 1959), was ionospheric absorption measurements. They were necessary for making reasonable ionospheric radio wave predictions. To separate deviative (E-region) and nondeviative (D-region) contributions, Karl Rawer invented the "Spider-Web" for the frequency dependent absorption measurements. Five scientists, including Karl Rawer, had to slave every third night on a 12-h shift. (allowing vacation and travel). At 02:00 a.m. we had to make absorption measurements: 5 s looking at the oscilloscope screen, 5 s to write the amplitude of all the echoes down, etc. for 10 min. We did this for several years to determine the variation in ionospheric absorption. (One of my colleagues made completely random entrées).

After installing a new Ionosonde in Djibouti, I explained to the station chief and officer that he had to make absorption measurements. He told me: at 12 o'clock noon, I eat dinner and at 2 o'clock in the night, I sleep. Thus I took a spare oscilloscope and a spare movie camera and recorded the amplitudes on film. After returning to my institute, I could persuade Karl Rawer that film can average and integrate amplitudes better than some scientists' brains. Automatic recording of absorption measurements was approved.

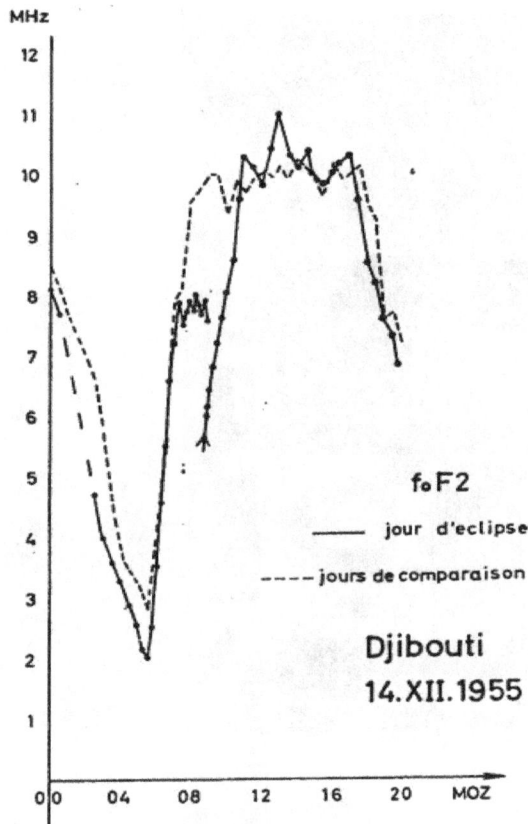

Fig. 2. Zero magnetic dip effect of the eclipse; a new F2-layer is generated independently of the old one.

3 Ionospheric Dynamics

Karl Rawer always was interested in the dynamics of the ionosphere. It was already mandatory for his first task of ionospheric predictions. He invented the use of quartiles and deciles to cover the substantial variability and statistical complexity of the ionosphere (Rawer,1951). I made the dynamics of the ionosphere the subject of my Ph.D. thesis (Bibl, 1964). But we worked even later on the subject using the capabilities of the Panorama-Ionosonde to make movies and to record continuously (Bibl and Rawer, 1959).

In respect to the need for standardization, Karl Rawer and I always were in agreement. Standardization allowed his team to build equipment inexpensively for other institutes. In support of the French communication network overseas, we built ionosondes for three African stations: Dakar, Lwiro, and Djibouti, as well as Nha Tran (Viet Nam) and Kerguelen (Antarctica). In Europe, we re-equipped Dourbes (Belgium), Darmstadt (Germany), and Genoa (Italy), in addition to our own station, first in Neuershausen and then in Breisach. This made the European network of ionosondes, left over from the geophysical year, the densest in the world.

In addition, we built a ground backscatter array of 12 rhombus antennas, connected in sequence to three high power fixed frequency transmitters. With the first continuous direct color recordings, we could monitor continuously the ionosphere in an area of 2000 km radius around the Breisach

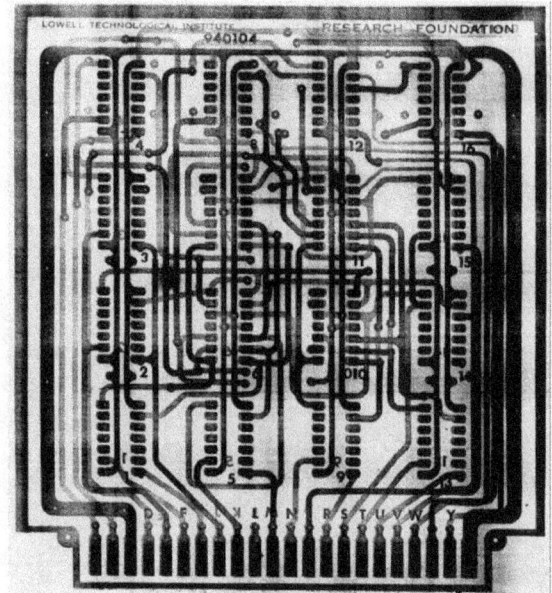

Fig. 3. Digital Memory card with 4 × 16 bits (1968).

station. Direct echoes from a distance of 500 km, arriving from the south, led to the discovery of a big hole in the ionosphere. Ionospheric maps, drawn from the dense network of stations, shows that this condition is relatively frequent, at least ten percent of the time (Bibl, 1964). It also verified the importance of the ionospheric station Sottens in Switzerland, which formerly was considered poor quality because sometimes it showed low critical frequencies.

Because I was unwilling to work for the German secret service and since any other interest in ionospheric research faded in Germany, I emigrated to the USA. How little did I imagine that the US, after the murder of President Kennedy, would become as belligerent as Germany was before?

4 Lowell: Center for Atmospheric Research

When I came to Lowell, MA, my first tasks were to complete an ionospheric drift station (Pfister and Bibl, 1972) and a sophisticated satellite signal recording station using multi-frequency dispersive Doppler and Faraday methods (Reinisch, 1970). These were the only subjects where I had no experience because my former colleague Ewald Harnischmacher had covered the first and Dr. Hess the second. After completion of the respective measuring equipment within a year, I thought I knew it all. Now I see that I know only ten percent of what I need to know to be successful. So fast is progress that I cannot keep up with it at all. It is absolutely necessary now to work in teams. For the first digital ionosonde, I developed digital frequency synthesis and a digital integrator with a comparator and a forward/reverse counter. Its printed circuit card with a 4×16 bit memory is presented as Fig. 3. But the first Digisonde, which was used successfully, already required co-workers specialized in programming and computers (Bibl and Reinisch, 1978).

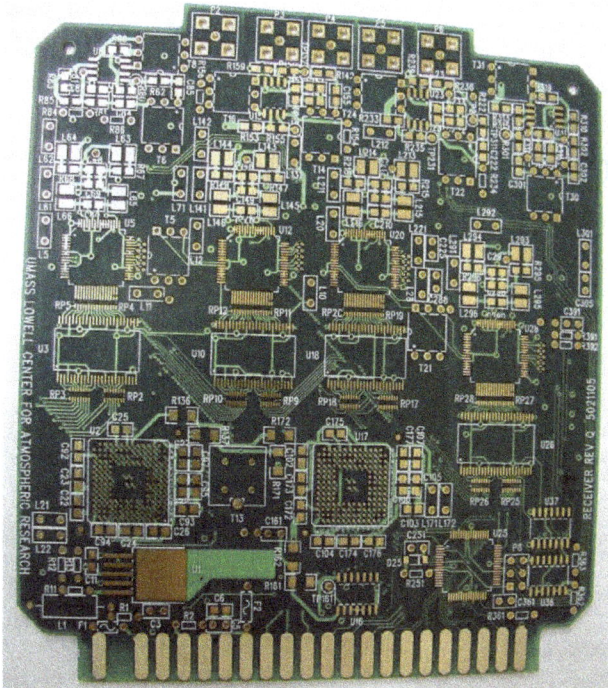

Fig. 4. Digital Receiver card replacing 16 analog cards of same size.

Completing Karl Rawer's legacy in helping other institutes to obtain the most advanced equipment and to promote international co-operation, we have built more than 70 Digisondes used everywhere in the world (Reinisch et al., 1997), one even in space (Reinisch et al., 2000).

Now, after 60 years, we are working on digital transmitters and receivers. They allow any shape for the pulse we want. For the transmitter pulse I have chosen a one-half sine pulse, for reasons explained below. All ionosondes I have designed have analog tuning on the front end because they should work under strongest interference conditions. For the first Digisonde I followed a recommendation by B. W. Reinisch to use the same circuit with switches for transmission first and then for reception. This scheme only works because the receiver was a Gaussian filter and could recover completely from the transmitter pulse very quickly. Even the digital receiver will have digital tuning in front with a bandwidth larger than the pulse to accommodate four adjacent frequencies. The Digital Receiver card (Fig. 4) will replace 16 analog card of the same size. This card has inputs and digitizers for four antennas. The digitizer outputs go to digital inputs of two Greychips. Each Greychip mixes the data digitally with a different frequency. With single-line half-cycle complex spectrum analysis of the received pulse, amplitudes and absolute phase differences of two frequencies are simultaneously analyzed, measuring dispersion or precision group height. Digital cleaning of coherent interferers (Bibl, 2000) and digital extraction of two frequencies from a single pulse will speed up the Digisonde frequency scan to make again 1/2 min ionograms, but with information content orders of magnitude higher than the analog ionosondes.

5 Conclusion

In trying to follow the professional ethics of my mentor, Professor Karl Rawer, although I don't have his range of capabilities, I have established a pattern of work that creates new knowledge in our field and allows other institutes and colleagues the same.

References

Bibl, K.: Zur Dynamik der Ionosphäre, Zeitschr. f. Geoph., Sonderheft 1, 1–33, 1958.

Bibl, K.: Does magnetic field aligned displacement govern the F2-region?, Ann. Geophysicae, 18, 294–297, 1962.

Bibl, K.: The detection of an important anomaly in the F-region ionization over central Europe, Ann. Geophysicae, 20, 447–453, 1964;

Bibl, K.:Die Sonnenfinsternis vom 14. December 1955 bei Djibouti, Kleinheubacher Berichte, 97–100, 1958.

Bibl, K.: Patent application # 09/728,846, 2000.

Bibl, K. and Delobeau, F.: Ionosphärische Beobachtungen während der totalen Sonnenfinsternis vom 30. Juni 1954, Zeitschr. f. Geophys., 21, 215–228, 1956.

Bibl, K. and Rawer, K.: Travelling disturbances originating in the outer ionosphere, J. Geoph. Res. 64, 2232–2238, 1959.

Bibl, K. and Reinisch, B. W.: The Universal Digital Ionosonde, Radio Science 13, 619–529, 1978.

Bibl, K., Paul, A., and Rawer, K.: Die Frequenzabhängigkeit der ionosphärischen Absorption, JATP 16, 324–339, 1959.

Huber, L. and Rawer, K.: Zur Frage des besten Impulsempfängers, AEÜ 4, 475–484 and 523–526, 1950.

Pfister, W. and Bibl, K.: A modernized technique for ionospheric drift with spectrum analysis, Space Res. XII, 975–982, 1972.

Rawer, K.: Ausbreitungsvorhersage für Kurzwellen mit Hilfe von Ionosphärenbeobachtungen, AEÜ 5, 154–167, 1951.

Reinisch, B. W.: Second Order Phase Path Calculations for the Trans-ionospheric Propagation, Proc. of the Symposium on the Future Application of Satellite Beacon Experiments, Lindau/Harz, W. Germany, 28–1 to 28–6, 1970.

Reinisch, B. W. and Sales, G. S.: Ionospheric sounding support of OTH Radar, Radio Sci. 13, 1681–1694, 1997.

Reinisch, B. W., Haines, D. M., Bibl, K., Cheney, G. P., Galkin, I. A., Huang, X., Myers, S. H., Sales, G. S., Benson, R. F., Fung, S. F., Green, J. L., Boardsen, S., Taylor, W. L., Bougeret, J.-L., Manning, R., Meyer-Vernet, N., Moncuquet, M., Carpenter, D. L., Galagher, D. L., and Reif, P.: The Radio Plasma Imager Investigation on the Image Spacecraft, Space Science Rev. 91, 319–359, 2000.

3

Long-period upper mesosphere temperature and plasma scale height variations derived from VHF meteor radar and LF absolute reflection height measurements

C. Jacobi[1] and D. Kürschner[2]

[1]Institute for Meteorology, University of Leipzig, Stephanstr. 3, 04103 Leipzig, Germany
[2]Institute of Geophysics and Geology, University of Leipzig, Collm Observatory, 04779 Wermsdorf, Germany

Abstract. The change of ionospheric absolute reflection heights h of low-frequency (LF) radio waves at oblique incidence in the course of the day is measured at Collm Observatory (51.3° N, 13.0° E) using 1.8 kHz sideband phase comparisons between the sky-wave and the ground wave of a commercial 177 kHz transmitter (Zehlendorf, reflection point at 52.1° N, 13.2° E). Plasma scale height estimates H are calculated from the decrease/increase of h in the morning/evening. The day-to-day variations of H are compared with those of daily mean temperatures at 90 km, measured with a VHF meteor radar (36.2 MHz) at Collm and using the amplitude decay of meteor reflections. A good qualitative correspondence is found between the two data sets. Since mesospheric long-period temperature variations are generally accepted to be the signature of atmospheric planetary waves, this shows that LF reflection height measurements can be used for monitoring the dynamics of the upper middle atmosphere.

1 Introduction

The dynamics of the mesosphere/lower thermosphere (MLT) region at 80–100 km altitude is forced through wave coupling, so that MLT mean circulation and temperature trends may indicate possible long-term trends of wave activity and vice versa. Usually, when atmospheric waves are considered, one distinguishes between short-period gravity waves, tides with periods of 24 h and harmonics, and other planetary waves in the so-called long-period range up to about 30 days.

Planetary wave analyses often describe either the total variability of winds or temperature in the period interval to about 30 days (Jacobi et al., 1998; Bittner et al., 2000), or they focus on one or few specific waves as, e.g. Rossby normal modes like the quasi 2-day wave (Chshyolkova et al.,

2005). Regarding long-term trends and possible indicators for climate variability, there is a tendency for a slight increase of overall day-to-day variability in the 1980's and 1990's (Jacobi et al., 1998, 2000), but this result is based on only few datasets, and thus far is neither confirmed by a hemispheric or global analysis, nor thus far explained through theoretical consideration. A more indirect analysis of planetary waves was performed using radio wave absorption variations (e.g. Lastovicka et al., 1994). These analyses showed a possible positive trend, but this is intermittent (Lastovicka, 2002).

To summarise, the current knowledge on planetary wave activity in the MLT is still incomplete, mainly because of the lack of long homogeneous time series. To contribute to the available data base and providing a potential for additional wave analysis, we propose here to use the diurnal change of lower E-region low-frequency (LF) reflection heights as an indicator for daily temperatures and to derive planetary waves from these analyses. Using LF radio waves as a tracer for synoptic measurements has been done since several decades (Entzian et al., 1976; Lauter et al., 1977). Phase-height measurements (von Cossart and Entzian, 1976) use LF transmitters located at a distance of several hundred kms from the receiver, with reflection heights well below 85 km during daytime. Here, we make the first attempt to analyse scales heights using LF waves on a short propagation path.

2 LF height measurements and analysis of scale height

Low frequency 177 kHz radio waves from a commercial radio transmitter are registered at Collm Observatory, Germany (51.3° N, 13.0° E, distance to transmitter 170 km). The virtual reflection heights h', referring to the reflection point at 52.1° N, 13.2° E, are estimated using measured travel time differences between the ground wave and the reflected sky wave through phase comparisons on sporadic oscillation bursts of the amplitude modulated LF radio wave in a small

Correspondence to: C. Jacobi (jacobi@uni-leipzig.de)

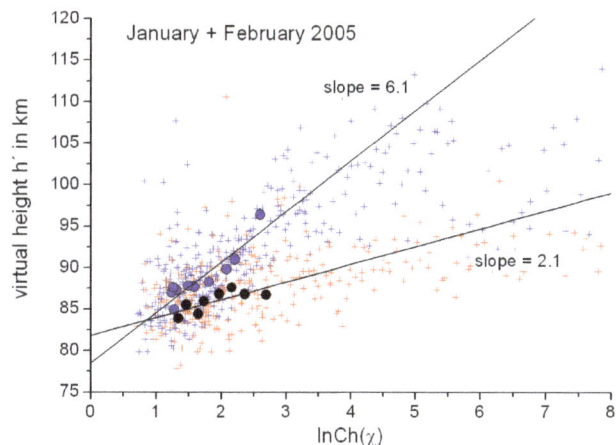

Fig. 1. Virtual height h' measured on 177 kHz vs. the logarithm of the Chapman function $Ch(\chi)$, for half-hourly mean data of all days in January and February 2005. Median values of h' are given for $Ch(\chi) < 3$, and for forenoon (blue) and afternoon (red) data separately. The slopes are calculated from the median values.

modulation frequency range around 1.8 kHz (Kürschner et al., 1987). The reflection height measurements have started in September 1982 and are carried out continuously since then, even after the temporal change to digital broadcasting in September 2005.

The height resolution of an individual reflection height measurement is nearly 2 km. The data are combined to half-hourly means that consist of approximately 6000 individual values on an average. Essentially caused by the variability of reflection heights the $1 \times \sigma$ variation of the half-hourly mean is in the order of 3 km below 95 km and slightly more than 5 km near an altitude of 100 km.

Half-hourly virtual reflection heights range between roughly 82 km during daytime and nearly 100 km during nighttime for the ordinary component. Especially in winter in the late night and early morning hours, and also in some midsummer nights, wave propagation is subject to stronger magnetoionic splitting into the ordinary and extraordinary component, with very large virtual reflection heights of the latter. In this case the half-hourly mean height consists of a mixture of both components with unknown weighting, so that these data will not be used here. During daylight hours, particularly in the summer months height measurements are not possible due to strong D-region absorption of sky waves then.

Assuming essentially equilibrium between ionisation and recombination, one single gas, an isothermal atmosphere, monochromatic radiation, and with that resulting a Chapman profile of the electron density, the height h of the layer maximum changes with zenith angle of the sun, χ, as:

$$h = h_0 + H \ln Ch(\chi), \tag{1}$$

with $h_0 = h(\chi = 0)$, $H = RT/Mg$ as the plasma scale height (T as absolute temperature, R as molar gas constant, g as the acceleration due to gravity and M as molecular mass) and the Chapman function $Ch(\chi)$. If $\chi < 75°$ then $Ch(\chi) \approx 1/\cos \chi$ is a suitable approximation in the case that the curvature of the earth is neglected. Experimentally, a relationship similar to Eq. (1) has been found between the virtual reflection height h' of LF waves below the layer maximum and the solar zenith distance as in the case of Chapman layer maximum:

$$h' = h'_0 + H \ln Ch(\chi). \tag{2}$$

Therefore, plotting the logarithm of the Chapman function against the virtual reflection height should provide, in a first approximation, a nearly linear correspondence, with the slope of the regression line as a scale height estimate:

$$H = \frac{\Delta h'}{\Delta \ln(Ch(\chi))}. \tag{3}$$

In the case of a delay of the reflection height curve with respect to the solar zenith distance, i.e. for the case of non-equilibrium, the slope differs from H. While von Cossart (1976) for mesospheric data used the average of forenoon and afternoon measurements and obtained a good correspondence with rocket measurements, the delay in the lower E-region may differ from that one at lower heights. Moreover, the difference between real and virtual reflection heights can be substantial and reach more than 2 km in the D-region and more than 5 km in the lower E-region. Therefore, the results of H estimations have to be considered as qualitative. In addition, at higher altitudes the derived H values are smaller, especially in winter (Entzian et al., 1976; Lauter et al., 1977), which is known as the so-called "winter anomaly". Caused by the comparatively short distance between transmitter and receiver, which is connected with a steep incidence angle, the reflection heights on 177 kHz lies above those in earlier works (75–95 km, Lauter et al., 1966). Thus, differences between forenoon and afternoon just as summer and winter behaviour on the H estimates are to be expected. As an example, for 2 month time interval in Fig. 1 h' vs. $\ln(Ch(\chi))$ is shown separately for the forenoon h' decrease (blue symbols) and the afternoon h' increase (red symbols). Median values for these two cases are given in the figure, and slopes are calculated. Clearly, the slopes are different and the change of slope for larger $\ln(Ch(\chi))$ values, i.e. for low elevation angles, can also be seen. Therefore, slopes have been calculated using the forenoon and afternoon median values only in the range of $\ln(Ch(\chi)) < 3$. For the afternoon the resulting slope is clearly too small, while the values for the forenoon are more realistic taking into account earlier results from other authors (e.g. von Cossart, 1976) and the temperature climatology of the mesopause region (e.g. CIRA86, Fleming et al., 1990). In the following we shall use forenoon data only.

Figure 1 also shows that sometimes at small values of $\ln(Ch(\chi)$ large values of h' are measured. These are owing

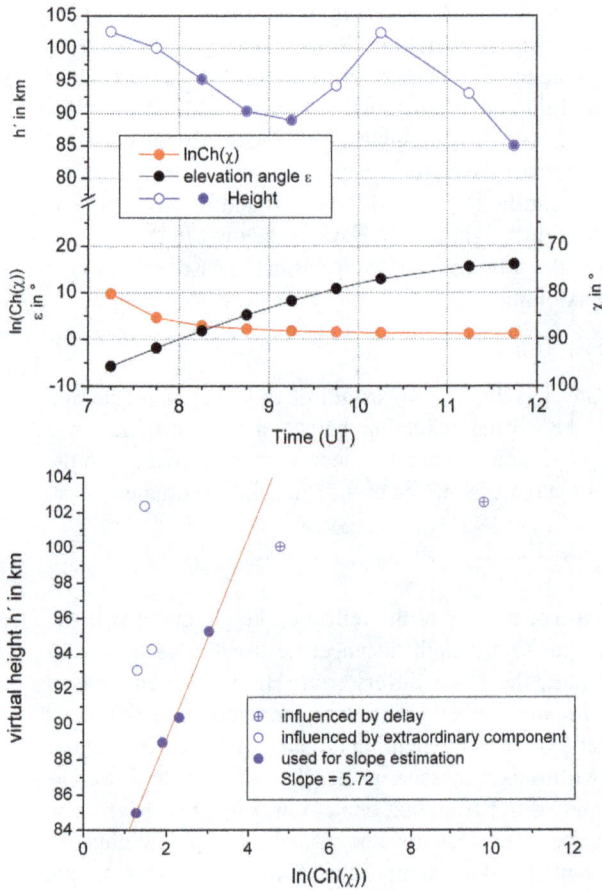

Fig. 2. Example of scale height H estimation for the forenoon of 10 January 2005. Upper panel: time series of logarithm of Chapman function $\ln Ch(\chi)$, elevation angle $\varepsilon = 90° \chi$, and virtual height h'. Lower panel: virtual height h' vs. $\ln Ch(\chi)$. Solid blue dots denote h' data points used for H estimation.

to reflections from the extraordinary component mixed with the ordinary component. These data cannot be used for the analysis. In Fig. 1 they are automatically discarded since we take the median of two months of data, but on a single day, as is exemplarily shown in Fig. 2 these data can completely alter the profile. As shown in the figure, data that are obviously dominated by the extraordinary component are discarded. It should be noted that, for single days, the analysis of the scale height my be done from only very few data points and large uncertainty resulting from that.

3   Mesopause temperatures derived from VHF meteor radar measurements

VHF meteor radars measure the radio wave reflection from the ionised trails of meteors entering the Earth's atmosphere. The decay time of the signal is detected from so-called underdense meteor trails, i.e. from those trails whose reflectiv-

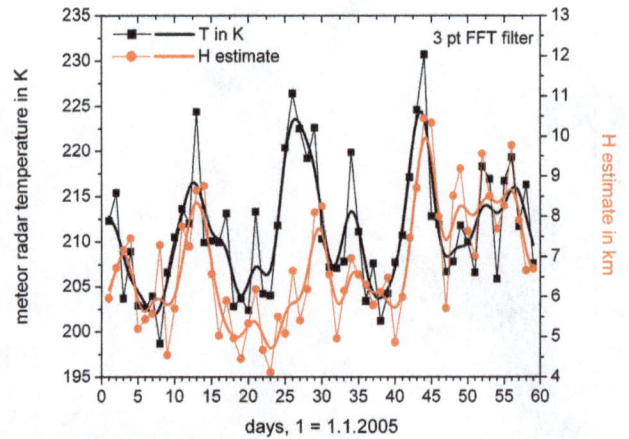

Fig. 3. Time series of scale height H estimates and meteor radar temperatures T during January and February 2005. The thick solid lines are smoothed data using a 3 point FFT filter.

ity is determined by their electron density, which decreases with time due to diffusion (Hocking, 1999, Hocking et al., 2001). The amplitude decreases exponentially depending on the ambipolar diffusion coefficients D_a:

$$A = A_0 \exp \left\{ -\frac{16\pi^2 D_a}{\lambda^2} t \right\}, \tag{4}$$

with A as the signal amplitude, λ as the radar wavelength and t as time. Measuring the decay of the amplitude allows to estimate D_a. The diffusion coefficient is proportional to the ratio of temperature squared and pressure. Introducing a temperature gradient from an empirical model and using exponential pressure decay with height, one obtains an equation for the temperature in dependence of the diffusion coefficient change with height:

$$\frac{d \ln D_a}{dz} = -2 \frac{dT/dz}{T_0} + \frac{mg}{kT_0}, \tag{5}$$

to be solved for T_0. Details can be found in Hocking (1999). Temperatures estimated with this method has been presented by Hocking et al. (2004) and Singer et al. (2004), who also showed validation results using rocket and ground-based optical methods. Temperature data are available as daily means at 90 km height, which is approximately the height where maximum meteor rates are found.

At Collm Observatory, a SKiYMET meteor radar is operated on 36.2 MHz since summer 2004. From meteor reflections mesopause region hourly wind profiles and daily temperatures are derived that roughly apply to a circle around Collm with 300 km diameter, and for heights between 80–100 km.

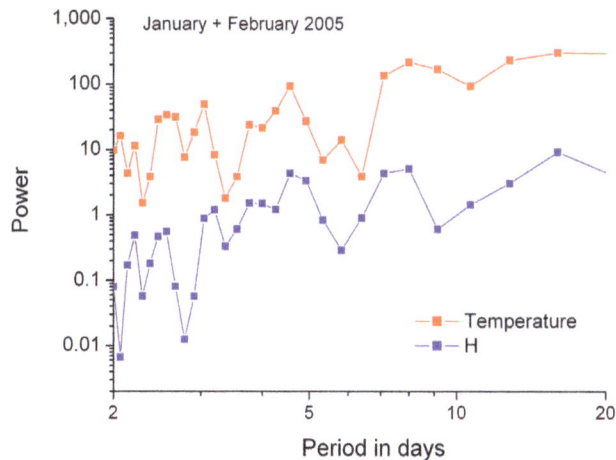

Fig. 4. Power spectra of scale height estimates and meteor radar temperatures.

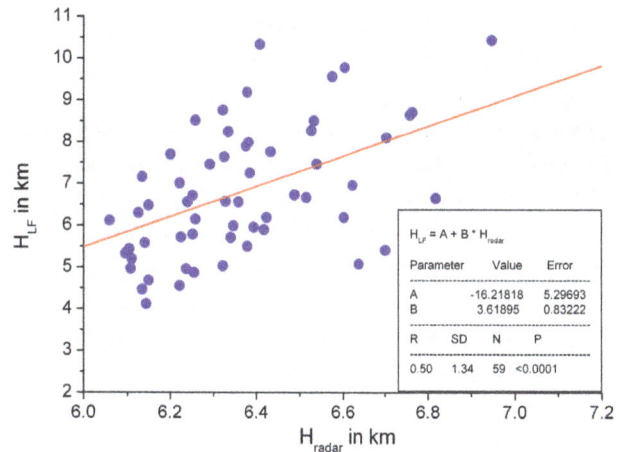

Fig. 5. Scale height estimates H_{LF} calculated from LF reflection height changes vs. scale heights H_{radar} calculated from meteor radar.

4   Results and discussion

The time series of meteor radar derived daily temperatures in early 2005 is shown in Fig. 3. The daily scale height estimates after Eq. (3) are added. Both curves have been subjected to a 3 point FFT filter (thick solid line) to suppress random irregularituies from the measurement and analysis uncertainty. The temperature and H curves show the same qualitative tendency. Obviously, the dominating day-to-day and long-period changes are visible in both datasets, indicating that their variability is of the same origin, i.e. the MLT temperature change. Power spectra of both parameters are shown in Fig. 4. Peaks are correspondingly found for periods of 3, 5, and 8 days.

Two months of LF absolute reflection height data has been used to derive scale height estimates in the lower E-region. These data has been compared to meteor radar temperatures. The results are clearly qualitative. In addition, the influence of the extraordinary component of the LF wave on the results has carefully to be analysed and removed from the data set. This, however, makes the analysis somewhat complicated, and increases the potential errors.

Nevertheless, the long-period variations of both parameters available after low pass filtering are in adequate agreement, indicating that the temperature effect on the diurnal scale height change is inferable using the 177 kHz measurements at Collm, with short propagation distances. Since LF height measurements are available for more than 2 decades, this provides the opportunity to analyse planetary waves and their long-term variations and trends.

Acknowledgements. This study has been partly supported by DFG under grant JA 836/19-1 (CPW-TEC) within the SPP 1176 "CAWSES - Climate And Weather of the Sun-Earth System".

References

Bittner, M., Offermann, D., and Graef, H.H.: Mesopause temperature variability above a midlatitude station in Europe, J. Geophys. Res., 105, 2045–2058, 2000.

Chshyolkova, T., Manson, A. H., and Meek, C. E.: Climatology of the quasi two-day wave over Saskatoon (52° N, 107° W): 14 Years of MF radar observations, Adv. Space Res., 35, 2011–2016, 2005.

Entzian, G., Lauter, E. A , and Taubenheim, J.: Synoptic monitoring of the mesopause region using D-region plasma as a tracer in different heights, Z. Meteorol., 26, 1–6, 1976.

Fleming, E. L., Chandra, S., Barnett, J. J., and Corney, M.: Zonal mean temperature, pressure, zonal wind and geopotential height as function of latitude, Adv. Space. Res. 10, 11–59, 1990.

Hocking, W. K.: Temperatures using radar-meteor decay times, Geophys. Res. Lett., 26, 3297–3300, 1999.

Hocking, W. K., Fuller, B., and Vandepeer, B.: Real-time determination of meteor-related parameters utilizing modern digital technology, J. Atmos. Solar-Terr. Phys., 63, 155–169, 2004.

Hocking, W. K., Singer, W., Bremer, J., Mitchell, N. J., Batista, P., Clemesha, B., and Donner, M.: Meteor radar temperatures at multiple sites derived with SKiYMET radars and compared to OH, rocket and lidar measurements, J. Atmos. Solar-Terr. Phys., 66, 585–593, doi:10.1016/j.jastp.2004.01.011, 2004.

Jacobi, Ch., Schminder, R., and Kürschner, D.: Planetary wave activity obtained from long-term (2-18 days) variations of mesopause region winds over Central Europe (52° N, 15° E), J. Atmos. Solar-Terr. Phys., 60, 81–93, 1998.

Kürschner, D., Schminder, R., Singer, W., and Bremer, J.: Ein neues Verfahren zur Realisierung absoluter Reflexionshöhenmessungen an Raumwellen amplitudenmodulierter Rundfunksender bei Schrägeinfall im Langwellenbereich als Hilfsmittel zur Ableitung von Windprofilen in der oberen Mesopausenregion, Z. Meteorol., 37, 322–332, 1987.

Lastovicka, J.: Long-term changes and trends in the lower ionosphere, Phys. Chem. Earth, 27, 497–507, 2002.

Lastovicka, J., Fiser, V., and Pancheva, D.: Long-term trends in planetary wave activity (2–15 days) at 80–100 km inferred from radio wave absorption, J. Atmos. Terr. Phys., 56, 893–899, 1994.

Lauter, E. A. and Entzian, G.: Überwachung der tiefen Ionosphäre mit Hilfe der Quasi-Phasenhöhenmessung im Langwellenbereich (100…200 kHz). Proceedings of the Summer School "Untere Ionosphäre", Kühlungsborn 1964, Academy of Sciences of the DDR, Berlin, 67–97, 1966.

Lauter, E. A., Entzian, G., von Cossart, G., Sprenger, K., and Greisiger, K. M.: Synoptische Erschließung von Prozessen in der winterlichen Mesopausenregion durch bodengebundene Beobachtungsverfahren, Z. Meteorol., 27, 75–84, 1977.

Singer, W., Bremer, J., Weiß, J., Hocking, W. K., Höffner, J., Don-

ner, M., and Espy, P.: Meteor radar observations at middle and arctic latitudes Part 1: Mean temperatures, J. Atmos. Solar Terr. Phys., 66, 607–616, doi:10.1016/j.jastp.2004.01.012, 2004.

Von Cossart, G.: Ein Beitrag zur synoptischen Untersuchung der atmosphärischen Struktur in der Mesopausenregion aus indirekten Phasenmessungen, PhD Thesis, Academy of Sciences of the DDR, 1976.

Von Cossart. G. and Entzian, G.: Ein Modell der Mesopausenregion zur Interpretation indirekter Phasenmessungen und zur Abschätzung von Ionosphären- und Neutralgasparametern, Z. Meteorol., 26, 220–230, 1976.

4

Investigations of long-term trends in the ionosphere with world-wide ionosonde observations[*]

J. Bremer

Leibniz-Institut für Atmosphärenphysik, Schloss-Str.6, D-18225 Kühlungsborn, Germany

[*]Dedicated to Prof. Dr. K. Rawer on the occasion of his 90th birthday.

Abstract. Basing on model calculations by Roble and Dickinson (1989) for an increasing content of atmospheric greenhouse gases in the Earth's atmosphere Rishbeth (1990) predicted a lowering of the ionospheric F2- and E-regions. Later Rishbeth and Roble (1992) also predicted characteristic long-term changes of the maximum electron density values of the ionospheric E-, F1-, and F2-layers. Long-term observations at more than 100 ionosonde stations have been analyzed to test these model predictions. In the E- and F1-layers the derived experimental results agree reasonably with the model trends (lowering of h'E and increase of foE and foF1, in the E-layer the experimental values are however markedly stronger than the model data). In the ionospheric F2-region the variability of the trends derived at the different individual stations for hmF2 as well as foF2 values is too large to estimate reasonable global mean trends. The reason of the large differences between the individual trends is not quite clear. Strong dynamical effects may play an important role in the F2-region. But also inhomogeneous data series due to technical changes as well as changes in the evaluation algorithms used during the long observation periods may influence the trend analyses.

1 Introduction

The estimation of ionospheric long-term trends is an important scientific topic for the investigation of possible anthropogenic changes in the Earth's atmosphere. Whereas an increasing content of atmospheric greenhouse gases (CO_2, CH_4, H_2O, ...) should cause increasing temperatures near the Earth's surface and in the troposphere (Hegerl et al., 1996) in the strato-, meso-, and thermosphere the temperatures should be reduced by an increasing cooling due to an enhanced infrared radiation of the greenhouse gases into space (Roble and Dickinson, 1989). Therefore, Rishbeth (1990) and Rishbeth and Roble (1992) predicted a shrinking of the ionosphere as well as characteristic changes of the maximum electron densities in the ionospheric E-, F1-, and F2-layers.

Data of ionosonde observations which are regularly derived at many different stations around the Earth since more than 40 or even 50 years can be used for trend analyses using different standard parameters. In this paper the following parameters have been analyzed: the maximum electron densities of different ionospheric layers characterized by their critical frequencies foE, foF1, and foF2 as well as the height parameters h'E and hmF2. The height of the maximum of the F2-layer, hmF2, has been derived from M(3000)F2 ionosonde values using the well-known simple formula derived by Shimazaki (1955).

Most of the data for more than 100 different ionosonde stations have been selected from CD-ROMs of NGDC, Boulder, USA, and from WDC-C at RAL, Chilton, UK.

1.1 Data analysis method

In Fig. 1 some examples of long-term variations are shown using foF2 and hmF2 data (noon values for June and December) observed at the station Juliusruh (54.6° N, 13.4° E) during the time period between 1957 and 2002. Both parameters are characterized by a marked 11-yearly variation caused by changes of the solar and geomagnetic activity shown in the lower part of Fig. 1. Here the solar sunspot number R is used as an index of the solar wave radiation and the geomagnetic A_p index as a proxy of the fluxes of precipitating high energy particles. These solar and geomagnetically induced variations of the plasma parameters are markedly stronger than possible long-term variations. Therefore, we used the following algorithm to eliminate these influences.

For monthly mean values at each full hour for $X=fo$E, h'E, foF1, foF2, or hmF2 a twofold regression equation has been derived:

$$X_{\text{th}} = a + b \cdot R + c \cdot A_p. \tag{1}$$

Correspondence to: J. Bremer
(bremer@iap-kborn.de)

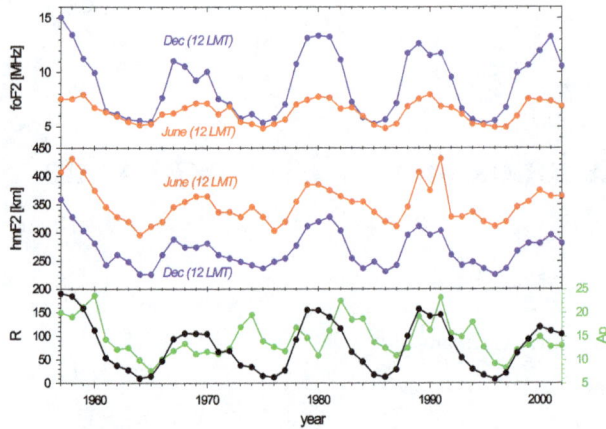

Fig. 1. Long-term variation of foF2 and hmF2 for June and December at noon condition from ionosonde observations at Juliusruh (54.6° N, 13.4° E) together with variations of yearly mean values of solar sunspot number R and geomagnetic A_p index.

For the solar activity index also other parameters than R can be used (e.g. solar radio flux F10.7), but the final trend results are not markedly influenced by this choice (Bremer, 2001). The solar and geomagnetically caused part has been subtracted from the corresponding observed values for each month and each hour.

$$\Delta X = X_{obs} - X_{th}. \tag{2}$$

From these hourly data yearly mean ΔX values have been estimated for the derivation of linear trends

$$\Delta X = d + e \cdot year \tag{3}$$

with the trend parameter b measured in MHz/year or km/year.

1.2 Experimental trends

Using the method shortly described in Sect. 1.1 the trends of different parameters observed at the ionosonde station Juliusruh (54.6° N, 13.4° E) are shown in Fig. 2. Both height parameters, hmF2 and h'E, have significant negative trends (red curves) in qualitative agreement with the model predictions of Rishbeth (1990), whereas the critical frequencies, foE, foF1, foF2, only slightly increase. Their trends are not significant (black curves). Compared with the strong variability of the original data shown in Fig. 1 the amplitudes of the long-term trends presented in Fig. 2 are markedly smaller, thus demonstrating that a careful elimination of the solar and geomagnetically induced variation is necessary to get reasonable trend results.

To get more information about mean global trends in the ionosphere analyses have been extended to data of different ionosonde stations all around the world. In Fig. 3 the results of foF1 trends are presented which have been derived from 51 different stations. In the upper part of this figure a histogram of the individual trends is shown for significant and non significant trends. The median trend is marked by an

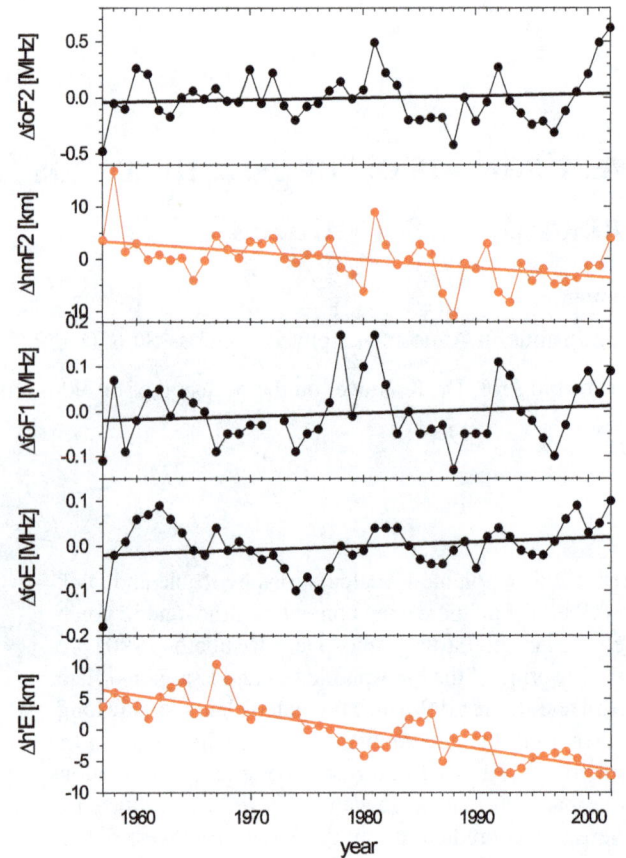

Fig. 2. Long-term trends of different ionospheric parameters observed at Juliusruh after elimination of the solar and geomagnetically induced variations.

arrow. In the lower part all ΔX values have been averaged and a mean trend was estimated. Both mean trend values derived by the two different methods are nearly identical. Also the estimated mean errors with 95% reliability level (for details of their estimation see Taubenheim, 1969) are smaller than the estimated mean values demonstrating that the derived global mean trend values are significant different from zero.

In Fig. 4 the histograms for all five investigated ionosonde parameters are shown together with the corresponding median values marked by arrows. The histograms are characterised by relatively broad distributions, mainly for the parameters of the F2-region. Figure 4 is an updated version of a similar picture shown earlier in Bremer (2001). Negative trends are presented by blue, positive trends by red colour. The number in brackets describe the number of individual stations used in the trend analyses. The median trend values of the histograms are collected in Table 1 together with their mean error values (Taubenheim, 1969). It can be seen that only the global mean trends in foE and foF1 are significant different from zero with a significance level of more than 95%. For all other cases (h'E, hmF2, foF2) the significance level is markedly lower (for foF2: 84%, h'E: 75%, hmF2: <50%).

Table 1. Mean experimental trends and error limits (95%) derived from trend analyses for different ionosonde parameters. N is the number of ionosonde stations used in the estimations of mean trends.

	Parameter	N	Mean trend	Error (95%)
F2-region	foF2	106	-0.0018 MHz/year	±0.0025 MHz/year
	hmF2	87	-0.009 km/year	±0.076 km/year
F1-region	foF1	51	0.0027 MHz/year	±0.0011 MHz/year
E-region	foE	72	0.0014 MHz/year	±0.0007 MHz/year
	h'E	31	-0.040 km/year	±0.070 km/year

Table 2. Mean experimental (exp) trends of different ionospheric parameters and expected changes of these data assuming a doubling of the atmospheric Greenhouse gases (CO_2*2). The model data (mod) are from Rishbeth (1990) and Rishbeth and Roble (1992).

	Parameter	Mean exp. Trend	CO_2*2 (exp)	CO_2*2 (mod)
F2-region	foF2	-0.0018 MHz/year	-0.36 MHz	$-0.2 \dots -0.5$ MHz
	hmF2	-0.009 km/year	-1.8 km	$-10 \dots -20$ km
F1-region	foF1	0.0027 MHz/year	0.54 MHz	$0.3 \dots 0.5$ MHz
E-region	foE	0.0014 MHz/year	0.28 MHz	$0.05 \dots 0.08$ MHz
	h'E	-0.040 km/year	-8.0 km	-2.5 km

From Fig. 4 and the mean results summarised in Table 1 it became clear that the variability of the individual trends in the F2-region is very strong. Especially the hmF2 trends differ markedly between the different stations analysed. In Fig. 5 these individual hmF2 trends are shown in dependence on latitude and longitude. Negative trends are marked by blue, positive by red symbols. Full dots represent significant trends (>95%), circles non significant trends. Strong regional differences can be observed e.g. with negative trends in Central and Western Europe and positive trends in Central Asia. Also in the foF2 trends some regional differences occur, but not so pronounced as in the hmF2 trends.

2 Discussion

Can the mean ionosonde trends shown in Table 1 be explained by an increasing atmospheric greenhouse effect? To answer this question the experimental trends have to be compared with model calculations of Rishbeth (1990) and of Rishbeth and Roble (1992). Their theoretical results have been derived for a doubling of the atmospheric greenhouse gases CO_2 and CH_4. The effective change of the greenhouse gases during the last 40 years where trends of the ionosonde data have been investigated is about 20% (Brasseur and de Rudder, 1987; Houghton et al., 2001). Assuming a linear dependence between the content of the atmospheric greenhouse gases and the ionospheric effect, the experimental trends can be extrapolated to a level of doubled greenhouse gases. These values called CO_2*2 (exp) are compared with the corresponding model values CO_2*2 (mod) of Rishbeth (1990) and Rishbeth and Roble (1992) in Table 2.

As to be seen from the data shown in Table 2 in the E-region the experimental and theoretical trend val-

ues agree qualitatively with a lowering of the height h'E and an increase of foE. However the experimental trends are markedly stronger than the model values. The derived positive foE trend is also in general agreement with rocket mass spectrometer measurements of the ion density ratio $[NO^+]/[O_2^+]$ in the E-region (Danilov and Smirnova, 1997). The observed negative trends of $[NO^+]/[O_2^+]$ cause increasing electron densities and therefore increasing foE values as the dissociative recombination coefficient of NO^+ is markedly larger than that of O_2^+.

The agreement of the mean experimental and model trends in the F1-region is surprisingly good as to be seen in Table 2.

Also in the F2-region the agreement between the experimental and model trends seems to be quite reasonable looking at the data in Table 2. However, the variability of the individual trends at different stations is very strong, and the derived mean trends are not statistically significant different from zero as can be seen in Table 1. Especially the variability of the hmF2 trends is very pronounced. Therefore, the agreement between model and experimental data is more accidental. The reason of the strong variability in the F2-region is not quite clear. After Fig. 5 there seem to be regional differences which could be caused by dynamical effects in the plasma of the F2-region. Such strong regional differences of the hmF2 trends have also earlier been detected by other authors using a more limited data volume (Ulich and Turunen, 1997; Bremer, 1998; Bencze at al., 1998). From satellite observations (Keating et al., 2000) it is known that the observed long-term neutral density reduction near 350 km altitude is in good agreement with model calculations of an increasing greenhouse effect (Akmaev, 2002). That means that the possible greenhouse effect in ionospheric data series is superposed by unknown dynamical processes which are

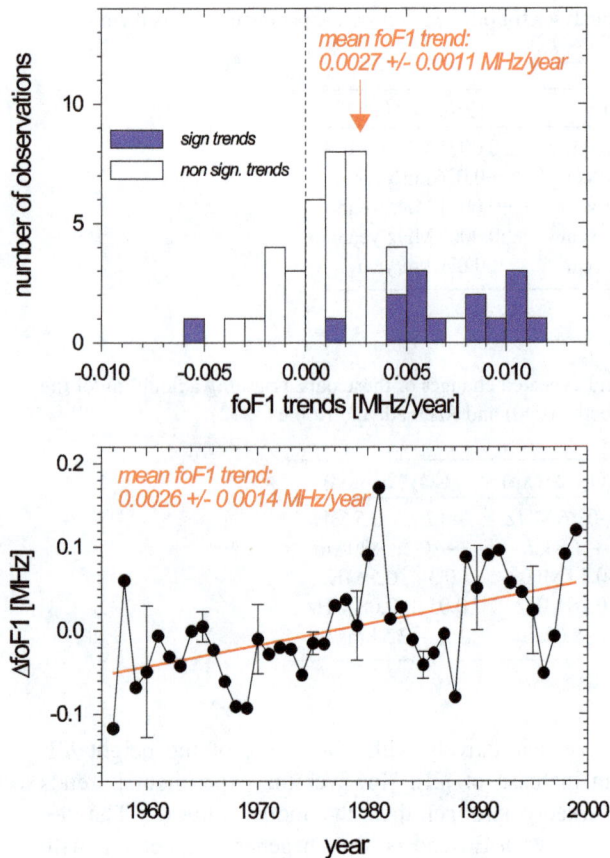

Fig. 3. Mean global foF1 trend deduced from world wide ionosonde observations. Upper part: Histogram of individual trends with median trend marked by an arrow. Lower part: Mean trend deduced from individual trends after elimination of solar and geomagnetic influences.

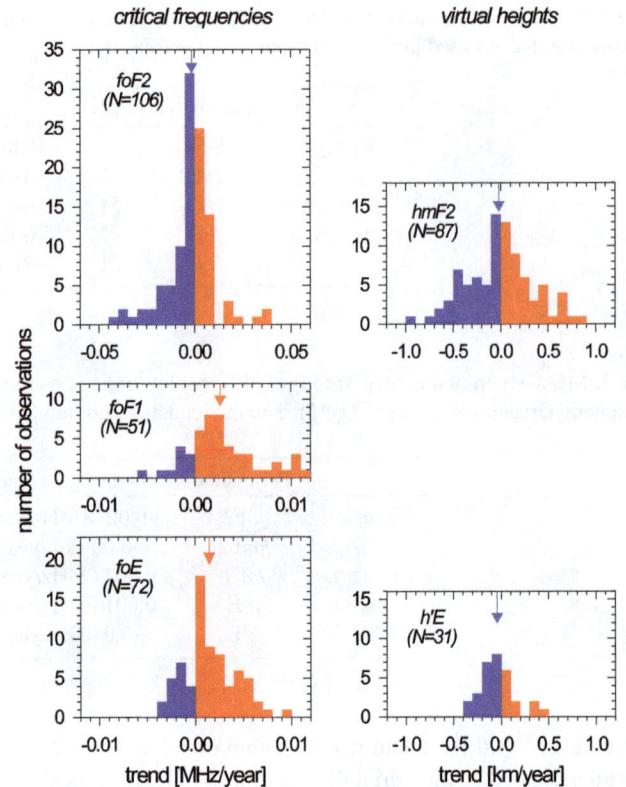

Fig. 4. Histograms of different ionospheric parameters (foF2, hmF2, foF1, foE, h'E) from world wide ionosonde observations. The median values are marked by arrows. The number of ionosonde stations used in the trend analyses are given in brackets.

3   Conclusions

Using data of world wide long-term ionosonde observations trends have been estimated for different characteristic ionospheric parameters in the E-, F1-, and F2-regions. The following conclusions can be given for the most important methodical, practical and scientific aspects:

– Long-term ionosonde data series have carefully to be checked concerning their homogeneity. Discontinuities caused by different technical changes can markedly influence the results of trend analyses.

– The solar and geomagnetically induced variations of the ionospheric parameters are essentially stronger than the long-term trends.

– The long-term trends are unimportant for practical ionospheric prediction models. In such models the influence of the solar variability is the most important external factor.

– The trends in the E-region (lowering of h'E, increase of foE) are in qualitative agreement with an increasing greenhouse effect. However, the experimental trends are stronger than the model results.

more important for the variability of the ionized component than for the neutral gas at F2-region heights.

Another reason of differences between trends of (partly neighbouring) stations may be caused by technical changes during the long ionosonde observation periods or by changes of the evaluation algorithms (one possible effect could be caused by changes from manual scaling to automatic scaling of the ionosonde observations). In Fig. 6 some examples of hmF2 data series are shown with discontinuities. Similar examples can be found in Bremer (2001) for h'E observations.

Taking into account the above mentioned increase of 20% of the greenhouse gases during the last 40 years, the doubling of the greenhouse gases would be expected for a time period of about 200 years. That means the different ionospheric parameters will change by the CO_2(exp) values of Table 2 during the next 200 years. Therefore, the expected mean changes of the different ionospheric parameters are not essential for practical purposes and have not to be considered in practical prediction models e.g. for ionospheric HF propagation.

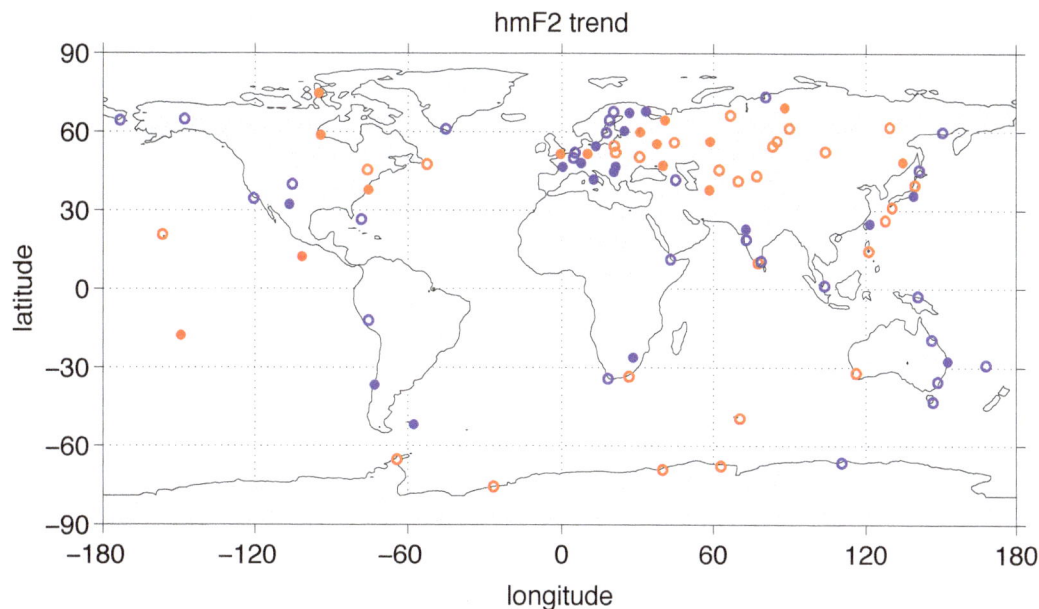

Fig. 5. Trends of *hm*F2 observations at different ionosonde stations in dependence on latitude and longitude. Positive trends: red, negative trends: blue, significant trends: full dots, non significant trends: circles.

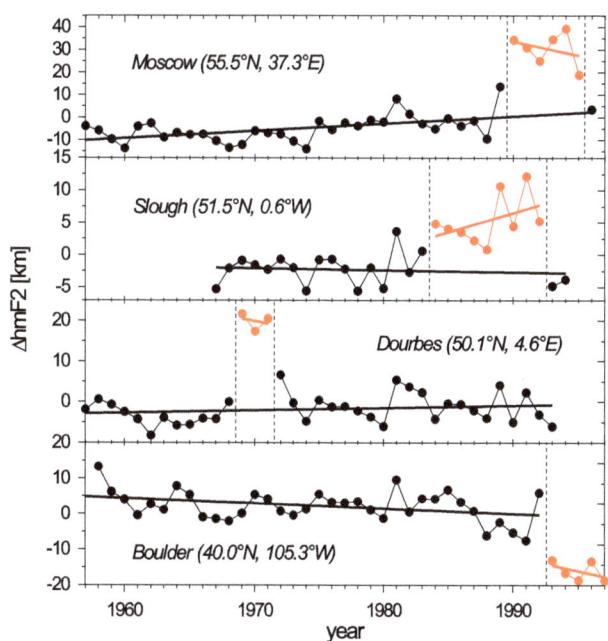

Fig. 6. Examples of individual *hm*F2 trends after elimination of the solar and geomagnetic influences with discontinuities which could be caused by technical changes.

– The mean trend in the F1-region (increase of *fo*F1) agrees quite well with model results of an increasing greenhouse effect.

– Due to a large variability of the individual trends in the F2-region no significant global trends could be derived for *fo*F2 and *hm*F2. Therefore, the relatively reasonable

agreement between the mean global experimental and model results is accidental. The regional differences of the trends hint to an unknown dynamical process which superpose a possible greenhouse effect in the F2-region.

References

Akmaev, R. A.: Modeling of the cooling due to CO_2 increases in the mesosphere and lower thermosphere, Phys. Chem. Earth, 27, 521–528, 2002.

Bencze, P., Sole, G., Alberca, L. F., and Poor, A.: Long-term changes of *hm*F2: possible latitudinal and regional variations, Proc. of the 2nd COST 251 Workshop, 30–31 March 1998, Side, Turkey, RAL UK, 107–113, 1998.

Brasseur, G. and de Rudder, A.: The potential impact on atmospheric ozone and temperature of increasing trace gas concentrations, J. Geophys. Res., 92, 10 903–10 920, 1987.

Bremer, J.: Trends in the ionospheric E- and F-regions over Europe, Ann. Geophys., 16, 986–996, 1998.

Bremer, J.: Trends in the thermosphere derived from global ionosonde observations, Adv. Space Res., 28, 7, 997–1006, 2001

Danilov, A. D. and Smirnova, N. V.: Long-term trends in the ion composition of the E-region (in Russian), Geomagn. Aeron., 37, 4, 35–40, 1997.

Hegerl, C. G., von Storch, H., Hasselmann, K., Sauter, B. D., Cubasch, U., and Jones, P. D.: Detecting greenhouse-gas-induced climate change with an optimal fingerprint method, J. Climate, 9, 2281–2306, 1996.

Houghton, J. T., Ding, Y., Groggs, D. J., Noguer, M., van der Linden, P. J., Dai, X., Maskell, K., and Johnson, C. A.: Climate Change: The Scientific Basis, Contribution of WG I to the 3rd Assessment Report of the IPCC, Cambridge, University Press, 2001.

Keating, G. M., Tolson, R. H., and Bradford, M. S.: Evidence of long term global decline in the Earth's thermospheric densities apparently related to anthropogenic effects, Geophys. Res. Lett., 27, 1523–1526, 2000.

Rishbeth, H.: A greenhouse effect in the ionosphere?, Planet. Space Sci., 38, 945–948, 1990.

Rishbeth, H. and Roble, R. G.: Cooling of the upper atmosphere by enhanced greenhouse gases – Modelling of the thermospheric and ionospheric effects, Planet. Space Sci., 40, 1011–1026, 1992.

Roble, R. G. and Dickinson, R. E.: How will changes of carbon dioxide and methane modify the mean structure of the mesosphere and thermosphere?, Geophys. Res. Lett., 16, 1441–1444, 1989.

Shimazaki, T.: World wide daily variations in the height of the maximum electron density in the ionospheric F2-layer, J. Radio Res. Labs., Japan, 2, 85–97, 1955.

Taubenheim, J.: Statistische Auswertung geophysikalischer und meteorologischer Daten, Akad. Verlagsgesellschaft Geest und Portig K.-G., Leipzig, 1969.

Ulich, Th. and Turunen, E.: Long-term behaviour of ionospheric F2-layer peak height on a global scale, Paper presented at Session 2.18 of the 8th Scientific Assembly of IAGA, Uppsala, 1997.

5

Measurements of Cosmic Magnetism with LOFAR and SKA

R. Beck

Max-Planck-Institut für Radioastronomie, Auf dem Hügel 69, 53121 Bonn, Germany

Abstract. The origin of magnetic fields in stars, galaxies and clusters is an open problem in astrophysics. The next-generation radio telescopes *Low Frequency Array (LOFAR)* and *Square Kilometre Array (SKA)* will revolutionize the study of cosmic magnetism. "The origin and evolution of cosmic magnetism" is a key science project for SKA. The planned all-sky survey of Faraday rotation measures (RM) at 1.4 GHz will be used to model the structure and strength of the magnetic fields in the intergalactic medium, the interstellar medium of intervening galaxies, and in the Milky Way. A complementary survey of selected regions at around 200 MHz is planned as a key project for LOFAR. *Spectro-polarimetry* applied to the large number of spectral channels available for LOFAR and SKA will allow to separate RM components from distinct foreground and background regions and to perform 3-D *Faraday tomography* of the interstellar medium of the Milky Way and nearby galaxies. – Deep polarization mapping with LOFAR and SKA will open a new era also in the observation of synchrotron emission from magnetic fields. LOFAR's sensitivity will allow to map the structure of weak, extended magnetic fields in the halos of galaxies, in galaxy clusters, and possibly in the intergalactic medium. Polarization observations with SKA at higher frequencies (1–10 GHz) will show the detailed magnetic field structure within the disks and central regions of galaxies, with much higher angular resolution than present-day radio telescopes.

contribute significantly to the total pressure of interstellar gas, are essential for the onset of star formation, and control the density and distribution of cosmic rays in the interstellar medium (ISM) and in the intracluster medium (ICM). In spite of their importance, the *evolution*, *structure* and *origin* of magnetic fields are all still open problems in fundamental physics and astrophysics. When and how were the first magnetic fields in the Universe generated? Was there is a connection between magnetic field formation and structure formation in the early Universe? Were the fields in young galaxies and clusters primordial or generated in the galaxies themselves? How did magnetic fields evolve as galaxies evolve? What are the strength and structure of the magnetic field of the intergalactic medium (IGM)?

Most of what we know about astrophysical magnetic fields comes through the detection of radio waves. *Synchrotron emission* measures the total field strength, while its *polarization* yields the orientation of the regular field in the sky plane and also gives the field's degree of ordering (Figs. 4–6). *Faraday rotation* of the polarization vector when the wave passes through magnetized plasma gives a measure of the regular field along the line of sight (Fig. 2). The combination yields a three-dimensional view of the regular field. The *Zeeman effect* provides an independent measure of field strength in cold gas clouds. However, measuring astrophysical magnetic fields is a difficult topic, restricted to nearby or bright objects when observing with present-day radio telescopes.

1 Introduction

Understanding the Universe is impossible without understanding magnetic fields. They fill interstellar and intracluster space, affect the evolution of galaxies and galaxy clusters,

2 LOFAR

The Low Frequency Array (LOFAR) is a new-generation *phased array* radio telescope which will observe in the frequency bands 30–80 MHz and 110–240 MHz. It is under construction in the Netherlands (see www.lofar.org and www.lofar.de). The first international station is under con-

Correspondence to: R. Beck
(rbeck@mpifr-bonn.mpg.de)

Fig. 1. SKA reference design (SKA Project Office and XILOSTUDIOS).

struction next to the Effelsberg 100-m telescope (Reich, this volume).

Low-frequency radio emission traces low-energy cosmic-ray electrons which suffer less from energy losses and hence can propagate further away from their sources into regions with weak magnetic fields. The lifetime of electrons in galaxies is generally limited by synchrotron losses and increases with decreasing frequency and decreasing field strength. In a 5 μG field the lifetime of electrons emitting in the LOFAR bands is $(2–5) \cdot 10^8$ yr. In magnetic fields weaker than $3.25\ \mu\text{G} \cdot (z+1)^2$ (where z is the redshift) the electron lifetime is limited by the Inverse Compton effect with the photons of the cosmic micowave background (CMB), so that the lifetime of electrons observed at frequency ν decreases weakly with decreasing field strength. Electrons with maximum lifetimes of $\sim 10^9$ yr can be observed at 50 MHz in fields of about 3 μG strength. Travelling through the hot gas in galactic halos or clusters with the Alfvén speed, they can propagate more than 100 kpc from their places of origin.

LOFAR will give access to the so far totally unexplored domain of weak magnetic field strengths. For a fixed signal-to-noise ratio of polarized intensity, the minimum detectable strength of the regular field is proportional to ν^2. Hence, ob-

serving diffuse polarized emission at low frequencies with LOFAR will reveal objects with weak fields such as galaxy halos and galaxy clusters. Faraday rotation increases with ν^{-2}, so that LOFAR will be able measure very small rotation measures (RM) and detect weak magnetic fields and low electron densities which are unobservable with present-day telescopes and even at the higher frequencies of SKA. Faraday rotation of polarized background sources with LOFAR may even allow to detect magnetic fields in the intergalactic medium (Sect. 9).

Four key science projects lead by Dutch scientists are under development on the following topics: The epoch of reionization, extragalactic surveys, transients and pulsars, and cosmic rays. The German Long Wavelength Consortium (GLOW) plans to perform further key projects on solar radio astronomy, jets, and cosmic magnetism.

3   SKA

The Square Kilometre Array (SKA) is an international project for the next decade to build a huge radio telescope with a collecting area of about one square kilometre, dis-

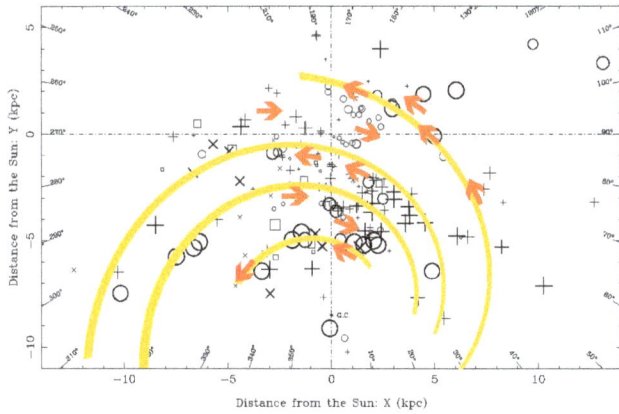

Fig. 2. Bird's eye view of the distribution of the Faraday rotation measures (RM) of pulsars within 8° of the Galactic plane. Positive RMs are shown as crosses and X, negative RMs as circles and open squares. The symbol sizes are proportional to the square root of $|RM|$, with the limits of 5 and 250 rad/m^2. The directions of a field model are given as arrows. The approximate location of four optical spiral arms is indicated as dotted lines (from Han et al., 1999).

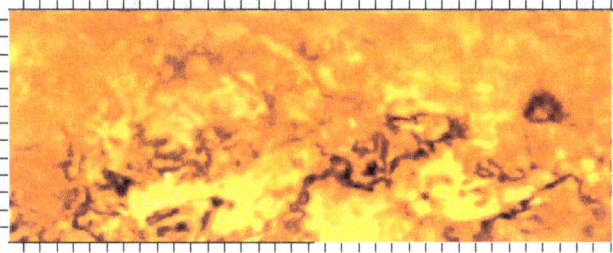

Fig. 3. Polarized intensity around the plane of the Milky Way ($l = 150° - 174°$, $b = -4.5° - +4.5°$) at 1.4 GHz, combined from data of the Effelsberg and Dwingeloo radio telecopes (from Reich et al., 2004).

Fig. 4. Total radio emission (contours) and magnetic field vectors at 4.8 GHz, combined from observations with the VLA and the Effelsberg telescope. The underlying image is from the Hubble Space Telescope (A. Fletcher and R. Beck / Hubble Heritage Team, graphics by the magazine "Sterne und Weltraum").

tributed over a large number of small antennas concentrated in several stations (Fig. 1). Requirements for the SKA design include a full frequency coverage of 0.3–20 GHz and a field of view at 1.4 GHz of at least 1 deg^2 which can be fully imaged at 1$''$ resolution. The frequency range 0.3–1 GHz will be covered by phased arrays, based on the experience with LOFAR and SKA prototypes, while classical parabolic dishes are considered at higher frequencies. A significant fraction of the collecting area will be concentrated into the central core of diameter ∼5 km, and longest baselines of ∼3000 km are planned. The site selection process has identified South Africa and Western Australia as the best sites for the SKA core and inner stations, with extensions to West and East Africa and to New Zealand, respectively. Construction of Phase 1 with 10% collecting area is planned for 2012, construction of the full array for 2015. SKA prototypes are un-

der construction in several countries. The European Community has started the SKA Design Study (SKADS) programme which involves 29 institutes from 12 countries.

Six SKA key science projects have been selected on the followings topics: The Dark Ages, galaxy evolution and large-scale structures, testing theories of gravitation, the Cradle of Life, cosmic magnetism, and exploration of the unknown. A science case book was published in 2004 (Carilli and Rawlings, 2004).

For more information see www.skatelescope.org.

4 Faraday rotation and spectro-polarimetry

Much of what LOFAR and SKA can contribute to our understanding of magnetic fields will come from their *polarimetric capabilities*. The crucial specifications are high polarization purity and multichannel spectro-polarimetric capability.

The former will allow detection of the relatively low linearly polarized fractions ($\leq$1%) from most astrophysical sources, while the latter will enable accurate measurements of Faraday rotation measures (RMs), intrinsic polarization position angles and Zeeman splitting.

With current instruments, the only way to simultaneously determine accurate values for both the RM and intrinsic polarization position angle is to make several observations at various frequencies across a broad frequency range. This is not only time-consuming, but the analysis must proceed cautiously, since various depolarizing effects have a strong frequency dependence (Sokoloff et al., 1998).

The high sensitivity and broad bandwidth of LOFAR and SKA eliminate these difficulties: a single spectro-polarimetric observation at a single IF can simultaneously provide good estimates of both RM and position angle, the limiting factor being the accuracy of ionospheric corrections to the observed Faraday rotation. For example, at an observing frequency of 1.4 GHz with a fractional bandwidth of 25%, a 1-min SKA observation of a source with a linearly polarized surface brightness of $\sim$8 μJy beam^{-1} will yield a RM determined to an accuracy $\Delta RM \approx \pm 5$ rad m^{-2}, sufficient to measure a regular field of 5 μG strength and 100 pc correlation length in a plasma of 0.05 cm^{-3} electron density.

Making use of its wide-band polarization facility, LOFAR can detect even smaller rotation measures, possibly down to about 0.1 rad m^{-2} (giving 36° total rotation at 120 MHz and 13° at 200 MHz), and hence will become the telescope to measure the weakest cosmic magnetic fields so far.

In the relatively strong magnetic fields of the interstellar medium in the Milky Way and other galaxies, low-frequency polarization is generally low due to various depolarization effects (Sokoloff et al., 1998). The method of *RM synthesis*, based on multichannel spectro-polarimetry, allows to measure a large range of RM values and to separate RM components from distinct regions along the line of sight (Brentjens and de Bruyn, 2005). This can be used for *Faraday tomography* of the interstellar medium in the Milky Way and in the disks and central regions of nearby galaxies.

5 Rotation measure surveys

Currently $\sim$1800 extragalactic sources and $\sim$300 pulsars have measured RM data (Kronberg, 1994; Brown et al., 2003; Han et al., 2006). These have proved useful probes of magnetic fields in the Milky Way, in nearby galaxies, and in clusters. However, the sampling of such measurements over the sky is very sparse.

With its low frequency range and wide-band polarization facility, LOFAR can detect weak rotation measures possibly down to about 0.1 rad m^{-2} (Sect. 4). An RM survey of selected fields around 200 MHz is the ideal project to measure weak magnetic fields and to develop the analysis tools needed for SKA. However, the RM contribution from the Galactic foreground in the Galactic plane are much larger than the RMs from galaxy halos, clusters, and the IGM. Application of RM synthesis (Sect. 4) with sufficiently large RM range and RM resolution will be essential to separate the RM components along the line of sight.

The key platform on which to base the SKA's studies of cosmic magnetism will be the *all-sky RM survey* at 1.4 GHz, in which spectro-polarimetric continuum imaging of 10 000 deg^2 of the sky can yield RMs for approximately 2×10^4 pulsars and 2×10^7 compact polarized extragalactic sources (see Gaensler et al., 2004 for a detailed description). This data set will provide a grid of RMs at a mean spacing of $\sim$30$'$ between pulsars and just $\sim$90$''$ between extragalactic sources.

The structure of the magnetic field in our Milky Way can be determined with help of pulsar RMs. The analysis of current data indicates a "bisymmetric" field with several field reversals (Fig. 2, Han et al., 1996, 2006). However, the uncertainties are still large. The large sample of pulsar RMs obtained with the SKA, combined with distance estimates to these sources from parallax or from their dispersion measures, can be inverted to yield a complete delineation of the magnetic field in the spiral arms and disk on scales $\geq$100 pc (Stepanov et al., 2002). Furthermore, the magnetic field geometry in the Galactic halo and the outer parts of the disk can be studied using the all-sky RM grid.

With the sensitivity of the SKA, deep observations of nearby galaxies and galaxy clusters can provide a huge number of background RMs and thus allow detailed maps of the magnetic structure. A large sample of objects could be studied in this way.

6 Polarization mapping of cosmic magnetic fields

The unprecedented sensitivity of LOFAR and SKA will allow to map the polarized synchrotron emission to much fainter levels and/or with much better resolution than with present-day telescopes. While LOFAR will concentrate on the weak, diffuse emission from galaxy halos and the intracluster medium of galaxy clusters, SKA will trace detailed field structures in the Milky Way and in galaxies.

Small-scale structures and turbulence in the interstellar medium of the Milky Way can be probed using *RM synthesis* (Sect. 4). Foreground ionized gas generates frequency-dependent Faraday features when viewed against diffuse Galactic polarized radio emission (Fig. 3, Wolleben et al., 2006; Reich, 2006).

In external galaxies and galaxy clusters, interstellar and intracluster magnetic fields can be directly traced by mapping the diffuse synchrotron emission and its polarization (Beck and Gaensler, 2004; Beck, 2005, Fig. 4). Combined with determinations of RMs for extended emission and background sources, this will allow us to derive detailed three-dimensional maps of the magnetic fields in these sources.

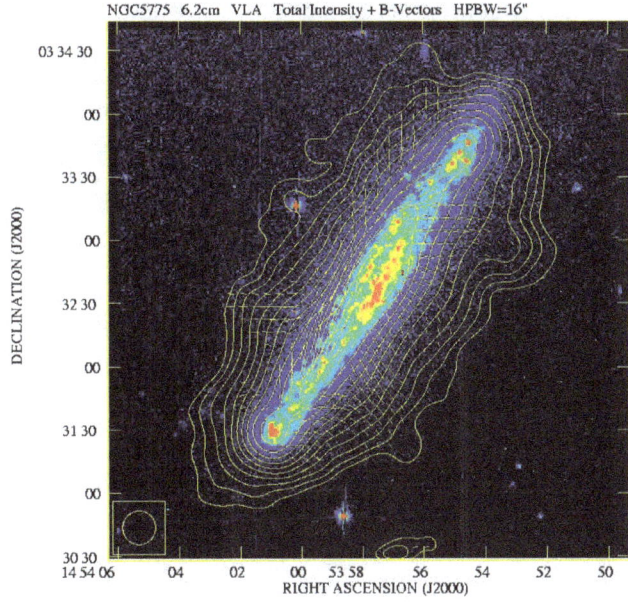

Fig. 5. Total radio emission (contours) and polarization B-vectors of the spiral galaxy NGC 5775, observed at 4.8 GHz with the VLA (from Tullmann et al., 2000).

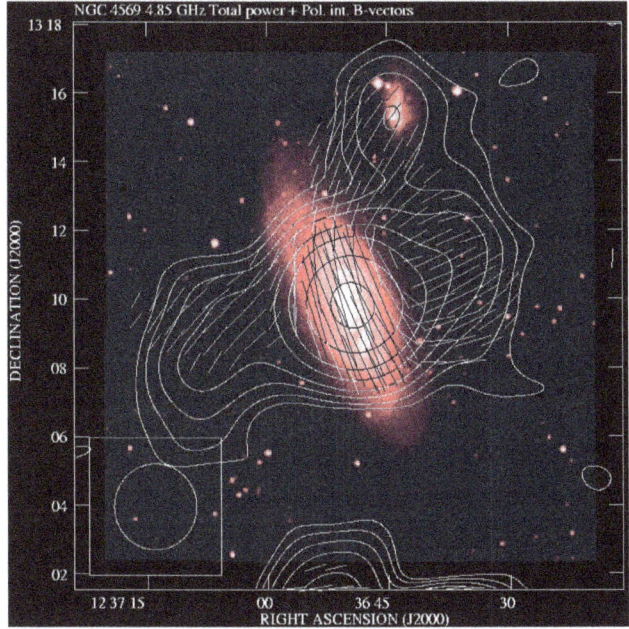

Fig. 6. Total radio emission (contours) and polarization B-vectors of the spiral galaxy NGC 4569 in the Virgo Cluster, observed at 4.8 GHz with the Effelsberg telescope (from Chyży et al., 2006).

SKA will provide sufficient resolution and sensitivity to identify individual features. Of special interest are the spectrum of magnetic turbulence, the number and location of magnetic reversals, the relation between magnetic and optical spiral arms, and the interaction between magnetic fields and gas flows.

LOFAR will show the extension of galaxies into the halo and the intergalactic space which emerge as a result of galactic winds or of interactions, phenomena which are observable only in a few cases with the most sensitive present-day telescopes (Figs. 5 and 6). LOFAR will also allow to trace the full extent of the magnetized halos of galaxy clusters.

7 The origin of magnetic fields in galaxies and clusters

The observation of large-scale patterns in RM in many galaxies (Beck, 2005) proves that some fraction of the magnetic field in galaxies has a coherent direction and hence is not generated by compression or stretching of irregular fields in gas flows. In principle, the *turbulent dynamo mechanism* is able to generate and preserve coherent magnetic fields, and they are of appropriate spiral shape (Beck et al., 1996). However, the physics of dynamo action is far from being understood. Primordial fields, on the other hand, are hard to preserve over a galaxy's lifetime due to diffusion, reconnection, and winding up by differential rotation. Even if they survive, they can create only specific field patterns.

The widely studied *mean-field $\alpha-\Omega$ dynamo* needs differential rotation and helical turbulence to operate. It gener-

ates a coherent magnetic field which can be represented as a superposition of modes of different azimuthal and vertical symmetries. The existing dynamo models predict that several azimuthal modes can be excited (Beck et al., 1996), the strongest being $m=0$ (an axisymmetric spiral field), followed by the weaker $m=1$ (a bisymmetric spiral field), etc. These generate a Fourier spectrum of azimuthal RM patterns. Observations of background RMs and of diffuse polarized synchrotron emission allow direct measurements of the azimuthal modes, but these measurements are limited at present to ~20 galaxies (Beck, 2000). With SKA, this sample can be increased by up to three orders of magnitude. These data can allow us to distinguish between different conditions for excitation of various dynamo modes. The presence and prevalence of reversals in the disk field structure, plus the structure of field in the halo, will together let us distinguish between dynamo and primordial models for field origin in galaxies (Beck, 2006).

Similarly, clarifying the origin of intracluster fields in galaxy clusters is a project of fundamental importance for LOFAR and SKA. Regular fields could be generated by dynamo action, cluster mergers, shock waves, or interactions between galaxies (Dolag, 2006; Shukurov et al., 2006).

8 Magnetic fields in distant galaxies

Measurements of magnetic fields in distant galaxies (at redshifts between $z\sim0.1$ and $z\sim2$) with LOFAR and SKA will provide direct information on how magnetized structures evolve and amplify as galaxies mature. The linearly polarized emission from galaxies at these distances will often be too faint to detect directly; Faraday rotation thus holds the key to studying magnetism in these sources. There are many distant, extended polarized sources (quasars and radio galaxies), providing the ideal background illumination for probing Faraday rotation in galaxies which happen to lie along the same line of sight. These experiments can deliver maps of magnetic field structures in galaxies more than 100 times more distant than discussed above.

At yet larger distances, we can take advantage of the sensitivity of the deepest SKA fields, in which we expect to detect the synchrotron emission from the youngest galaxies and proto-galaxies. Since standard dynamos need a few rotations or about 10^9 yr to build up a coherent galactic field (Beck et al., 1996), the detection of synchrotron emission in young galaxies would put constraints on the seed field which may call for alternative models.

9   Intergalactic magnetic fields

Fundamental to all the issues discussed above is the search for magnetic fields in the intergalactic medium (IGM). All of "empty" space in the Universe may be magnetized. Its role as the likely seed field for galaxies and clusters, plus the prospect that the IGM field might trace structure formation in the early Universe, places considerable importance on its discovery. A magnetic field already present at the epoch of re-ionization or even at the recombination era might have affected the processes occurring at those epochs (Subramanian, 2006). To date there has been no detection of magnetic fields in the IGM; current upper limits on the average strength of any such field suggest $|B_{IGM}| \lesssim 10^{-8}-10^{-9}$ G (Kronberg, 1994).

With LOFAR it will be possible to search for synchrotron radiation at the lowest possible levels in intergalactic space. Its detection will allow to probe the existence of magnetic fields in rarified regions of the intergalactic medium, measure their intensity, and investigate their origin and their relation to the structure formation in the early Universe. Fields of $B \simeq 10^{-9}-10^{-8}$ G are expected along filaments of 10 Mpc length with $n_e \simeq 10^{-5}$ cm^{-3} electron density (Kronberg, 2006) which yield Faraday rotation measures of RM=0.1–1 rad m^{-2}. Their detection is a big challenge, but possible. LOFAR has a realistic chance to measure **intergalactic magnetic fields for the first time**.

If this all-pervading magnetic field will turn out to be even weaker, it may still be identified through the all-sky RM grid with the SKA (Sect. 5). The correlation function of the RM distribution provides the magnetic power spectrum of the IGM as a function of cosmic epoch (Blasi et al., 1999). Such measurements will allow us to develop a detailed model of

the magnetic field geometry of the IGM and of the overall Universe.

Primordial fields existing already in the recombination era would induce Faraday rotation of the polarized CMB signals of the cosmic microwave background (CMB) (Kosowsky and Loeb, 1996) and generate a characteristic peak in the CMB power spectrum at small angular scales. The detection is challenging but possible with an instrument of superb sensitivity like SKA.

References

Beck, R.: Magnetic fields in normal galaxies, Phil. Trans. R. Soc. Lond. A, 358, 777–796, 2000.

Beck, R.: Magnetic fields in galaxies, in: Cosmic Magnetic Fields, edited by: Wielebinski, R. and Beck, R., Springer, Berlin, 41–68, 2005.

Beck, R.: The origin of magnetic fields in galaxies, Astr. Nachr., 327, 512–516, 2006.

Beck, R. and Gaensler, B.M.: Observations of magnetic fields in the Milky Way and in nearby galaxies with a Square Kilometre Array, in: Science with the Square Kilometre Array, edited by: Carilli, C. and Rawlings, S., New Astr. Rev., 48, 1289–1304, 2004.

Beck, R., Brandenburg, A., Moss, D., Shukurov, A., and Sokoloff, D.: Galactic magnetism: Recent developments and perspectives, Ann. Rev. Astron. Astrophys., 34, 155–206, 1996.

Blasi, P., Burles, S., and Olinto. A.V.: Cosmological magnetic field limits in an inhomogeneous Universe, Astrophys. J., 514, L79–L82, 1999.

Brentjens, M.A., and de Bruyn, A.G.: Faraday rotation measure synthesis, Astron. Astrophys., 441, 1217–1228, 2005.

Brown, J.C., Taylor, A.R., and Jackel, B.J.: Rotation measures of compact sources in the Canadian Galactic Plane Survey, Astrophys. J. Suppl., 145, 213–223, 2003.

Carilli, C., and Rawlings, S.: Science with the Square Kilometre Array, New Astr. Rev., 48, 2004.

Chyży, K.T., Soida, M., Bomans, D.J., et al.: Large-scale magnetized outflows from the Virgo cluster spiral NGC 4569, Astron. Astrophys., 447, 465–472, 2006.

Dolag, K.: Simulating large-scale structure formation with magnetic fields, Astr. Nachr., 327, 575–582, 2006.

Gaensler, B.M., Beck, R., and Feretti, L.: The origin and evolution of cosmic magnetism, in: Science with the Square Kilometre Array, edited by: Carilli, C. and Rawlings, S., New Astr. Rev., 48, 1003–1012, 2004.

Han, J.L., Manchester, R.N., and Qiao, G.J.: Pulsar rotation measures and the magnetic structure of our Galaxy, Mon. Not. Royal Astr. Soc., 306, 371–380, 1999.

Han, J.L., Manchester, R.N., Lyne, A.G., Qiao, G.J., and van Straten, W.: Pulsar rotation measures and the large-scale structure of the Galactic magnetic field, Astrophys. J., 642, 868–881, 2006.

Kosowsky, A. and Loeb, A.: Faraday rotation of microwave background polarization by a primordial magnetic field, Astrophys. J., 469, 1–6, 1996.

Kronberg, P.P.: Extragalactic magnetic fields, Rep. Prog. Phys., 57, 325–382, 1994.

Kronberg, P. P.: Extragalactic radio sources, IGM magnetic fields, and AGN-based energy flows, Astr. Nachr., 327, 517–522, 2006.

Reich, W.: Galactic polarization surveys, astro-ph/0603465, 2006.

Reich, W., Fürst, E., Reich, P., Uyanıker, B., Wielebinski, R., and Wolleben, M.: The Effelsberg 1.4 GHz medium Galactic latitude survey (EMLS), in: The Magnetized Interstellar Medium, edited by: Uyanıker et al., Copernicus, Katlenburg, p. 45, 2004.

Shukurov, A., Subramanian, K., and Haugen, N. E. L.: The origin and evolution of cluster magnetism, Astr. Nachr., 327, 583–586, 2006.

Sokoloff, D. D., Bykov, A. A., Shukurov, A., Berkhuijsen, E. M., Beck, R., and Poezd, A. D.,: Depolarization and Faraday effects in galaxies, Mon. Not. Royal Astr. Soc., 299, 189–206, 1998, and Mon. Not. Royal Astr. Soc., 303, 207–208, 1999 (Erratum).

Stepanov, R., Frick, P., Shukurov, A., and Sokoloff, D.: Wavelet tomography of the Galactic magnetic field, Astron. Astrophys., 391, 361–368, 2002.

Subramanian, K.: Primordial magnetic fields and CMB anisotropies, Astr. Nachr., 327, 403–409, 2006.

Tüllmann, R., Dettmar, R.-J., Soida, M., Urbanik, M., and Rossa, J.: The thermal and non-thermal gaseous halo of NGC 5775, Astron. Astrophys., 364, L36–L41, 2000.

Wolleben, M., Landecker, T. L., Reich, W., and Wielebinski, R., An absolutely calibrated survey of polarized emission from the northern sky at 1.4 GHz, Astron. Astrophys., 448, 411–424, 2006.

6

Numerical modeling of solar wind influences on the dynamics of the high-latitude upper atmosphere

M. Förster[1], **B. E. Prokhorov**[1,2], **A. A. Namgaladze**[3], **and M. Holschneider**[2]

[1]GFZ German Research Centre for Geosciences, Helmholtz Centre Potsdam, Germany
[2]University Potsdam, Institute for Applied Mathematics, Potsdam, Germany
[3]Murmansk State Technical University, Murmansk, Russia

Correspondence to: M. Förster (mfo@gfz-potsdam.de)

Abstract. Neutral thermospheric wind patterns at high latitudes obtained from cross-track acceleration measurements of the CHAMP satellite above both polar regions are used to deduce statistical neutral wind vorticity distributions and were analyzed in their dependence on the Interplanetary Magnetic Field (IMF). The average pattern confirms the large duskside anticyclonic vortex seen in the average wind pattern and reveals a positive (cyclonic) vorticity on the dawnside, which is almost equal in magnitude to the duskside negative one. The IMF dependence of the vorticity pattern resembles the characteristic field-aligned current (FAC) and ionospheric plasma drift pattern known from various statistical studies obtained under the same sorting conditions as, e.g., the EDI Cluster statistical drift pattern. There is evidence for hemispheric differences in the average magnitudes of the statistical patterns both for plasma drift and even more for the neutral wind vorticity. The paper aims at a better understanding of the globally interconnected complex plasma physical and electrodynamic processes of Earth's upper atmosphere by means of first-principle numerical modeling using the Upper Atmosphere Model (UAM). The simulations of, e.g., thermospheric neutral wind and mass density at high latitudes are compared with CHAMP observations for varying IMF conditions. They show an immediate response of the upper atmosphere and its sensitivity to IMF changes in strength and orientation.

1 Introduction

More than fifty years since the start of the space era with the first satellites, the efforts to understand our space plasma environment and the complexities of its link to solar activities,

propagating by the solar wind past the Earth, increase continuously. The dependencies are highly complex and many of them are still not understood in spite of many decades of research with several generations of space missions.

Understanding this complexity is a fundamental problem in physics. This is the subject of space weather science, a relatively new field of research. The term space weather is defined as the conditions on the Sun and in the solar wind, Earths magnetosphere, ionosphere, and thermosphere that can influence the performance and reliability of space-borne and ground-based technological systems and endanger human life or health. As our modern society inexorably increases its dependence on space, the necessity of predicting and mitigating space weather becomes ever more acute (Eastwood, 2008).

This study aims at a better understanding of the coupled magnetosphere-ionosphere-thermosphere (M-I-T) system including its electrodynamics. For the present paper, we confine to investigations of solar wind and IMF forcing processes at the high-latitude upper atmosphere, conveyed by the complex magnetospheric current system and their concomitant electric fields. Beside the dayside thermal heating due to solar EUV and X-ray illumination, the auroral energy sources of particle precipitation and Joule heating, they constitute a main driver for the high latitude upper thermosphere.

The thermospheric mass density and wind have been measured with a relatively new technique, namely an accelerometer as part of the CHAllenging Minisatellite Payload (CHAMP) mission. CHAMP operated throughout the last decade in a near-polar and almost circular orbit near the F-region maximum height. The satellite was developed and managed at the GFZ German Center of Geosciences in Potsdam and turned out to be an exceedingly

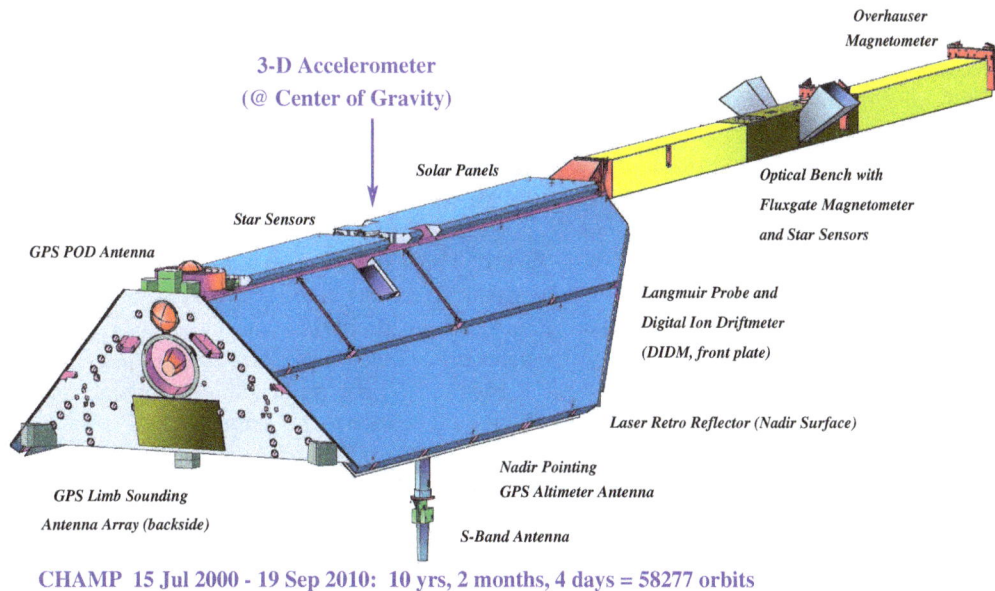

Fig. 1. CHAMP satellite and its equipment (backside view). It wrote a track record of more than one decade in near-Earth orbit.

successful near-Earth geoscience mission. It was launched in summer 2000 into an orbit at $\sim$460 km with an inclination of $\sim$87.3° (Reigber et al., 2002) and completed its active measurement lifetime in September 2010 at approximately 260 km orbit height. It was equipped with a payload suitable for monitoring many parameters relevant for space weather research and application (see Fig. 1). The forthcoming multi-satellite constellation mission Swarm with a similar diagnostics payload will allow even more profound studies of the Earth's magnetic field and the near-Earth environment including space weather aspects and the upper thermosphere dynamics (Friis-Christensen et al., 2006, 2009).

One key scientific instrument onboard CHAMP was the triaxial accelerometer, located at the spacecraft's center of mass. It effectively sampled the in situ acceleration with an accuracy of $\sim$3 × 10^{-9}ms^{-2}. From the air drag observations, thermospheric mass density and cross-track neutral wind data have been obtained (Doornbos et al., 2010).

First-principle physical-numerical models are nowadays mandatory tools for solving forecasting challenges and for a deeper understanding of the complex interrelated processes in Earth's environment. Modern first-principle numerical models are time dependent, three dimensional (3-D), and global. In this study we will use the Upper Atmosphere Model (UAM) of the Polar Geophysical Institute (PGI) and the Murmansk State Technical University (MSTU).

We had made use of this model already during the preparatory phase of the CHAMP mission as part of the theoretical grounding studies for CHAMP's observational facilities and their applicability for space weather monitoring tasks, as well as for optimal measurement ranges and precision require-

ments (Förster et al., 1999a). On that occasion, we studied a moderate storm event and its global-scale implications for neutral air density, composition and wind effects at subauroral to low latitudes as well as the electromagnetic changes with regard to their observability by the instrumentation of CHAMP (Förster et al., 1999b; Namgaladze et al., 2000).

Within the last decade the actual observational outcome of the CHAMP mission is tremendous and the abundance of its data wealthiness is far from being exhausted yet. The CHAMP data provide comprehensive evidences both for influences by solar wind and magnetospheric processes from above (e.g. Förster et al., 2008, 2011) and for upward propagating disturbances from below (e.g. Häusler and Lühr, 2009; Lühr et al., 2011). All these coupling processes are crucial to our understanding of the system's dynamics and variability.

In this study we will focus on global M-I-T coupling effects at high-latitudes. First we present some statistical results of CHAMP's neutral wind accelerometer measurements, which show the systematic dependence of upper thermospheric winds on IMF conditions in the solar wind (Sect. 2). Then we describe the UAM model (Sect. 3) and the specific challenges for modeling the near-Earth space plasma environment and the coupling of magnetospheric processes within the high-latitude ionosphere and thermosphere (Sect. 4). We present our first attempts for systematic analyses of M-I-T coupling processes under various stable IMF conditions (Sect. 5) and conclude the paper with a discussion and an outlook for future modeling efforts (Sect. 6).

2 Average neutral wind vorticity at high latitudes

In the classical M-I-T coupling theory (as reviewed, e.g., by Cowley, 2000), the ionosphere/thermosphere end is treated as a height-integrated boundary of the magnetosphere, where field-aligned current (FAC) systems mediate stress and energy with the magnetosphere-magnetosheath current generation system. It is common view that this coupling occurs by means of ion drag between ionospheric ions set in motion by solar wind induced magnetospheric convection and thermospheric neutrals. Likewise it has long been known that the high-latitude thermospheric wind is stongly influenced by solar wind conditions, in particular by direction and strength of the IMF. This has been shown by satellite missions like Dynamics Explorer (DE-2) back in the early 80-ies already (Hays et al., 1984; Killeen and Roble, 1988; Rees and Fuller-Rowell, 1989; Killeen et al., 1995).

An indirect mechanism of energy transfer from the solar wind to the upper atmosphere was recently brought up by Siscoe and Siebert (2006). It involves the Region–1 FAC system and the geomagnetic field configuration in such a way that the $J \times B$ force acting against the solar wind at the high-altitude end of the Region–1 current loop is transmitted to the Earth as a $J \times B$ force acting on the thermosphere (Siscoe, 2006). Vasyliunas (2007) compares this Lorentz force mechanism as an analog to mechanical leveraging and quotes it the "mechanical advantage" of the magnetosphere in the M-I-T system, resulting in an increased thermospheric drag by nearly an order of magnitude over the direct drag mechanism.

The thermospheric wind moves, on the other hand, the conducting ionized layers across the geomagnetic field lines, resulting in a neutral wind dynamo effect that contributes to the overall electrodynamics of the coupled system (Blanc and Richmond, 1980). Moreover, the inertia of the upper atmosphere's motion can help to maintain the ionospheric convection independently of the magnetospheric driver processes which is also known for long time as "fly-wheel effect" of the upper atmosphere (Banks, 1972; Coroniti and Kennel, 1973). The coupled M-I-T system will adjust the various driving processes to result in the plasma convection and thermospheric wind pattern, which we observe.

The complex dynamic response of the coupled processes can be addressd by first-principle global numerical modelings as exemplified later on using the UAM (Sect. 3–5). But first we will show average patterns of the high-latitude upper atmosphere dynamics (Figs. 2–4), in particular its vorticity, which is directly related to these external drivers (see also Förster et al., 2008, 2011). Statistical patterns, like in this study, represent some average behaviour, which might be far from equilibrium conditions within the highly dynamic M-I-T system.

The comprehensive CHAMP accelerometer data set and the novel methodology of its data analysis (Doornbos et al., 2010) allow detailed studies of the high-latitude thermospheric wind behaviour at F region heights. The method to obtain statistically averaged full vector patterns of the high-latitude thermospheric wind from the cross-track (one-component) accelerometer measurements has been explained by Förster et al. (2008). In short, the reconstruction of the full horizontal wind vector is done with a Singular Value Decomposition (SVD) method which combines the multiple component measurements of a given bin with their known directions to get the best-fit wind vector estimation for each bin individually. The novelty of our study, based on the well-founded statistics, consists in the first systematic analysis of the thermospheric vorticity in polar regions of both hemispheres in dependence on the IMF orientation.

Figure 2, upper panels, shows the statistically averaged high-latitude horizontal thermospheric wind patterns both for the Northern (left side) and Southern Hemisphere (right) during the years 2002/03 of moderate to high solar activity. We used magnetic coordinates, specifically Altitude Adjusted Corrected Geomagnetic (AACGM) coordinates (e.g., Weimer, 2005, Appendix A), because of the strong geomagnetic control of high-latitude thermospheric dynamics, that has been shown already by early satellite observations (e.g. Hays et al., 1984), and we binned the neutral wind data in the same way as the plasma drift measurements obtained from the Cluster/EDI data over the same time period (Haaland et al., 2007; Förster et al., 2007), in order to make them well comparable. The MLT/magnetic latitude projection of the averaged thermospheric neutral wind vectors is presented here in the same viewing direction for both polar regions, namely such as looking from atop the northern pole downward.

The bottom panels of Fig. 2 present the corresponding neutral wind vorticity, deduced from the circulation patterns by applying Stoke's theorem to each bin, similar to the study of Sofko et al. (1995). It approximates the vorticity ω across each bin's surface with the integrated flow along the closed path at its boundary to the neighbouring grid cells. Here we define the horizontal wind vorticity with respect to the radial or locally upward direction similar to, e.g., Thayer and Killeen (1991) with positive values (blue) representing cyclonic (counter-clockwise) and negative (red) for anticyclonic (clockwise) rotation. Further we applied spherical harmonic functions (again like in the analogous study for the plasma convection, Haaland et al., 2007) with the same order and degree of the associated Legendre polynomials to represent the vorticity pattern in a similarly smoothed fashion.

The high-latitude neutral wind vorticity in Fig. 2 shows as expected very similar patterns for both the Northern (bottom left) and Southern Hemisphere (bottom right panel). They comprise the well-known large duskside anticyclonic rotation cell with minimum values of -0.67 mHz and -0.52 mHz, respectively, i.e. with about 25 % larger absolute magnitudes on average at the Northern Hemisphere compared with the Southern Hemisphere. The dawnside cyclonic vorticity is slightly smaller at both hemispheres (about 20 %) for the overall averages with maximum values of 0.53 mHz

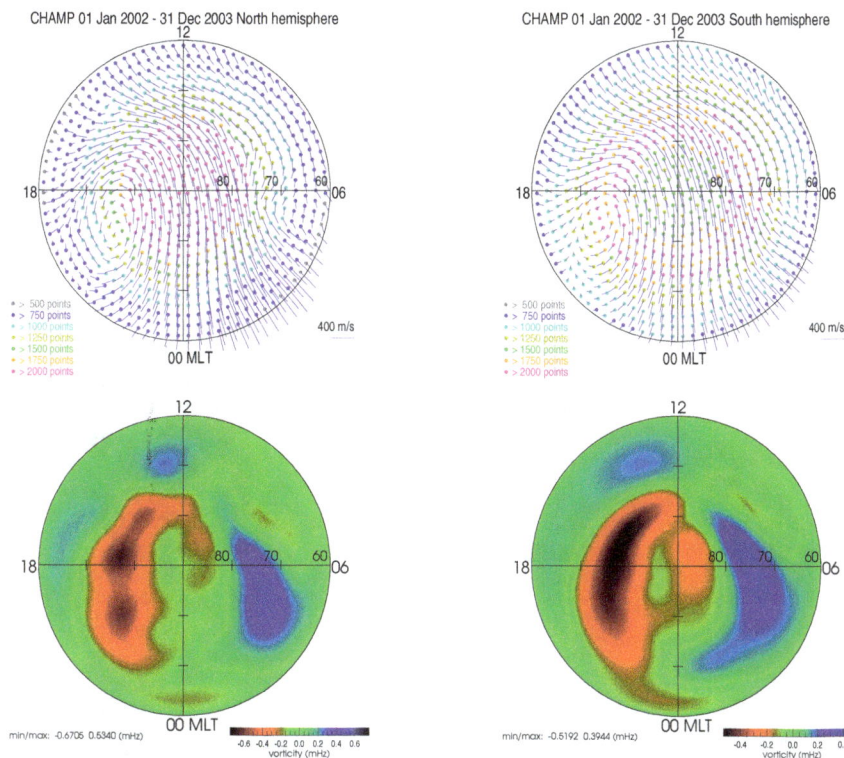

Fig. 2. Average statistic thermospheric wind pattern at ∼400 km height over the Northern (left side panels) and Southern polar hemisphere (right side) obtained from CHAMP accelerometer data of the whole years 2002 and 2003. The MLT versus magnetic latitude dials have an outer boundary of 60°and sun direction is on top, dusk to the left and dawn to the right side. The neutral wind direction and magnitude are shown in the upper panels by small vectors with the origin in the dots at the bin's position. The bottom panels show the vorticity pattern of the horizontal wind vectors. Negative values (red) indicate a clockwise, positive (blue) a counterclockwise circulation according to the colour bar at the bottom. Minimum and maximum values are indicated there as well (note the different ranges of the colour scales).

Table 1. Minimum and maximum values of high-latitude thermospheric vorticity [mHz] in both hemispheres at dusk and dawn, respectively, according to the smoothed IMF-dependent patterns in Figs. 3 and 4. The bottom line shows the average behaviour (Fig. 2).

IMF Sector	Northern Hemisphere			Southern Hemisphere		
	min	max	Δ	min	max	Δ
0	−0.630	0.353	0.983	−0.503	0.346	0.849
1	−0.661	0.445	1.106	−0.664	0.342	1.006
2	−0.603	0.613	1.216	−0.682	0.430	1.112
3	−0.742	0.813	1.555	−0.718	0.570	1.288
4	−0.917	0.700	1.617	−0.579	0.562	1.141
5	−0.835	0.766	1.601	−0.591	0.581	1.172
6	−0.845	0.611	1.456	−0.663	0.407	1.070
7	−0.735	0.409	1.144	−0.626	0.426	1.052
av.	−0.674	0.533	1.207	−0.519	0.392	0.911

and 0.40 mHz, respectively, and shows about the same discrepancy between North and South. The outer border of the vorticity cells follows in shape the statistical equatorward boundary of Region–1 FACs (cf., Förster et al., 2011). The two symmetric crescent-shaped vorticity areas are slightly

turned clockwise with respect to the noon-midnight meridian and a view to the wind vector patterns in the upper panels shows their different quality. The duskside vorticity is due to shear motion forming the large dusk vortex, while the dawnside vorticity represents a slight curvature on the large background wind circulation.

Figures 3 and 4 specify the vorticity patterns in dependence on IMF orientation for the Northern and Southern Hemispheres, respectively. The eight sectors represent average neutral wind vorticities for separate IMF ranges, each 45° wide, around IMF directions indicated on top of each dial. Table 1 summarizes all the minimum and maximum values of the statistic pattern shown in Figs. 3 and 4. Generally, the average vorticity magnitude increases with B_z turning southward and becomes largest for southward IMF (sector 4/5 at the Northern, sector 3 at the Southern Hemisphere), while the vorticity magnitudes are smallest and least in latitudinal extent for B_z+ (sector 0). The dynamic range of the vorticity maxima of the average pattern with respect to the IMF B_z component is larger at the Northern Hemisphere.

These IMF-dependencies and their two-cell pattern shapes remind of corresponding patterns for characteristic FAC (cf.,

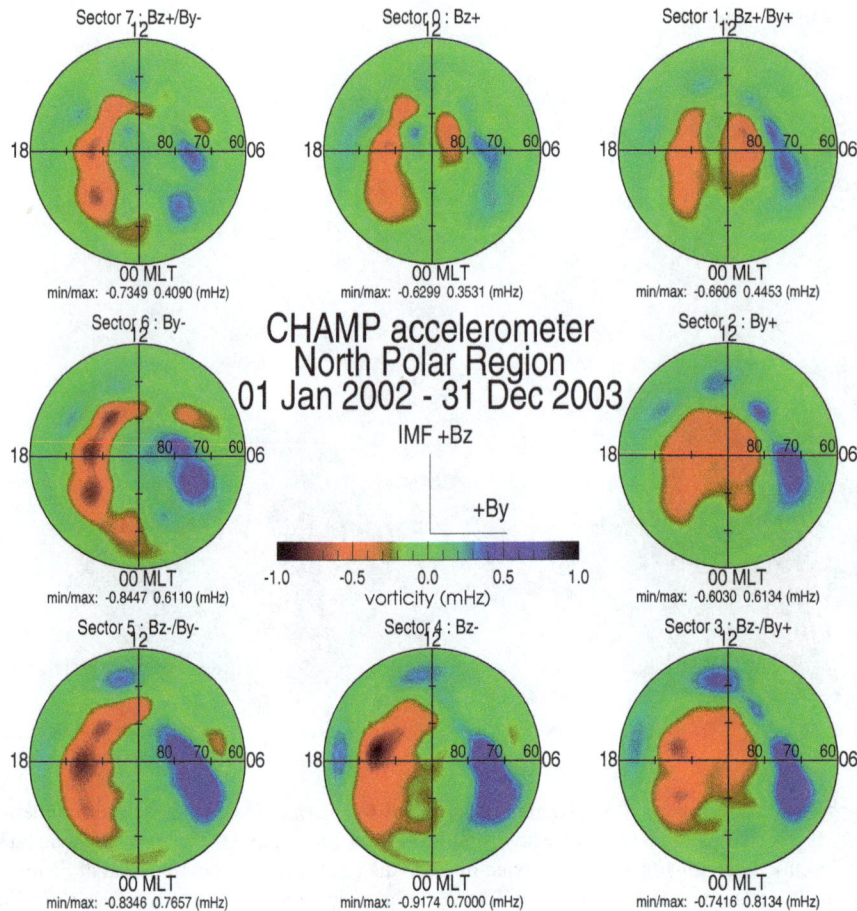

Fig. 3. Average Northern Hemisphere thermospheric vorticity patterns, sorted for 8 distinct sectors of IMF orientation. Each sector comprises 45° of IMF clock angle range centered around the direction indicated on top of each panel. The data are for the same time interval, presented in the same magnetic latitude/MLT coordinates and colour scale as in Fig. 2, bottom panel. Minimum and maximum values are indicated below each dial (see also Table 1).

e.g., Weimer, 2001; Papitashvili et al., 2002) and plasma drift dependencies on IMF orientation (Ruohoniemi and Greenwald, 2005; Haaland et al., 2007; Förster et al., 2007, 2009). Large circular vorticity cells ("round-shaped") form on the duskside of the Northern Hemisphere for B_y+ (sectors 1–3), while "crescent-shaped" cells appear there under B_y- conditions (sectors 5–7). At the same time the dawnside cell is occupied by "crescent-shaped" or larger "round-shaped" regions, respectively. The patterns are mirror-symmetric with respect to IMF B_y for the Southern Hemisphere, where the average magnitudes are generally somehow smaller. For northward IMF (sector 0), there is even an indication of a four-cell structure at high dayside latitudes both at North and South.

3   The Upper Atmosphere Model (UAM) concept

The global numerical model of the Earth's upper atmosphere (UAM) has originally been developed at the Kaliningrad Observatory (now West Department) of IZMIRAN (Namgaladze et al., 1988, 1990, 1991). Later it was modified at the Polar Geophysical Institute and the Murmansk State Technical University, in particular for applications at high latitudes (Namgaladze et al., 1996a,b; Volkov and Namgaladze, 1996). The model describes the thermosphere, ionosphere, plasmasphere and inner magnetosphere of the Earth as a single system by means of numerical integration of the corresponding time-dependent three-dimensional continuity, momentum and heat balance equations for neutral, ion and electron gases as well as the equation for the electric field potential. It covers the height range from 60 km up to 15 Earth radii of geocentric distance, using a fixed dipolar geomagnetic field configuration. The tilt between the geomagnetic and geographic axes of the Earth is taken into account.

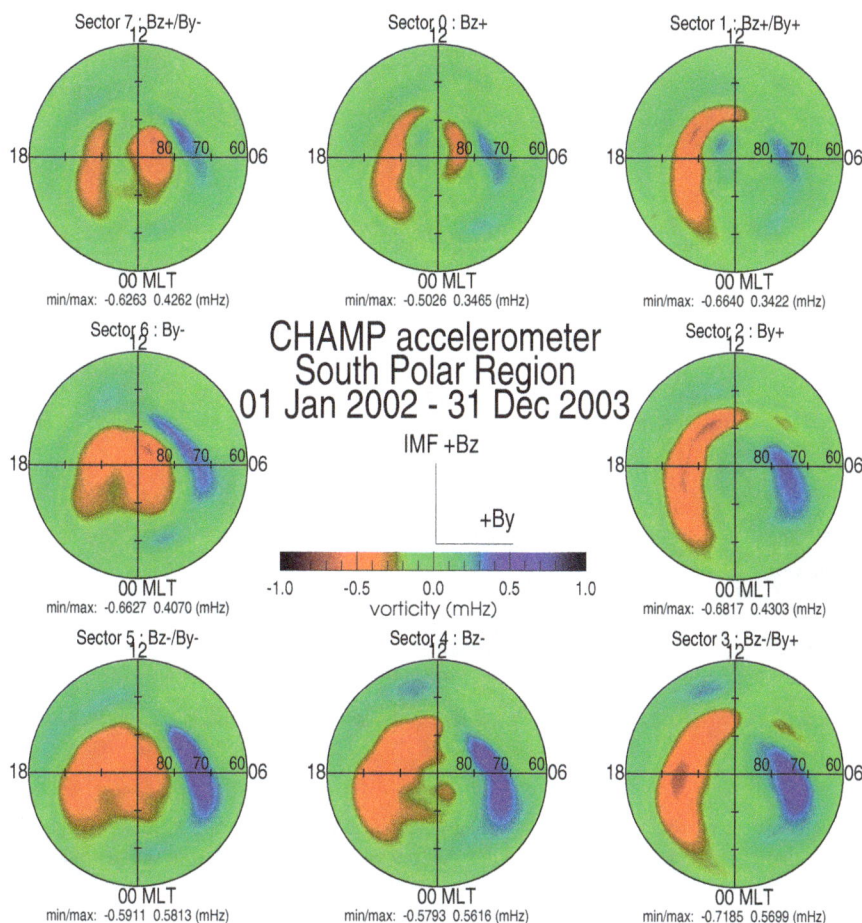

Fig. 4. Same as Fig. 3, but for the Southern Hemisphere thermospheric vorticity patterns.

The UAM solves numerically the well-known hydrodynamical continuity, momentum, and heat balance equations for the corresponding constituents by means of finite difference methods (Namgaladze et al., 1988, 1991). It consists of four main blocks (code units) as schematically illustrated in Fig. 5. These blocks describe specific subsets of the complex system in its own appropriate coordinate system as described in detail in previous model descriptions as, e.g., Namgaladze et al. (1996b, 1998c, 2000). The magnetospheric block, in particular, has been incorporated into the model in the 90-ies (Volkov and Namgaladze, 1996). This block calculates the density, pressure and drift velocity of the hot ions in the magnetospheric plasma sheet as well as the Region–2 FACs. The exchange of information between the blocks is realized by interpolation schemes from the grid points of one coordinate system to the other, while the solutions are found in an iterative manner. Improved versions of the model (Namgaladze et al., 1996b, 1998a,b) allow variable grid schemes with higher resolutions in certain regions as, e.g. the auroral and cusp zones, in dependence of the physical problem to be solved. The time steps have to be adopted accordingly to

avoid undesirable numerical effects (Ridley et al., 2010).

The time-dependent transport equations of the model are second-order parabolic ones; the model formulation has to comprise therefore both initial and boundary conditions for all the constituents. An inherent part of the model formulation is also the state-of-the-art compilation of a multitude of transport coefficients, cross sections, chemical reaction rates, heating efficiencies etc., which can have a devastating effect on the results, if they are not properly chosen.

Model inputs are (1) the solar UV and EUV spectra, (2) the precipitating particle fluxes (electrons and energetic ions), and (3) the cross-polar cap potential drop ($\Delta\Phi_{PC}$) or, alternatively, suitable Region–1 (and, when indicated, Region–0) FACs, which represent the primary external drivers for the M-I-T system (reconnection processes).

4   Model challenges (open questions)

The necessary input values, mentioned at the end of the previous section, require the use of empirical models. Such

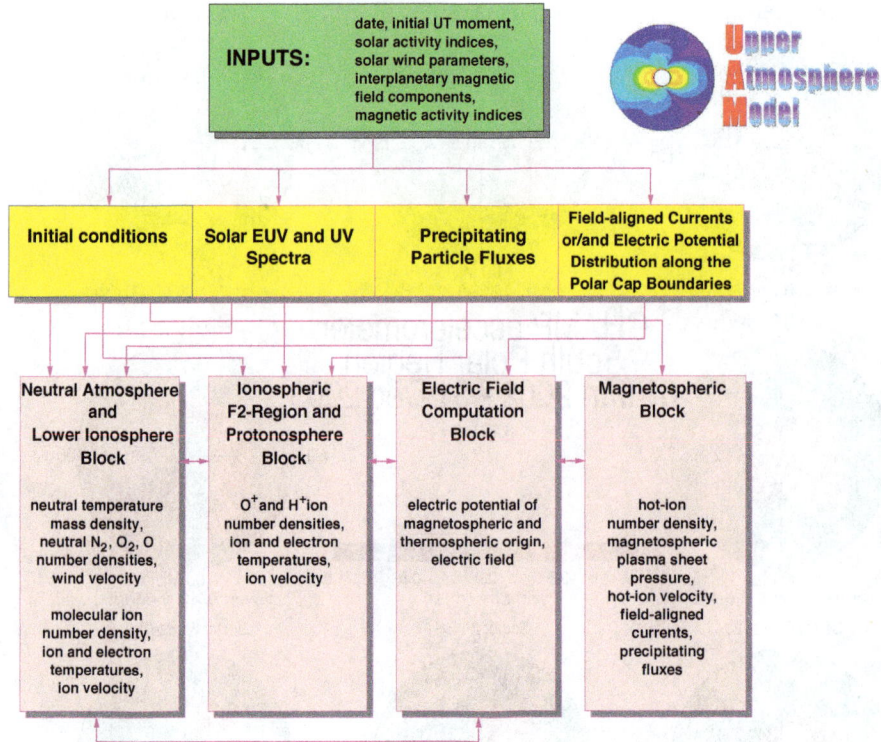

Fig. 5. Schematic structure of the Upper Atmosphere Model (UAM Namgaladze et al., 1988, 1991, 1996b, 1998a,b) with its inputs (initial and boundary conditions), the four main computational blocks, and their output parameters (taken from Förster et al., 1999a, Fig. 3).

Table 2. Selected time intervals with extended stable IMF conditions.

Date	timing	MJD	IMF	Average IMF values [nT]				3-hourly								Daily
	UT, h	2000	Sector	$\|B\|$	Bx	By	Bz	Kp indices								F10.7
14 Apr 2001	15–11*	469	6	7.7	−2.5	−7.2	−1.3	4+	4−	3+	3+	3+	3+	3−	2o	181.0
25 May 2002	18–12*	875	1	9.8	−6.3	6.3	4.0	1−	0+	0+	0+	1+	2o	1+	1o	187.4
18 Nov 2002	00–24	1052	5	8.8	2.1	−6.4	−5.6	3o	2o	2−	1+	2+	2o	3o	3o	174.8
01 Feb 2003	14–14*	1127	6	9.8	3.2	−9.2	−1.2	3o	2o	1−	1−	1+	3−	5−	4+	122.1
23 Oct 2003	00–21	1391	0	7.0	−1.9	−0.6	6.7	2−	3+	3−	3−	3o	4−	3o	4o	141.7
30 Aug 2004	08–22	1703	4	12.6	6.8	3.0	−10.2	2+	2o	3+	4o	4o	6−	5o	7o	91.6
13 Sep 2004	22–21*	1717	6	10.6	6.0	−8.4	2.2	0o	0o	0o	0+	0o	0+	4+	4+	119.1
06 Oct 2004	13–22	1740	2	4.1	0.1	4.0	−1.0	1o	2−	1+	1o	0o	0o	0+	0+	92.0
10 Dec 2004	09–21	1805	2	8.1	−1.0	7.6	−2.4	2o	2+	2o	2o	2+	3o	2+	3o	82.3
13 Jun 2005	00–14	1990	3	12.2	−1.9	9.8	−6.9	6o	6−	4+	3o	4o	3−	3+	2−	94.7
30 Apr 2006	10–24	2311	2	3.4	−3.2	1.1	0.5	0o	0o	0o	0o	0o	0o	0o	0o	101.4
02 Nov 2006	21–08*	2497	2	5.5	3.5	4.0	−1.3	2+	2o	1o	1o	2−	3−	3o	2+	86.8

* Interval starts at the previous day already.

models usually describe the magnitude, spatial distributions and temporal variations of certain input values in dependence of either solar, IMF parameters or geomagnetic indices. These models are as a rule generalizations of large observational data sets and describe often so-called climatological variations, but not necessarily the actual situation.

To overcome this difficulty and to skip the "ditch" between generalized or "climatological" modeling and actual simulations of real situations, one has to seek for as much as possible real, actual measurements for all kind of input values from solar EUV fluxes, electrodynamic parameters, till high-energy precipitations. Any real observation of the environmental plasma situation, which can be acquired for comparison, can therefore improve the modeling performance by constraining or optimizing the possible solutions.

In this study we are concerned with complex processes at high-latitudes, which constitute particularly interesting matters for the externally driven M-I-T interlinking. The correct matching of the various inputs (fluxes, FACs, boundary positions) within relatively small distances (which might even reach to sub-grid dimensions) is here of particular importance. In contrast to global studies, which focus on sub-auroral to mid- and low latitudes (like, e.g., the study of Förster et al., 1999b; Namgaladze et al., 2000), where the high-latitude energy input can be handled as a far-distance standard-shaped oval source, we have to treat these inputs very carefully within and near the auroral oval.

Such input values are for instance the solar EUV and x-ray spectra with newly developed solar indices (Tobiska et al., 2006) and electron precipitation fluxes given by Hardy et al. (1985) and their latitudinal and longitudinal distribution. Most of such models are parametrized with respect to solar or global geomagnetic indices like the $F_{10.7}$ proxy of solar EUV flux intensities and the three-hourly Kp values. Somehow exceptional are models that describe environmental conditions with respect to solar IMF strength and orientation. They exist for such input quantities like the ionospheric convection electric field (Ruohoniemi and Greenwald, 2005; Haaland et al., 2007; Förster et al., 2009) and FAC distributions (Weimer, 2001; Papitashvili et al., 2002).

Any misfit between input models can have tremendous consequences for UAM results of, e.g., the electric field pattern and therefore also for the plasma distribution and neutral wind pattern. It is well known that the electric field and current distributions depend strongly on the ionospheric conductance (Ridley et al., 2004). This concerns in particular the current carrying layers, which are formed at auroral latitudes primarily by energetic particle precipitations. But also the terminator position plays a role, which depends on season (Ridley, 2007), but also on UT (longitude) due to the tilt between the geomagnetic and geographic axes for a dipolar field configuration like in the UAM (the real geomagnetic field with differences of the configurations between the hemispheres can complicate the situation even more).

For this model study with its focus on M-I-T coupling processes in dependence of the IMF strength and orientation, we used the FAC model of Papitashvili et al. (2002), including its extended auxiliary material. The idea was to have an external driver representation, which comprises explicitly the IMF dependencies. Alternatively, we could use also IMF-dependent models of the magnetospheric drift pattern like that of Förster et al. (2009) or the ground-based SuperDARN observations of the ionospheric drift pattern as summarized by Ruohoniemi and Greenwald (2005). All these models represent parametrized statistical averages, i.e. they show smoothed, quasi-static patterns in a "climatological" sense. They show generally smaller amplitudes as a multitude of real distribution patterns, all the second-order dependencies and the natural variabilities are "smeared" into spherical harmonic functions of finite order and degree.

The FAC model of Papitashvili et al. (2002) was derived from high precision magnetic field measurements from the Ørsted and Magsat satellites, being parameterized by the IMF strength and direction for summer, winter, and equinox as well as explicitly for both polar regions separately, resolving the seasonal dependence of interhemispheric asymmetries. The FAC distributions are fitted with associated Legendre functions, limited to degree $n = 21$ and order $m = 3$ to obtain the FAC patterns with the fewest terms, and continued downward to the ionospheric altitude of 115 km using current conservation (Papitashvili et al., 2002).

First UAM simulation attempts with these patterns revealed promptly that the resulting drift pattern and the cross-polar cap potential (CPCP) show unrealistic small values, obviously due to the outspread FAC distributions of the spherical harmonics. We therefore repeated the experiments with a modified FAC distribution, which limits the actual FAC belts in latitude ranges corresponding to the most likely auroral oval positions, keeping the main characteristics of the FAC model patterns with its IMF and seasonal dependence. This was done by integrating the model current densities along each meridional cut and assigning these values to the belt's position, both for Region–1 and Region–0 (within the auroral oval) FACs. The oval position, including B_y^{IMF}-dependent turning and B_z^{IMF}-dependent widening or shrinking, has to be harmonized with other model's boundary positions like, e.g., the open-closed boundary (OCB) and the precipitation patterns. The cusp position is of particular importance because of its crucial role in the M-I-T dynamics as observed by CHAMP (e.g., Schlegel et al., 2005; Rentz and Lühr, 2008) and modeled with UAM (Namgaladze et al., 1996b, 1998c).

The choice of the suitable coordinate system is an important issue for a meaningful model–observation comparison, in particular at high-latitudes. The model operates with a tilted dipol magnetic field constellation, while most of empirical models for the high-latitude parameters and the observations of CHAMP are provided in corrected geomagnetic coordinates (or even AACGM for a correct mapping of data along geomagnetic field lines). While the model results in the subsequent presentations (Figs. 8–9) are plotted in geomagnetic coordinates (MLT versus geomagnetic latitude), the overplotted CHAMP orbital track and the measurements are shown according to their AACGM coordinates. We think that comparisons come hence as close as possible to reality.

A slightly different, but related question is the correct positioning of the various high-latitude boundaries within the dipolar model formulation and how well it can reflect the observations. In future, it might be necessary to replace UAM realizations of the geomagnetic field configuration – at least for the electric field computation and the magnetospheric block – by more realistic approximations of the actual field, as it is supplied, e.g., by the Tsyganenko model of the magnetosphere (Tsyganenko, 2002a,b). This model

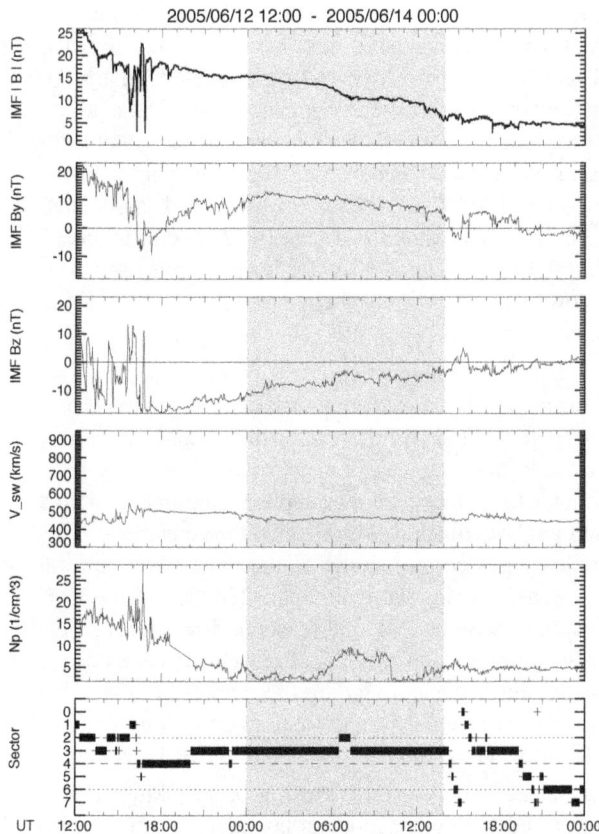

Fig. 6. Solar wind and IMF conditions during one of the intervals with stable IMF conditions (shaded, # 10 according to Table 2), obtained from ACE observations at the Earth-sun L1 libration point upstream of Earth and time-shifted to the magnetopause position according to a procedure, which is described in detail in Haaland et al. (2007). From top to bottom, the panels show the IMF magnitude, the IMF y- and z-components in GSM coordinates, the solar wind velocity, the solar wind plasma density, and the corresponding sector of the IMF in the GSM-y-z plane (see text).

is also parametrized with respect to solar wind and IMF conditions.

5   First comparisons for stable IMF conditions

Magnetospheric behaviour at any given time moment depends strongly on the actual forcings and its pre-history. In other words, the M-I-T system is very dynamic and highly time-dependent. This is reproduced by the UAM as a time-dependent 3-D realization with initial and boundary conditions. The temporal development constitutes a key question for realistic modeling, in particular for storm events. As we focus in this study on IMF dependencies and hemispheric differences, we primarily tested the UAM performance under relatively stable IMF conditions. As listed in Table 2, we selected a series of time intervals of the CHAMP lifetime with

various IMF conditions, each of which persisted for longer time intervals of at least several hours. It is supposed that the M-I-T system then attains some kind of dynamic equilibrium state that can more easily be compared with observations.

Figure 6 shows the solar wind and IMF conditions around one of the intervals with (more or less) stable IMF conditions, namely the 10th of the 12 listed in Table 2. The IMF values obtained by the Advanced Composition Explorer (ACE) spacecraft, are time shifted to the frontside magnetopause (at $X_{GSM} = 10 \, R_E$), using the phase front propagation technique of Weimer et al. (2003) in a slightly modified version, which is based on constrained minimum variance analyses of the IMF (Haaland et al., 2006). The procedure of sorting for specified IMF directions (clock angle sectors) and the bias value filtering of solar wind data for stable IMF conditions has been described in the companion papers of Haaland et al. (2007) and Förster et al. (2007). Here, we use the same methodology with the same bias value (≥ 0.96) threshold as for these Cluster/EDI plasma convection analyses and previous high-latitude neutral wind studies (Förster et al., 2008, 2011).

The grey-shaded period in Fig. 6 is supposed to be sufficiently stable – in this case for about 14 h. This interval is insofar exceptional, as it proceeds during the recovery phase of a moderate storm that occured during the preceding day. The geomagnetic indices Dst, AE, and Kp in Fig. 7 illustrate this situation. The 3-hourly Kp values are still at a high level during the begin of the interval and relax slowly, while the AE values fluctuate around 900 nT indicating considerable auroral activity. The IMF orientation is stable with considerable B_z- and B_y+ values corresponding to sector 3 conditions, while the magnitude $|B^{IMF}|$ declines steadily during the interval.

Fig. 8 shows UAM simulation results for high-latitude circumpolar regions in geomagnetic coordinates with $|\phi_m| > 60°$ for both the Northern (left panel) and Southern (right) Hemisphere together with CHAMP cross-track neutral wind measurements along two subsequent passages at almost ~400 km altitude over the respective polar regions. The modeled parameters shown are the electric potential in the background with the black-white scale and the horizontal neutral wind field represented by coloured arrows with the scale on top. The model results refer to an altitude level that is indicated in the lower right corner; it corresponds to the satellite height at the moment indicated by the black arrow and written in the middle of the bottom line. Minimum and maximum values along the CHAMP orbital section are shown in the lower left corner of each panel both for the observed and the modeled values.

During this time interval, CHAMP crosses the large dusk vortex and the dawn cell approximately in the dawn-dusk plane. A large "round-shaped" dusk convection cell and simultaneously a large dusk side neutral wind vortex can be noticed at the Northern Hemisphere (Fig. 8, left panel) with maximum wind speeds across the polar cap region on the

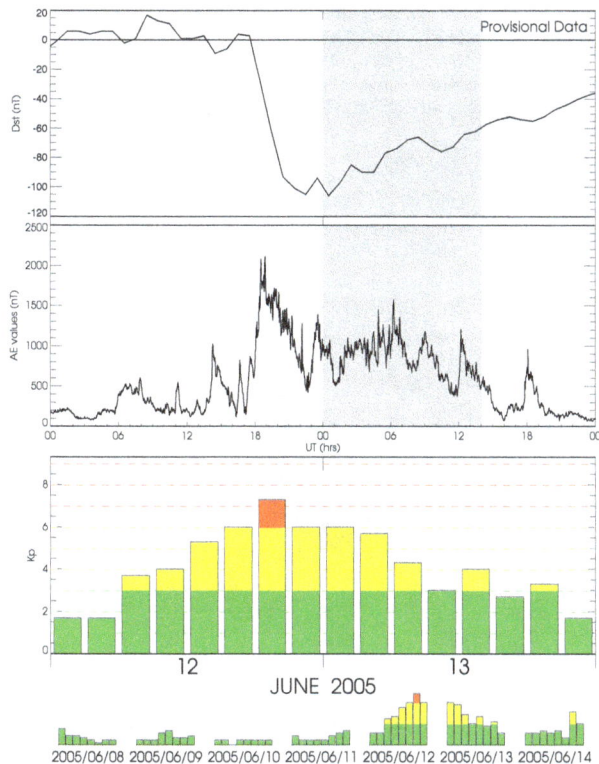

Fig. 7. Geomagnetic conditions during the same interval (# 10 of Table 2) as in Fig. 6 with stable IMF conditions (shaded). The panels from top to bottom show the hourly Dst index, the one-minute AE values, and the 3-hourly Kp index for the current and the encircling days. The interval appears to happen during a minor storm recovery phase.

dawnward side. This is in accord with the statistical average pattern of sector 3 IMF orientation both for plasma convection (cf. Fig. 7 of Haaland et al. (2007)) and the neutral horizontal wind (Fig. 3 above).

The modeled neutral wind component in the cross-track direction along the satellite's orbit agrees closely with the observed wind profile. The position of the dusk vortex center, as it is accidentally almost centrally crossed by the satellite, seems to be closely reproduced as it is indicated by the change of the wind direction from sunward in the subauroral region ($< 70°$) to anti-sunward at the poleward side. The modeled wind velocity shows smooth variations in contrast to the observations, which reveal superposed smaller-scale minor variations.

The plasma convection and neutral wind circulation pattern during the crossing over the Southern polar region (Fig. 8, right panel) shows a more symmetric two-cell pattern with largest wind velocities in antisolar direction over the central polar cap. The dusk vortex center is shifted to lower latitudes and to slightly earlier local time hours (~ 17 MLT). The dusk vortex shrinked in its latitudinal extent, so that both

vortices have now about the same size. This corresponds to the statistically expected pattern for Sect. 3 IMF conditions at the Southern Hemisphere (cf. Fig. 8 of Haaland et al. (2007) and Fig. 4, respectively).

The modeled neutral wind profile along the orbit reproduces the latitudinal variation, but the amplitude does not everywhere match the observations. It differs on the duskward side inside the polar cap region up to about a factor of two, while it coincides in the region of the dawnside vortex. The maximum wind amplitude of the CHAMP observations of more than $900 \, \text{m s}^{-1}$ in this case appears to be quite large.

Figure 9 shows UAM-CHAMP comparisons for the neutral mass density in the same manner and for the same time moments as in Fig. 8. Here, the modeled parameter is shown in the background with the colour scale on top and the minimum/maximum values of the model area indicated just below the scale on its begin and end. The corresponding min/max values of both CHAMP and UAM values along the orbital trace are written in the lower left corner of each panel.

The agreement between model and observation for the Northern traverse of the polar region (Fig. 9, left panel) is good, except for two mass density peaks on the dawn side. They probably indicate the position of the cusp in the pre-noon sector near $\sim 74°$ and 08:30 MLT (cf. Schlegel et al., 2005) and an other auroral feature at lower latitudes ($\sim 66°$).

The mass density profile of the CHAMP observations over the Southern Hemisphere (Fig. 9, right panel) clearly differs from the modeled ones both in amplitude and latitudinal trend. This misfit over the southern winter polar region is not seen at the beginning of this stable IMF interval (not shown), but started around 03:00 UT and continued with a steady increase toward the interval's end for unclear reasons. At the Northern Hemisphere the density differences between model results and observations starts later on this day ($\sim 12:00$ UT), is less distinctive and does not show differences in trend. There is obviously some thermospheric heating process, which is not yet correctly reproduced by the UAM modeling.

6 Discussion and conclusions

This paper presents statistical neutral wind vorticity patterns based on accelerometer data of the CHAMP satellite. It shows the strong IMF dependence of the high-latitude upper atmosphere neutral wind at F-region heights (~ 400 km). These patterns confirm previous findings based on early satellite data in the 1980s (e.g., Killeen and Roble, 1988; Killeen et al., 1995), but can now rest upon a much broader statistical base due to the good coverage of the CHAMP data set (cf. also Förster et al., 2011).

Averaging over seasonal effects, these patterns provide an observational evidence for hemispheric differences in the amplitudes of the neutral wind vorticity, which amounts in

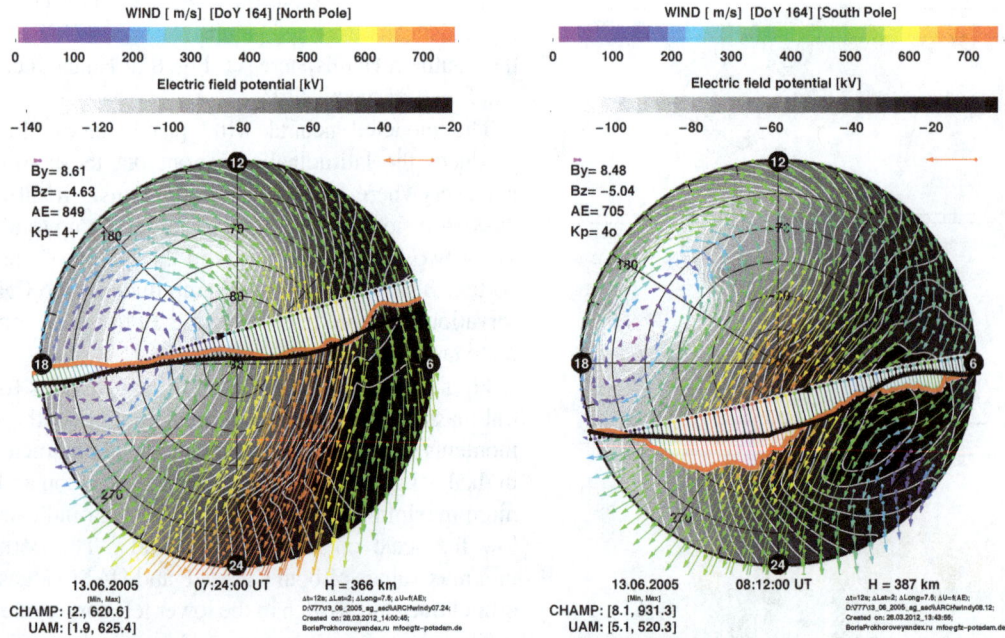

Fig. 8. UAM model results compared with cross-track accelerometer measurements of the CHAMP satellite for overflights over the Northern (left panel) and Southern Hemispheres (right panel) in polar geomagnetic coordinates at latitudes higher than 60° during the interval # 10 according to Table 2 (see also Figs. 6 and 7 for solar and geophysical conditions). The dials show sun direction (12 MLT) on top, dusk side (18 MLT) to the left and dawn (06 MLT) to the right, respectively. Magnetic longitude coordinates are indicated with the black crosslines. The dotted line follows the orbital track from ~18 MLT to ~08 MLT at North (left) and from ~20 MLT to ~06 MLT at South (right panel) with larger arrows at the actual spacecraft positions at 07:24 UT and 08:12 UT, respectively. The cross-track fine black arrows correspond to the calculated UAM crosswind speed, while the colour arrows show cross-track wind component measurements of CHAMP. The model wind vector pattern (coloured arrows according to the scale on top) and the electric potential (b/w colour scale) are shown in the background.

the statistical average to about 25 % – with larger values at the Northern Hemisphere compared to the Southern. A small difference between North and South in the statistical sense was already found for the plasma drift pattern (or CPCP) with ~7 % larger average drift velocities at the Northern Hemisphere compared to the Southern (Förster et al., 2007).

These differences are most likely due to the different geographic-geomagnetic offsets between the hemispheres (Förster et al., 2008). Alternatively, it might be conceivable that the differences of statistical averages could be ascribed to unequal weighting effects due to seasonal or UT dependent conductance differences between the hemispheres. Finally, they could even be due to different patterns of geomagnetic flux densities at the opposite polar regions, i.e. due to higher order geomagnetic field components, which are known to exist to an appreciable extent. Further UAM modeling studies should help to clarify this question. In this regard it will be interesting, whether the present symmetric model configuration with the tilted dipolar field is sufficient to explain the hemispheric differences or whether a more realistic geomagnetic field configuration has to be invoked.

It was noted already in earlier studies that the anticyclonic dusk side thermospheric neutral wind vortex is almost invari-

ably stronger than the cyclonic vortex associated with the dawn side ion convection cell (Killeen and Roble, 1988). These observations were accompanied by comprehensive theoretical-numerical studies (e.g. Rees and Fuller-Rowell, 1989), which ascribe this difference to the effects of the Coriolis and centrifugal forces. They tend to maintain the dusk side anticyclonic vorticity, while partly compensating each other on the dawn side.

A previous CHAMP study by Lühr et al. (2007) on average high-latitude neutral thermospheric wind pattern came, bye the way, to different results than those shown in Fig. 2. They obtained distributions of mean thermospheric wind vectors in both polar regions by means of a minimum variance technique (Fig. 4 and equ. 4 of Lühr et al., 2007, respectively). Because of the sometimes shallow minima, they guess that the correct wind direction could also be obtained from the maximum values of that function, when turning it by 90°. This methodological error results in different wind pattern, mainly in the early morning and afternoon sectors.

The variation of the vorticity pattern with IMF B_y and B_z (Fig. 3 and 4) strongly resembles the corresponding pattern of the plasma drift (e.g., Ruohoniemi and Greenwald, 2005; Haaland et al., 2007) and FACs (e.g., Papitashvili et al.,

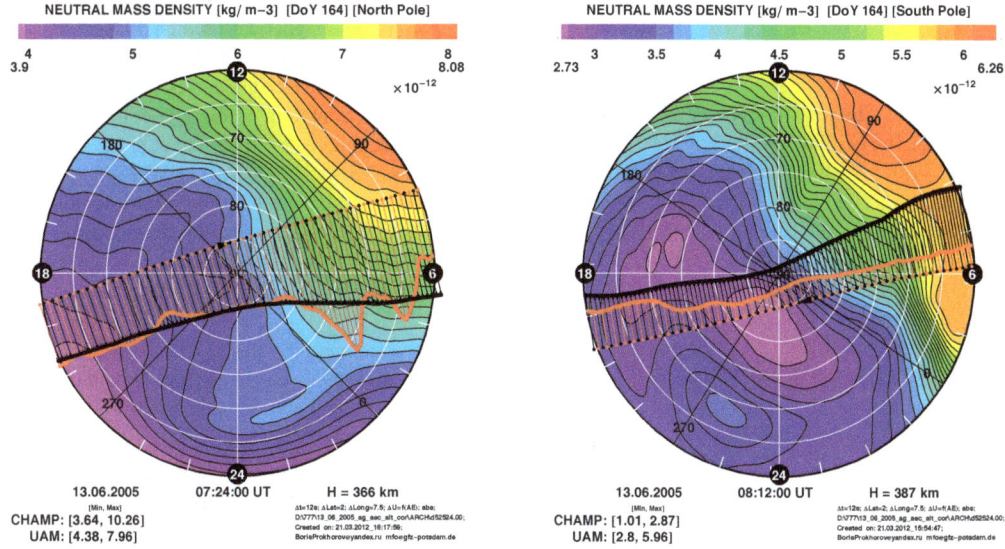

Fig. 9. UAM model results for the same interval and in a similar fashion like in Fig. 8, but here compared with CHAMP accelerometer mass density measurements for overflights over the Northern (left panel) and Southern Hemispheres (right panel) in polar geomagnetic coordinates at latitudes higher than 60°. The absolute mass density values along the orbital track are shown as cross-track distances with dashed black lines for the UAM model results and red coloured ones for the CHAMP observations. Mimimum and maximum values along this orbital section are listed in the left bottom corner both for CHAMP and UAM model data (in units of the top colour scale). The mass density pattern at the height level of the satellite as indicated in the bottom line is shown in the background with the colour scale at the top.

2002). Additional to the strong changes of extent, shape and position of the dusk and dawn circulation cells with the IMF orientation, there is also a systematic variation of their maximum amplitudes. They are largest for sector 3 at South and for sector 4/5 at the North Hemisphere. This comes along with the largest cross-polar thermospheric wind amplitudes over the central polar cap for these IMF orientations at the respective hemisphere, as shown by Förster et al. (2008).

Statistical patterns of high-latitude magnetic field-aligned plasma vorticity with the same sorting for 8 sectors of IMF orientation have been deduced recently also from Super-DARN measurements (Chisham et al., 2009). They show the close affinity to corresponding FAC pattern, despite of smaller variations due to seasonal (ionospheric conductance) effects. In a subsequent study, Chisham and Freeman (2010) investigated the probability density function of plasma vorticity measurements and showed that it is best modeled by the q-exponential distribution across most of the polar ionosphere, except in the dayside Region-1 current region, where the Weibull distribution provides the best model. They interpret this as an indication for the relative importance of different physical mechanisms affecting the plasma drift vorticity in the various regions with a dominance of convective (barotropic) effects within the dayside Region-1 and baroclinic vorticity elsewhere (Chisham and Freeman, 2010). The clear-cut neutral wind vorticity regions and their obvious IMF dependence indicate a strong relevance of the neutral wind for this plasma vorticity behaviour.

For a deeper understanding of the M-I-T system, adequate numerical models are indispensable due to the complexity and nonlinearity of the inherent coupling processes. This concerns first of all global, time-dependent first-principal MHD models like the UAM in this study. We presented our first attempts for an adequate modeling of the high-latitude CHAMP measurements to obtain new physical insights into complex M-I-T processes. The forthcoming Swarm satellite mission will allow much more detailed studies of the high-latitude thermosphere dynamics and its complex interaction with magnetospheric configurations under varying solar wind conditions.

Acknowledgements. Work at GFZ German Research Centre for Geosciences Potsdam and the University Potsdam was supported by Deutsche Forschungsgemeinschaft (DFG). We thank the ACE SWEPAM and MAG instrument teams and the ACE Science Center for providing the ACE data. The CHAMP mission is sponsored by the Space Agency of the German Aerospace Center (DLR) through funds of the Federal Ministry of Economics and Technology, following a decision of the German Federal Parliament (grant code 50EE0944). The data retrieval and operation of the CHAMP satellite by the German Space Operations Center (GSOC) of DLR is acknowledged.

References

Banks, P. M.: Magnetospheric processes and the behavior of the neutral atmosphere, Space Res., 12, 1051–1067, 1972.

Blanc, M. and Richmond, A. D.: The ionospheric disturbance dynamo, J. Geophys. Res., 85, 1669–1686, 1980.

Chisham, G. and Freeman, M. P.: On the non-Gaussian nature of ionospheric vorticity, Geophys. Res. Lett., 36, L12103, doi:10.1029/2010GL043714, 2010.

Chisham, G., Freeman, M. P., Abel, G. A., Bristow, W. A., Marchaudon, A., Ruohoniemi, J. M., and Sofko, G. J.: Spatial distribution of average vorticity in the high-latitude ionosphere and its variation with interplanetary magnetic field direction and season, J. Geophys. Res., 114, A09301, doi:10.1029/2009JA014263, 2009.

Coroniti, F. V. and Kennel, C. F.: Can the ionosphere regulate magnetospheric convection?, J. Geophys. Res., 78, 2837–2851, 1973.

Cowley, S. W. H.: Magnetosphere–Ionosphere Interactions: A Tutorial Review, in: Magnetospheric Current Systems, edited by: Ohtani, S.-I., Fujii, R., Hesse, M., and Lysak, R. L., 118, Geophysical Monograph, 91–106, American Geophysical Union, 2000.

Doornbos, E., van den IJssel, J., Lühr, H., Förster, M., and Koppenwallner, G.: Neutral density and crosswind determination from arbitrarily oriented multi-axis accelerometers on satellites, J. Spacecraft Rockets, 47, 580–589, doi:10.2514/1.48114, 2010.

Eastwood, J. P.: The science of space weather, Phil. Trans. R. Soc. A, 366, 4489–4500, doi:10.1098/rsta.2008.0161, 2008.

Förster, M., Lühr, H., Reigber, C., König, R., Namgaladze, A. A., and Yurik, R. Y.: The CHAMP satellite and its space weather monitoring capability, in: Proceedings of the 14th ESA Symposium on European Rocket and Balloon Programmes and Related Research, SP–437, 255–259, Noordwijk, Netherland, 1999a.

Förster, M., Namgaladze, A. A., and Yurik, R. Y.: Thermospheric composition changes deduced from geomagnetic storm modeling, Geophys. Res. Lett., 26, 2625–2628, 1999b.

Förster, M., Paschmann, G., Haaland, S. E., Quinn, J. M., Torbert, R. B., McIlwain, C. E., Vaith, H., Puhl-Quinn, P. A., and Kletzing, C. A.: High-latitude plasma convection from Cluster EDI: Variances and solar wind correlations, Ann. Geophys., 25, 1691–1707, 2007, http://www.ann-geophys.net/25/1691/2007/.

Förster, M., Rentz, S., Köhler, W., Liu, H., and Haaland, S. E.: IMF dependence of high-latitude thermospheric wind pattern derived from CHAMP cross-track measurements, Ann. Geophys., 26, 1581–1595, 2008, http://www.ann-geophys.net/26/1581/2008/.

Förster, M., Feldstein, Y. I., Haaland, S. E., Dremukhina, L. A., Gromova, L. I., and Levitin, A. E.: Magnetospheric convection from Cluster EDI measurements compared with the ground-based ionospheric convection model IZMEM, Ann. Geophys., 27, 3077–3087, 2009, http://www.ann-geophys.net/27/3077/2009/.

Förster, M., Haaland, S. E., and Doornbos, E.: Thermospheric vorticity at high geomagnetic latitudes from CHAMP data and its IMF dependence, Ann. Geophys., 29, 181–186, 2011, http://www.ann-geophys.net/29/181/2011/.

Friis-Christensen, E., Lühr, H., and Hulot, G.: Swarm: A constellation to study the Earth's magnetic field, Earth Planets Space, 58, 351–358, 2006.

Friis-Christensen, E., Lühr, H., Hulot, G., Haagmans, R., and Purucker, M.: Geomagnetic Research From Space, EOS Trans. AGU, 90, 213, doi:10.1029/2009EO250002, 2009.

Haaland, S. E., Paschmann, G., and Sonnerup, B. U. Ö.: Comment on "A new interpretation of Weimer et al.'s solar wind propagation delay technique" by Bargatze et al., J. Geophys. Res., 111, A06102, doi:10.1029/2005JA011376, 2006.

Haaland, S. E., Paschmann, G., Förster, M., Quinn, J. M., Torbert, R. B., McIlwain, C. E., Vaith, H., Puhl-Quinn, P. A., and Kletzing, C. A.: High-latitude plasma convection from Cluster EDI measurements: Method and IMF-dependence, Ann. Geophys., 25, 239–253, 2007, http://www.ann-geophys.net/25/239/2007/.

Hardy, D. A., Gussenhoven, M. S., and Holeman, E. A.: A statistical model of auroral electron precipitation, J. Geophys. Res., 90, 4229–4248, 1985.

Häusler, K. and Lühr, H.: Nonmigrating tidal signals in the upper thermospheric zonal wind at equatorial latitudes as observed by CHAMP, Ann. Geophys., 27, 2643–2652, 2009, http://www.ann-geophys.net/27/2643/2009/.

Hays, P. B., Killeen, T. L., Spencer, N. W., Wharton, L. E., Roble, R. G., Emery, B. A., Fuller-Rowell, T. J., Rees, D., Frank, L. A., and Craven, J. D.: Observations of the dynamics of the polar thermosphere, J. Geophys. Res., 89, 5597–5612, 1984.

Killeen, T. L. and Roble, R. G.: Thermosphere Dynamics: Contributions from the First 5 Years of the Dynamics Explorer Program, Rev. Geophys., 26, 329–367, 1988.

Killeen, T. L., Won, Y.-I., Niciejewski, R. J., and Burns, A. G.: Upper thermosphere winds and temperatures in the geomagnetic polar cap: Solar cycle, geomagnetic activity, and interplanetary magnetic field dependencies, J. Geophys. Res., 100, 21,327–21,342, 1995.

Lühr, H., Rentz, S., Ritter, P., Liu, H., and Häusler, K.: Average thermospheric wind pattern over the polar regions, as observed by CHAMP, Ann. Geophys., 25, 1093–1101, 2007, http://www.ann-geophys.net/25/1093/2007/.

Lühr, H., Park, J., Ritter, P., and Liu, H.: In-situ CHAMP observation of ionosphere-thermosphere coupling, Space Sci. Rev., 168, 237–260, doi:10.1007/s11214-011-9798-4, 2012.

Namgaladze, A. A., Korenkov, Y. N., Klimenko, V. V., Karpov, I. V., Bessarab, F. S., Surotkin, V. A., Glushchenko, T. A., and Naumova, N. M.: A global numerical model of the thermosphere, ionosphere, and protonosphere of the Earth, Geomagn. Aeron. (Engl. translation), 30, 515–521, 1990.

Namgaladze, A. A., Korenkov, Y. N., Klimenko, V. V., Karpov, I. V., Surotkin, V. A., and Naumova, N. M.: Numerical modelling of the thermosphere–ionosphere–protonosphere system, J. Atmos. Terr. Phys., 53, 1113–1124, 1991.

Namgaladze, A. A., Korenkov, Y. N., Klimenko, V. V., Karpov, I. V., Bessarab, F. S., Surotkin, V. A., Glushchenko, T. A., and Naumova, N. M.: Global model of the thermosphere–ionosphere–protonosphere system, Pure Appl. Geophys., 127, 219–254, 1988.

Namgaladze, A. A., Martynenko, O. V., Namgaladze, A. N., Volkov, M. A., Korenkov, Y. N., Klimenko, V. V., Karpov, I. V., and Bessarab, F. S.: Numerical simulation of an ionospheric disturbance over EISCAT using a global ionospheric model, J. Atmos. Terr. Phys., 58, 297–306, 1996a.

Namgaladze, A. A., Namgaladze, A. N., and Volkov, M. A.:

Numerical modelling of the thermospheric and ionospheric effects of magnetospheric processes in the cusp region, Ann. Geophys., 14, 1343–1355, 1996b.

Namgaladze, A. A., Martynenko, O. V., and Namgaladze, A. N.: Global model of the upper atmosphere with variable latitudinal integration step, International Journal of Geomagnetism and Aeronomy, 1, 53–58, 1998a.

Namgaladze, A. A., Martynenko, O. V., Volkov, M. A., Namgaladze, A. N., and Yurik, R. Y.: High-latitude version of the global numerical model of the Earth's upper atmosphere, Proceedings of the Murmansk State Technical University, 1, 23–84, 1998b.

Namgaladze, A. A., Namgaladze, A. N., and Volkov, M. A.: Seasonal effects in the ionosphere–thermosphere response to the precipitation and field–aligned current variations in the cusp region, Ann. Geophys., 16, 1283–1298, 1998c.

Namgaladze, A. A., Förster, M., and Yurik, R. Y.: Analysis of the positive ionospheric response to a moderate geomagnetic storm using a global numerical model, Ann. Geophys., 18, 461–477, 2000,
http://www.ann-geophys.net/18/461/2000/.

Papitashvili, V. O., Christiansen, F., and Neubert, T.: A new model of field-aligned currents derived from high-precision satellite magnetic field data, Geophys. Res. Lett., 29, 1683, doi:10.1029/2001GL014207, 2002.

Rees, D. and Fuller-Rowell, T. J.: The response of the thermosphere and ionosphere to magnetospheric forcing, Phil. Trans. R. Soc. London, A328, 139–171, 1989.

Reigber, C., Lühr, H., and Schwintzer, P.: CHAMP mission status, Adv. Space Res., 30, 129–134, 2002.

Rentz, S. and Lühr, H.: Climatology of the cusp-related thermospheric mass density anomaly, as derived from CHAMP observations, Ann. Geophys., 26, 2807–2823, 2008,
http://www.ann-geophys.net/26/2807/2008/.

Ridley, A. J., Gombosi, T. I., and De Zeeuw, D. L.: Ionospheric control of the magnetosphere: conductance, Ann. Geophys., 22, 567–584, 2004,
http://www.ann-geophys.net/22/567/2004/.

Ridley, A. J.: Effects of seasonal changes in the ionospheric conductances on magnetospheric field-aligned currents, Geophys. Res. Lett., 34, L05101, doi:10.1029/2006GL028444, 2007.

Ridley, A. J., Gombosi, T. I., Sokolov, I. V., Tóth, G., and Welling, D. T.: Numerical considerations in simulating the global magnetosphere, Ann. Geophys., 28, 1589–1614, 2010,
http://www.ann-geophys.net/28/1589/2010/.

Ruohoniemi, J. M. and Greenwald, R. A.: Dependencies of high-latitude plasma convection: Consideration of interplanetary magnetic field, seasonal, and universal time factors in statistical patterns, J. Geophys. Res., 110, A09204, doi:10.1029/2004JA010815, 2005.

Schlegel, K., Lühr, H., St.-Maurice, J.-P., Crowley, G., and Hackert, C.: Thermospheric density structures over the polar regions observed with CHAMP, Ann. Geophys., 23, 1659–1672, 2005,
http://www.ann-geophys.net/23/1659/2005/.

Siscoe, G. L.: Global force balance of region 1 current system, J. Atmos. Sol.-Terr. Phys., 68, 2119–2126, 2006.

Siscoe, G. L. and Siebert, K. D.: Bimodal nature of solar wind-ionosphere-thermosphere coupling, J. Atmos. Sol.–Terr. Phys., 68, 911–920, 2006.

Sofko, G. J., Greenwald, R. A., and Bristow, W. A.: Direct determination of large-scale magnetospheric field-aligned currents with SuperDARN, Geophys. Res. Lett., 22, 2041–2044, 1995.

Thayer, J. P. and Killeen, T. L.: Vorticity and divergence in the high-latitude upper thermosphere, Geophys. Res. Lett., 18, 701–704, 1991.

Tobiska, K. W., Bouwer, S. D., and Bowman, B. R.: The development of new solar indices for use in thermospheric density modeling, in: AIAA/AAS Astrodynamics Specialist Conference and Exhibition, American Institute of Aeronautics and Astronautics (AIAA), 21–24 Aug 2006, Keystone, Colorado, 2006.

Tsyganenko, N. A.: A model of the near magnetosphere with a dawn-dusk asymmetry 1. Mathematical structure, J. Geophys. Res., 107, 1179, doi:10.1029/2001JA000219, 2002a.

Tsyganenko, N. A.: A model of the near magnetosphere with a dawn-dusk asymmetry 2. Parameterization and fitting to observations, J. Geophys. Res., 107, 1176, doi:10.1029/2001JA000220, 2002b.

Vasyliunas, V. M.: The mechanical advantage of the magnetosphere: solar-wind-related forces in the magnetosphere-ionosphere-Earth system, Ann. Geophys., 25, 255–269, 2007,
http://www.ann-geophys.net/25/255/2007/.

Volkov, M. A. and Namgaladze, A. A.: Models of field–aligned currents needful to simulate the substorm variations of the electric field and other parameters observed by EISCAT, Ann. Geophys., 14, 1356–1361, 1996,
http://www.ann-geophys.net/14/1356/1996/.

Weimer, D. R.: Maps of ionospheric field–aligned currents as a function of the interplanetary magnetic field derived from Dynamics Explorer 2 data, J. Geophys. Res., 106, 12889–12902, 2001.

Weimer, D. R., Ober, D. M., Maynard, N. C., Collier, M. R., McComas, D. J., Ness, N. F., Smith, C. W., and Watermann, J.: Predicting interplanetary magnetic field (IMF) propagation delay times using the minimum variance technique, J. Geophys. Res., 108, 1026, doi:10.1029/2002JA009405, 2003.

Weimer, D. R.: Improved ionospheric electrodynamic models and application to calculating Joule heating rates, J. Geophys. Res., 110, A05306, doi:10.1029/2004JA010884, 2005.

7

Advantages of the new model of IRI (IRI-Plas) to simulate the ionospheric electron density: case of the European area

O. A. Maltseva, G. A. Zhbankov, and N. S. Mozhaeva

Institute for Physics, Southern Federal Universiy, Russia

Correspondence to: O. A. Maltseva (mal@ip.rsu.ru)

Abstract. Satellite telecommunications, positioning and navigation systems require knowledge of the electron distribution in height Ne(h) to high-altitude orbits of satellites. One of the possibilities to construct such profiles is associated with the use of the ionospheric total electron content TEC. This paper is devoted to three advantages of the IRI-Plas model. They include introduction of the topside basis scale height Hsc, expansion of the IRI model to the plasmasphere, ingestion of experimental values of TEC. Testing of this model according to different satellite experiments (CHAMP, DMSP) shows the high efficiency of this model. The method of adaptation of the IRI-Plas model to the plasma frequency at altitudes of CHAMP and DMSP satellites allows us to produce behavior of Ne(h)-profiles during the disturbances, as well as to refine the values of TEC, which determine the accuracy of positioning. Results were obtained using data of the European area.

1 Introduction

The operation of the various satellite communications, navigation, positioning systems depends on the state of the ionosphere and needs to know the electron distribution in height Ne(h) in near space. In application of radio and satellite communications, the empirical model of the ionosphere IRI (Bilitza, 2001) is most widely used, but it determines the Ne(h)-profile to a height of 2000 km. Ability to determine the profiles at higher altitudes is associated with the total electron content TEC. However, the IRI model gives a large discrepancy when compared with the experimental TEC because of the profile shape of the topside ionosphere, so that the model has been modified several times in this century (IRI2001, IRI2007, Bilitza, 2001; Bilitza, Reinisch, 2008) and the modification is continuing (in 2010, a new version of IRI2010 (Bilitza et al., 2010) was proposed). The next step

can be connected with the IRI-Plas model based on papers of T. Gulyaeva (e.g. Gulyaeva, 2003, 2011). The main advantages of this model are introduction of the topside basis scale height Hsc, expansion of the IRI model to the plasmasphere, ingestion of experimental values of TEC. Previous versions allowed us to adapt the model to two experimental parameters of the ionosphere: the critical frequency foF2 and the maximum height hmF2. Now there are three parameters. This should allow to determine the more precise shape of Ne(h)-profiles. The aim of this work is: (1) determination of the behavior of Ne(h)-profiles when adapting the model to three parameters, (2) clarification of the values of TEC by means of adaptation of the model to the plasma frequency fne(sat) at altitudes of satellites CHAMP and DMSP. The IRI-Plas model was taken from site ftp://ftp.izmiran.rssi.ru/pub/izmiran/SPIM/. The values of the parameters foF2 and hmF2 are taken from the database (http://spidr.ngdc.nasa.gov/spidr/index.jsp), TEC values are taken from IONEX files of global maps of JPL, CODE, UPC, ESA (ftp://cddis.gsfc.nasa.gov/pub/gps/products/ionex/). Data of fne(sat) was taken from http://cindispace.utdallas.edu/DMSP/dmsp_data_at_utdallas.html (Langmuir Probe Data), http://isdc.gfz-potsdam.de/. Juliusruh is chosen as a reference station. In addition, the results for other stations of the European area are presented.

2 Testing the model IRI-Plas according to different experiments

Testing by means of satellite data is carried out for 4 cases used in various applications: (1) the initial model IRI, and models adapted to: (2) the experimental value of foF2, (3) the experimental value of the TEC, (4) experimental values of foF2 and TEC, to demonstrate the difference between the results for these methods. Option 1 is used when there is no current information, and determines the average (quiet)

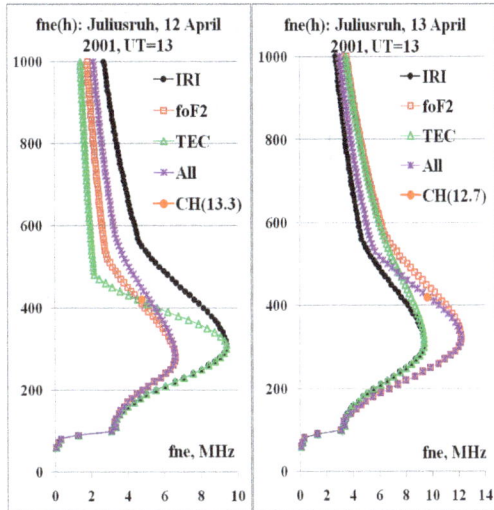

Fig. 1. Comparison of Ne(h)-profiles corresponding to the initial IRI model (black circles) with Ne(h)-profiles of the model adapted to experimental values of foF2 (red squares), experimental values of TEC (green triangles) and joint values of foF2 and TEC (violet asterisks). The red circles show values of the plasma frequency of the CHAMP satellite. The left panel illustrates case of the positive disturbance, the right panel applies to negative one.

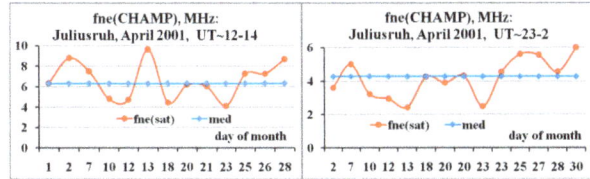

Fig. 2. The behavior of the plasma frequency fne(CHAMP) near noon (left panel) and midnight (right panel) along with medians of these values to detect periods of positive and negative disturbances in the topside ionosphere.

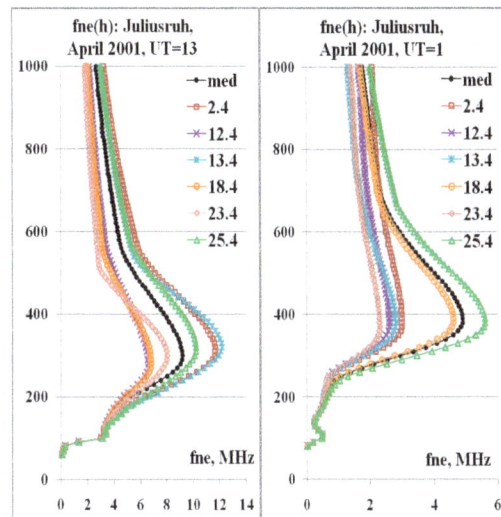

Fig. 3. The behavior of Ne(h)-profiles during disturbances in April 2001. Black circles display the median profiles obtained by the model adapted to experimental medians of foF2 and TEC. The other profiles apply to days indicated by corresponding marks. The left panel concerns to time near noon (UT = 13), the right panel – time near midnight (UT = 1).

ionospheric state. It is a standard for comparison with other options. Option 2 uses the current value of foF2 and completely defines the bottom side of the profile. Option 3 is widely used in connection with the TEC measurements. The advantages of this option before the second one are in a continuous global monitoring. Adapting the model to the current values of the TEC allows us to obtain new values of foF2. Option 4 is one of the main differences between the IRI-Plas model and previous IRI versions. It allows to determine the Ne(h)-profile at the location of ionosondes. Testing of these options is to compare the plasma frequency at altitudes of satellites calculated for the model with the experimental values of fne. A comparison was carried out for two satellites CHAMP (hsat ∼ 400 km) and DMSP (hsat ∼ 840 km). Results are given for April 2001, including two strong and two weak disturbances. Figure 1 displays cases related to the negative and positive perturbations.

A substantial difference between foF2(obs) and foF2(IRI) is seen. The profiles shown by black dots present the profiles for the initial model. The profiles shown by green triangles correspond to the traditional adaption techniques when the construction of N(h)-profile uses the experimental value of the TEC and the model value of foF2 (McNamara, 1985; Houminer and Soicher, 1996, Gulyaeva, 2003), as well as assimilation techniques involving TEC (e.g., Khattatov et al., 2005). The difference can be large. The best agreement corresponds to the new version. It allows us to illustrate how can change N(h)-profiles, adapted to the foF2 and TEC during the disturbances of various types.

3 Examples of the behavior of Ne(h)-profiles during the disturbances

Figure 2 shows the behavior of the plasma frequency fne(sat) according to CHAMP satellite in April 2001 over the Juliusruh station, along with the median to detect cases of disturbances in the upper ionosphere. In the left panel are cases that fall near noon, on the right side – near midnight. Figure 3 shows Ne(h)-profiles for the Juliusruh station together with Ne(h)-profiles calculated for the experimental medians of foF2 and TEC (the lines shown by black dots).

A positive deviation is seen for the daytime. At night, negative deviations dominate with lower gradients in the topside. Another example refers to a period of quiet Sun, which attracted a great attention of the scientific community because of the unusual duration (e.g. Liu et al., 2011; Zakharenkova et al., 2011). In terms of the perturbation this

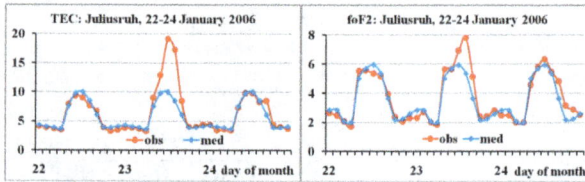

Fig. 4. Typical example of the TEC burst and corresponding variation of foF2: "obs" presents experimental values, "med" shows medians.

period is characterized by the absence of strong and moderate disturbances. The disturbances were observed in the form of positive bursts of TEC (Cander and Haralambous, 2011) synchronously covering the European zone from Chilton to Athens. Example of the TEC burst and corresponding variation of foF2 for the Juliusruh station for 22–24 January 2006 is displayed in Fig. 4. Behavior of Ne(h)-profiles for these days and UT = 8–18 is given in Fig. 5 along with profiles for medians. At the time UT = 8, Ne(h)-profiles are still little different from the medians. At the time UT = 10, only ionization of the topside part increased. At the time UT = 12, a gain propagated to lower ionosphere, reaching a maximum at UT = 14. At the time UT = 16, decrease in the ionization of the topside part occurred most rapidly. At the time UT = 18, peaks of TEC and foF2 have completely disappeared. This illustrates that development of disturbance is connected with transition of ionization from plasmasphere to the region near the F2 maximum. This behavior is characteristic for other mid latitude stations (Chilton, Ebre, Athens, Leningrad, Moscow), and for other cases. Different behavior of Ne(h)-profiles is visible for the Tromso station, as shown in Fig. 6. The perturbation starts earlier (at UT = 0) and passes the maximum phase in UT = 12. Some profiles seem unusual, but a comparison of profiles in the times when measurements of fne were in the satellite CHAMP (Fig. 7) shows a complete coincidence for adapted profiles. Since it is difficult to obtain the moments of coincidence between measurements of the satellite and the TEC, Fig. 7 shows the results for the nearest moments that are indicated in parentheses. The difference between profiles for four versions is great.

The above examples show that the profiles adapted to both of the foF2 and TEC parameters can be an important tool for studying the behavior of the ionosphere. N(h)-profiles adapted to one of the parameters may be far away from the real profiles.

4   Refinement of the TEC values from satellite experiments

There is a variety of the TEC values, which can be used to adapt the model. These are: (1) measurement at individual receivers, the values of which can differ up to ~ 10 TECU

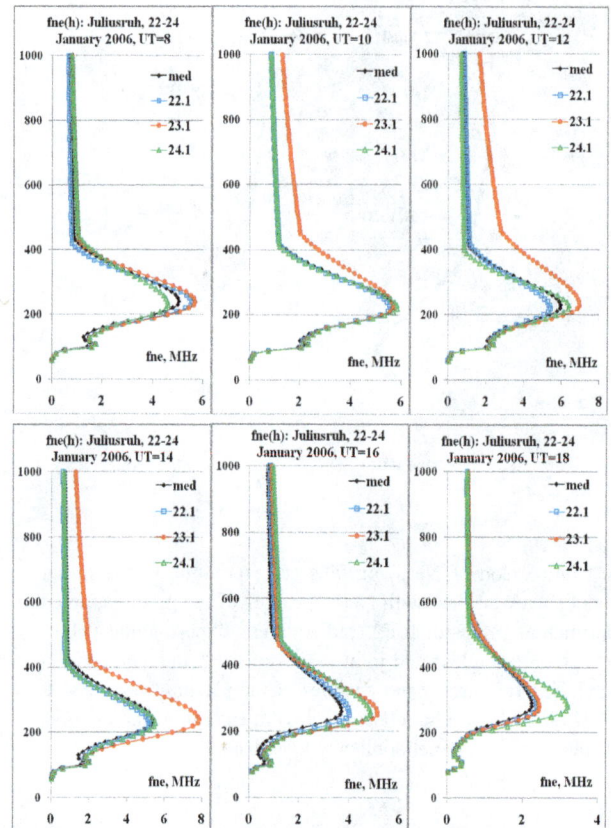

Fig. 5. The behavior of Ne(h)-profiles during the burst of TEC on 22–24 January 2006 over the Juliusruh station. The Ne(h)-profiles shown by black circles were obtained by the model adapted to experimental medians of foF2 and TEC. The Ne(h)-profiles shown by red circles apply to the day of the burst (23 January 2006).

of each other (e.g. Choi et al., 2010), (2) the global maps, the values of which can differ up to 2–3 times (Arikan et al., 2003), (3) tomographic measurements (Chartier et al., 2012) and others. A typical example of differences was done in Fig. 7 of the paper (Arikan et al., 2003). This Figure compares TEC values obtained by various methods: authors of Arikan et al. (2003), Rutherford Appleton Laboratory (Ciraolo and Spalla, 1997), Canadian maps and four IGS maps for the Kiruna station on 25 and 28 April 2001 (difference in 2–3 times). We had only the values of global maps of JPL, CODE, UPC, ESA. The values of these maps for the Juliusruh station on 25 April are shown in our Fig. 8 (the left panel): the difference for four maps may lie in the range of 10–30 TECU. This set of values leads to variety of N(h)-profiles that differ greatly from each other. It is difficult to specify the criterion to select the TEC close to the true value. But there is a criterion for choosing N(h)-profiles. This is measurement of the plasma frequency fne(sat) on the satellites. Of two satellites, DMSP should be preferred due to a high-altitude orbit and the values of CHAMP sometimes meet several profiles. Ne(h)-profiles adapted to fne(DMSP)

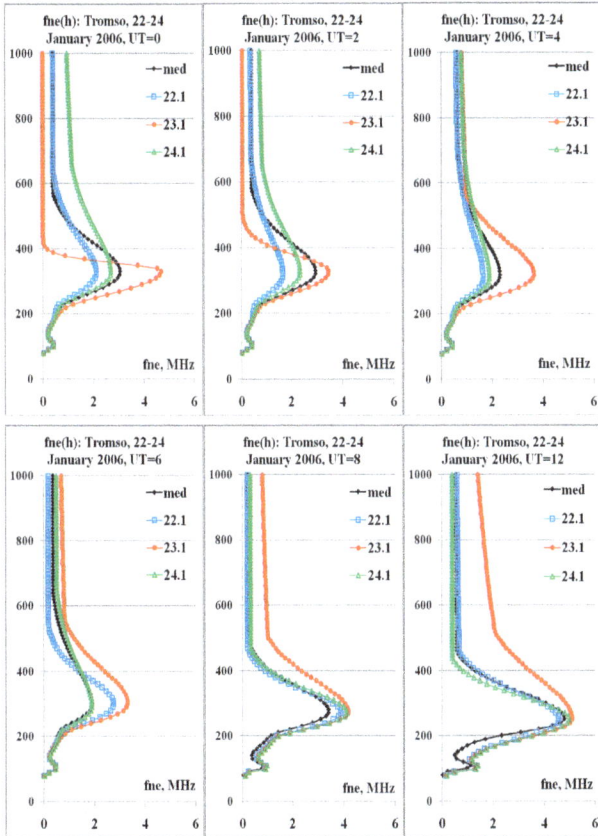

Fig. 6. The behavior of the Ne(h)-profiles during the burst of TEC on 22–24 January 2006 for the Tromso station. Designations are the same as on Fig. 5.

provide new values of the TEC, which are different from the values of all global maps. The right panel of Fig. 8 shows these values for two days (25, 28 April 2001) and time of satellite passages indicated in parenthesis. For example, 25(9) means 25 April 2001 and UT = 9. Another example is presented in Fig. 9 for the Moscow station and other disturbed month (July 2004).

These examples and statistics from many stations and geophysical conditions show that the values of TEC corresponding fne(DMSP) may lie within the range of maps below the weighted average IGS values (Hernandez-Pajares et al., 2009), but may be below that range.

5   Conclusions

TEC is the most important parameter of the ionosphere used in scientific and engineering applications, so there is a need of its model. One such model is the IRI. This paper deals with the IRI-Plas version of this model, which has advantages of a rigorous account of the plasmaspheric part of the profile, the introduction of Hsc parameter to determine the shape of the upper part of the profile, ingestion of the TEC. We used

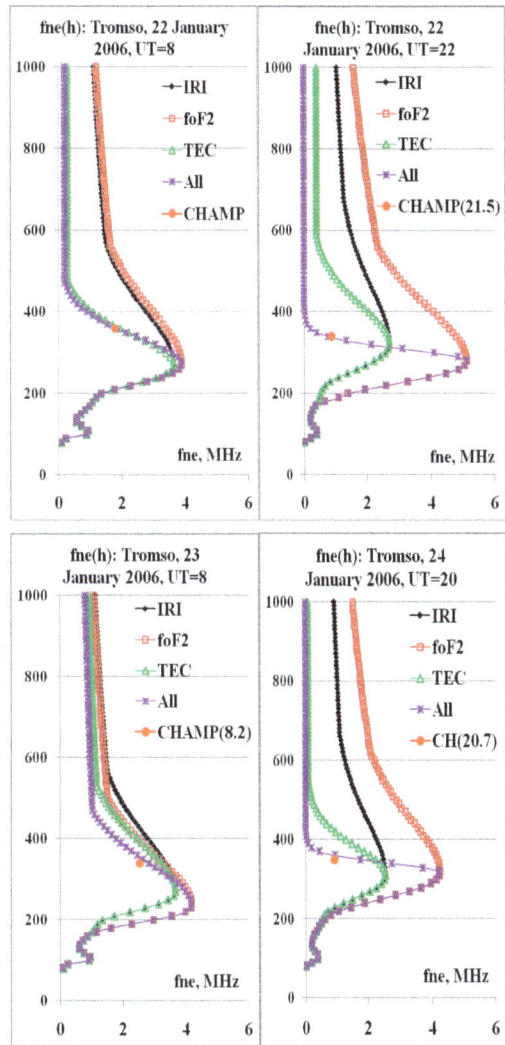

Fig. 7. Ne(h)-profiles during the burst of TEC on 22–24 January 2006 in the Tromso station at moments, close to passages of the satellite.

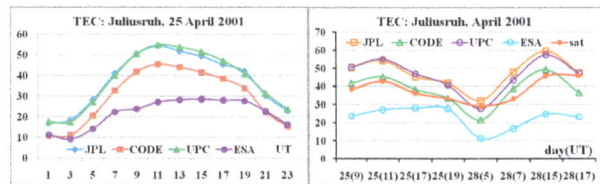

Fig. 8. Differences of TEC calculated from the various global maps (left panel) and TEC obtained by adaptation of the model to the plasma frequency fne of the DMSP satellite (right panel).

these advantages to show: (1) the adaptation of the model to the TEC has provided Ne(h)-profiles, very different from the profiles of the original model, and allows to assess the behavior of the ionosphere during the disturbances, (2) due to the variety of values of TEC, adaptation of the model to

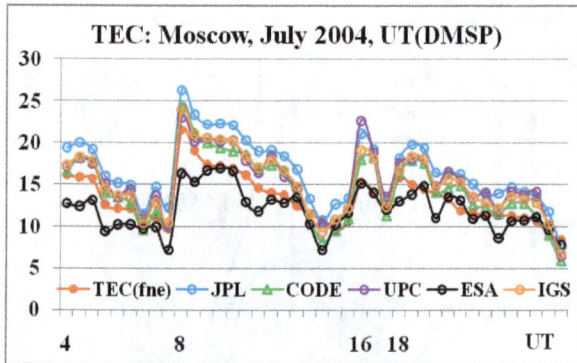

Fig. 9. New TEC values in comparison with data of global maps of JPL, CODE, UPC, ESA and weighted average IGS values. Results are given for days (not shown in the figure) and time UT of the satellite passages indicated on the x-axis.

one of the parameters, or even three parameters (foF2, hmF2, TEC) does not always lead to the Ne(h)-profile corresponding plasma frequencies fne(sat), and (3) an adaptation of the model to fne(sat) can lead to new values of TEC, which can lie within the existing set of values, and may be below that set. We can point to two possible ways to use these values. If we can specify a value of "true" TEC, the difference can be attributed to the plasmaspheric part and thus we get the way of correction of this part, which is still of some interest (e.g., Cherniak et al., 2012). If this value cannot be specified, then the values of TEC for profiles adapted to fne(sat), can be considered as an improved value of the TEC. For applications in navigation and positioning, an empirical ionospheric model should be used to minimize the ionospheric errors. The Klobuchar model (e.g., Klobuchar, 1987) widely used for this aim provides a 50 % correction on a global basis. It is possible to suppose that the IRI-Plas model can compensate the ionospheric error with much higher accuracies. These results and conclusions were obtained for the European region. It is not difficult to obtain such results for any other region of the globe.

Acknowledgements. Authors thank scientists provided data of SPIDR, global maps of TEC, operation and modification of the IRI model, A. Karpachev for CHAMP data, two reviewers for very important comments.

References

Arikan, F., Erol, C. B., and Arikan, O.: Regularized estimation of vertical total electron content from Global Positioning System data, J. Geophys. Res., 108, 1469, doi:10.1029/2002JA009605, 2003.

Bilitza, D.: International Reference Ionosphere, Radio Sci., 36, 261–275, 2001.

Bilitza, D. and Reinisch, B. W.: International Reference Ionosphere 2007: Improvements and New Parameters, Adv. Space Res., 42, 599–609, 2008.

Bilitza, D., Reinisch, B. W., and Gulyaeva, T.: ISO technical specification for the ionosphere – IRI recent activities, in: Report presented for COSPAR Scientific Assembly, Bremen, Germany, C01-0004-10, 2010.

Cander, L. R. and Haralambous, H.: On the importance of TEC enhancements during the extreme solar minimum, Adv. Space Res., 47, 304–311, 2011.

Chartier, A., Mitchell, C. N., and Jackson, D. R.: A 12 year comparison of MIDAS and IRI2007 ionospheric total electron content, Adv. Space Res., 49, 1348–1355, 2012.

Cherniak, Iu. V., Zakharenkova, I. E., Krankowski, A., and Shagimuratov, I. I.: Plasmaspheric electron content from GPS TEC and FORMOSAT-3/COSMIC measurements: solar minimum conditions, Adv. Space Res., 50, 427–440, 2012.

Choi, B.-K., Chung, J.-K., and Cho, J.-H.: Receiver DCB estimation and analysis by types of GPS receiver, J. Astron. Space Sci., 27, 123–128, 2010.

Ciraollo, L. and Spalla P.: Comparison of ionospheric total electron content from the Navy Navigation Satellite System and the GPS, Radio Sci., 32, 1071–1080, 1997.

Gulyaeva, T. L.: International standard model of the Earth's ionosphere and plasmasphere, Astron. and Astrophys. Transaction, 22, 639–643, 2003.

Gulyaeva, T. L.: Storm time behavior of topside scale height inferred from the ionosphere-plasmasphere model driven by the F2 layer peak and GPS-TEC observations, Adv. Space Res., 47, 913–920, 2011.

Hernandez-Pajares, M., Juan, J. M., Orus, R., Garcia-Rigo, A., Feltens, J., Komjathy, A., Schaer, S. C., and Krankowski, A.: The IGS VTEC maps: a reliable source of ionospheric information since 1998, J. Geod., 83, 263–275, 2009.

Houminer, Z. and Soicher, H.: Improved short-term predictions of foF2 using GPS time delay measurements, Radio Sci., 31, 1099–1108, 1996.

Khattatov, B., Murphy, M., Gnedin, M., Sheffel, J., Adams, J., Cruickshank, B., Yudin, V., Fuller-Rowell, T., and Retterer J.: Ionospheric nowcasting via assimilation of GPS measurements of ionospheric electron content in a global physics-based time-dependent model, Q. J. Roy. Meteorol. Soc., 131, 3543–3559, 2005.

Klobuchar, J. A.: Ionospheric time-delay algorithm for single-frequency GPS users, IEEE Transactions on aerospace and electronic systems, AES-23, 325–331, 1987.

Liu, L., Chen, Y., Le, H., Kurkin, V. I., Polekh, N. M., and Lee, C.-C.: The ionosphere under extremely prolonged low solar activity, J. Geophys. Res., 116, A04320, doi:10.1029/2010JA016296, 2011.

McNamara, L. F.: The use of total electron density measurements to validate empirical models of the ionosphere, Adv. Space Res., 5, 81–90, 1985.

Zakharenkova, I. E., Krankowski, A., Bilitza, D., Cherniak, Yu. V., Shagimuratov, I. I., and Sieradzki, R.: Comparative study of foF2 measurements with IRI-2007 model predictions during extended solar minimum, Adv. Space Res., 51, 620–629, 2011.

8

Numerical modeling of the equatorial electrojet UT-variation on the basis of the model GSM TIP

M. V. Klimenko[1], **V. V. Klimenko**[2], **and V. V. Bryukhanov**[1]

[1]Kaliningrad State Technical University, Kaliningrad, Russia
[2]West Department of IZMIRAN, Kaliningrad, Russia

Abstract. In the presented work the results of numerical modeling of the UT-variation of the equatorial electrojet, executed on the basis of the model GSM TIP are presented, taking into account the dynamo electric fields generated by thermospheric winds in a current-carrying layer of the ionosphere at heights 80–175 km above a surface of the Earth. To the Global Self-consistent Model of the Thermosphere, Ionosphere and Protonosphere (GSM TIP), developed in WD IZMIRAN, a new block for the calculation of electric fields in the ionosphere has been added. In this block the solution of the three-dimensional equation describing the conservation law of the full current in the Earth's ionosphere is reduced to the solution of the two-dimensional equation by integration along geomagnetic field lines. Calculations of parameters of the near-Earth space plasmas have been executed for quiet equinoctial conditions on 22 March 1987 during the minimum of solar activity.

It has been shown, that there is a distinct semidiurnal harmonic in the diurnal behavior of the linear density of the equatorial electrojet with maxima at 23:00 UT and 15:00 UT, as well as with minima at 06:00 UT and 20:00 UT. The greatest and smallest values of the peak intensity of the equatorial electrojet with respect to the diurnal behavior can differ by a factor of two. The longitudinal extent of the area of the equatorial electrojet does hardly show any UT-variation, but the greatest longitudinal extent is at 06 UT. With the growth of the peak intensity of the equatorial electrojet its latitudinal extent also increases (on ∼5–10°) a little. At the same time the equatorial electrojet in the maxima of intensity has approximately an identical width, whereas in the minima the electrojet is narrow in the morning and wide in the afternoon.

As for the surface density of the equatorial electrojet, its UT-variation is much weaker and equals ∼1–3 A/km² and the peak intensity is equal ∼15–20 A/km². The latitudinal extent of the surface density of the equatorial electrojet is maximal at 23:00 UT and 15:00 UT and minimal at 06:00 UT and 20:00 UT.

Correspondence to: M. V. Klimenko
(maksim.klimenko@mail.ru)

1 Introduction

The equatorial ionosphere is unique in many respects. Global scale dynamo action, i.e. the generation of currents by electromotive forces due to tidal winds results in the generation of planetary scale east-west electric fields at low latitudes around the geomagnetic equator. These electric fields, in combination with the north-south magnetic fields, cause different geophysical phenomena near the geomagnetic equator. The most important equatorial ionospheric phenomena are: the equatorial ionization anomaly (or Appleton anomaly), the equatorial electrojet, and the generation of plasma density irregularities. An outstanding problem is the cause for the day-to-day variability in the intensity of the electric fields that are responsible for the equatorial anomaly and the electrojet. The enhanced variations of the Earth's magnetic field over the equator were explained by Egedal (1947) to be due to an enhanced east-west current flow in a narrow latitudinal belt ±3° around the geomagnetic equator. This was later named equatorial electrojet by Chapman (1951). The explanation for the electrojet was offered on the basis of the electrodynamics of a horizontally stratified ionosphere with anisotropic conductivities in a horizontal magnetic field. In a magnetoplasma with mutually perpendicular electric E and magnetic B fields, Pedersen currents flow parallel to E and Hall currents flow perpendicular to both E and B. In the presence of the almost non-conducting boundaries above and below the dynamo region (80–175 km), the flow of Hall currents is inhibited. Under such circumstances, the east-west conductivities are enhanced, increasing the flow of currents, namely, the equatorial electrojet. Cowling (1933), Martyn et al. (1948), Baker and Martyn (1953), and Sugiura and Cain (1966) made important contributions to the explanation of this phenomenon.

Connected with the equatorial electrojet are several phenomena occurring in the equatorial ionosphere during quiet and disturbed geomagnetic conditions, such as equatorial spread F and equatorial plasma bubbles which influence radio waves propagation in the vicinity of geomagnetic equator.

2 Statement of the problem and a brief description of the model

In this work results of numerical modeling of the UT-variation of the equatorial electrojet, executed with the Global Self-consistent Model of the Thermosphere, Ionosphere and Protonosphere (GSM TIP) are presented. The calculations were carried out taking into account only the dynamo electric fields generated by winds of the regular thermospheric circulation, but neglecting thermospheric tides on the bottom boundary of thermosphere at a height of 80 km and in the current-carrying layer of the ionosphere over the altitude range of 80 to 175 km above the Earth's surface. The GSM TIP was developed in the West Department of IZMIRAN. For given input data (possibly time dependent) the model calculates the time-dependent global three-dimensional structure of temperature, composition (O_2, N_2, O) and mass velocity vector of the neutral gas; and densities, temperatures and vector velocities of atomic (O^+, H^+) ions, molecular (N_2^+, O_2^+, NO^+) ions and electrons, and the two-dimensional distribution of the electric field potential both of dynamo and magnetospheric origin. Additionally the mismatch of the geographic and geomagnetic Earth's axes is taken into account.

The solution is performed numerically on a global grid with a resolution of 5° in geomagnetic latitude for the neutral atmospheric and ionospheric equations. The integration of the latter equations is executed along geomagnetic field lines. The geomagnetic field is presented in the model by a tilted centred dipole. Field lines are considered to be open at $L > 14.9$, where $L = \frac{r}{R_E}$ is the McIlwain parameter. The longitudinal resolution step is 15° in a spherical geomagnetic coordinate system. In the vertical dimension, the thermospheric code uses 30 height grid points between 80 and 520 km altitude above the Earth's surface.

The ionospheric part of the code (F2-region and above) has variable spatial steps along the magnetic field lines from a base altitude 175 km to a maximum distance of 15 Earth's radii. The paper by Namgaladze et al. (1988) is devoted to the detailed description of the general statement of the problem of modeling parameters of the thermosphere-ionosphere-protonosphere system as a whole. The statement of the problem is also given in papers by Namgaladze et al. (1991), Korenkov et al. (1998).

To the model a new block of calculation of electric fields in the ionosphere is added (Klimenko et al., 2005, 2006a, b). The modelled physical principles, the mathematical structure of the new block of calculation of electric field and zonal current in the Earth's ionosphere of the model GSM TIP and the used algorithm of the calculations has been described in detail by Klimenko et al. (2006a, b). The distribution of the quasi-stationary large-scale electric field in the Earth's ionosphere is described by the current density conservation law:

$$div\, \boldsymbol{j} = 0. \tag{1}$$

where $\boldsymbol{j} = \widehat{\sigma} \cdot (\boldsymbol{E} + \boldsymbol{V}_n \times \boldsymbol{B}) = \widehat{\sigma} \cdot \boldsymbol{E'}$ is the surface density of the current, $\widehat{\sigma}$ is the ionospheric conductivity tensor, $\boldsymbol{E}$ is the electric field of polarization, $\boldsymbol{V}_n \times \boldsymbol{B}$ is the dynamo field, $\boldsymbol{V}_n$ is the velocity of the average mass motion of the neutral gas, and $\boldsymbol{B}$ is the geomagnetic field induction.

The three-dimensional Eq. (1) was reduced to two-dimensional integration over the height of the current-conducting layer. The transition to the two-dimensional equation is performed by taking the integral along the geomagnetic field lines assuming that the electric field is constant in the current-conducting ionospheric layer along these field lines, which are expected equipotential. Such an approach to modeling the electric field in the Earth's ionosphere has long been known and is used by many researchers. This approach has been described in many monographs and manuals, e.g., in (Gurevich et al., 1976; Richmond, 1982; Volland, 1984; Singh and Cole, 1987; Heelis, 2004).

Assuming that the geomagnetic field is a dipole, we introduce the dipole (dipolar) coordinate system (q, v, u), where:

$$q = \rho^2 \cdot \cos \Theta, \quad v = \Lambda, \quad u = \rho \cdot \sin^2 \Theta.$$

Here $\rho = \frac{R_E}{r} = \frac{R_E}{R_E + h}$, R_E is the Earth's radius, h is the height above the Earth's surface, (r, Θ, Λ) is a spherical geomagnetic coordinate system, r is the radius vector, Θ is the geomagnetic colatitude (polar angle), and Λ is the geomagnetic longitude. The conductivity tensor shape in the dipole (dipolar) and spherical geomagnetic coordinate systems can be found, e.g., in (Gurevich et al., 1976).

In the dipole (dipolar) coordinate system (q, v, u), the equation of current density conservation in the Earth's ionosphere Eq. (1) has the form:

$$\frac{1}{h_q \cdot h_v \cdot h_u} \cdot \left(\frac{\partial}{\partial v} \left(h_q \cdot h_u \cdot (\sigma_P \cdot E_v + \sigma_H \cdot E_u + \right. \right.$$

$$+ (\sigma_P \cdot V_{nu} - \sigma_H \cdot V_{nv}) \cdot B \,)) + \frac{\partial}{\partial u} \left(h_q \cdot h_v \cdot (\sigma_P \cdot E_u - \right.$$

$$\left. \left. - \sigma_H \cdot E_v - (\sigma_P \cdot V_{nv} + \sigma_H \cdot V_{nu}) \cdot B \,))) = 0, \right.$$

where $h_q = \frac{R_E}{\rho^3 \cdot k}$, $h_v = \frac{R_E}{\rho} \cdot \sin \Theta$, and $h_u = \frac{R_E}{\rho^2 \cdot k \cdot \sin \Theta}$ are the Lame coefficients, and $k = \sqrt{1 + 3 \cdot \cos^2 \Theta}$

If the electric field potential in the Earth's ionosphere is taken into account, we obtain:

$$\frac{\partial}{\partial u} \left(\frac{h_q \cdot h_v}{h_u} \cdot \sigma_P \cdot \frac{\partial \Phi}{\partial u} \right) + \frac{\partial}{\partial v} \left(h_q \cdot \sigma_H \right) \cdot$$

$$\cdot \frac{\partial \Phi}{\partial u} + \frac{\partial}{\partial v} \left(\frac{h_q \cdot h_u}{h_v} \cdot \sigma_P \cdot \frac{\partial \Phi}{\partial v} \right) -$$

$$- \frac{\partial}{\partial u} \left(h_q \cdot \sigma_H \right) \cdot \frac{\partial \Phi}{\partial v} = \frac{\partial}{\partial v} \left(h_q \cdot h_u \cdot (\sigma_P \cdot V_{nu} - \sigma_H \cdot V_{nv}) \cdot B \right) -$$

$$- \frac{\partial}{\partial u} \left(h_q \cdot h_v \cdot (\sigma_P \cdot V_{nv} + \sigma_H \cdot V_{nu}) \cdot B \right), \tag{2}$$

where Φ is the electric field potential; $B = |\boldsymbol{B}|$, σ_P and σ_H are the Pedersen and Hall conductivities of the ionosphere, the expressions of which can be found in (Gershman, 1974; Heelis, 2004).

Let us integrate Eq. (2) along the field line segment in the current-conducting layer from q_1 to q_2. If the field line completely lies in the ionospheric current-conducting layer, the integration is performed from the bottom of this line in the given hemisphere to the top.

$$\frac{\partial}{\partial u} \int\limits_{q_1}^{q_2} \frac{h_q \cdot h_v}{h_u} \cdot \sigma_P \cdot \frac{\partial \Phi}{\partial u} \cdot dq + \int\limits_{q_1}^{q_2} \frac{\partial}{\partial v}(h_q \cdot \sigma_H) \cdot \frac{\partial \Phi}{\partial u} \cdot dq +$$

$$+\frac{\partial}{\partial v} \int\limits_{q_1}^{q_2} \frac{h_q \cdot h_u}{h_v} \cdot \sigma_P \cdot \frac{\partial \Phi}{\partial v} \cdot dq - \int\limits_{q_1}^{q_2} \frac{\partial}{\partial u}(h_q \cdot \sigma_H) \cdot \frac{\partial \Phi}{\partial v} \cdot dq = \psi,$$

where

$$\psi = \frac{1}{R_E^3} \cdot \frac{\partial}{\partial v} \int\limits_{q_1}^{q_2} (\sigma_P \cdot V_{nu} - \sigma_H \cdot V_{nv}) \cdot B \cdot$$

$$\cdot \frac{r^5}{\sin \Theta \cdot (1 + 3 \cdot \cos^2 \Theta)} \cdot dq -$$

$$-\frac{1}{R_E^2} \cdot \frac{\partial}{\partial u} \int\limits_{q_1}^{q_2} (\sigma_P \cdot V_{nv} + \sigma_H \cdot V_{nu}) \cdot B \cdot \frac{r^4 \cdot \sin \Theta}{\sqrt{1 + 3 \cdot \cos^2 \Theta}} \cdot dq$$

Assuming that inflowing and outflowing currents are absent at the lower boundary of the ionospheric current conducting layer at a height of 80 km, we have the boundary condition for the three-dimensional modeling Eq. (1): $j_q = 0$ (everywhere except at the equator) and $j_u = 0$ (at the equator) at a height of $h = 80$ km, which is used as the lower limit during the integration over the thickness of the ionospheric current-conducting layer.

We now determine the zonal electrojet as a zonal current surface density integrated over the thickness of the ionospheric current-conducting layer, i.e., as a zonal current linear density. In this case the positive and negative signs of the zonal electrojet will correspond to the direction toward east and west, respectively.

In the model where the integration is performed along the geomagnetic field lines, the zonal electrojet is calculated using the formula:

$$J_v = \sum\nolimits_P \cdot E_v + \sum\nolimits_H \cdot E_u + \int\limits_{q_1}^{q_2} B(\sigma_P \cdot V_{nu} - \sigma_H \cdot V_{nv}) \cdot h_q \cdot dq,$$

where $\sum_P = \int\limits_{q_1}^{q_2} \sigma_P \cdot h_q \cdot dq$; $\sum_H = \int\limits_{q_1}^{q_2} \sigma_H \cdot h_q \cdot dq$

In papers by Klimenko et al. (2006a, b) the modelled physical principles and the mathematical structure of the new

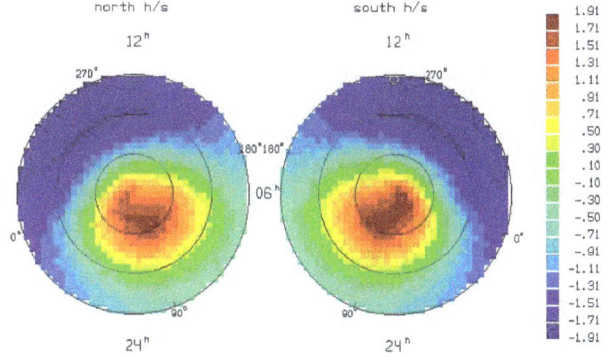

Fig. 1. Dynamo electric field potential distribution in polar geomagnetic coordinate system (latitude-longitude) obtained in the model GSM TIP for 00:00 UT.

block of calculation of electric field and zonal current in the Earth's ionosphere of the model GSM TIP, and the used algorithm of the calculations are explained in every detail.

The equatorial electrojet is essentially a LT (local time) phenomenon with the UT component resulting from longitudinal differences arising from the differences in conductivity due to various factors, e.g. the fact that the dipole axis and the geographic axis do not coincide, or that the ambient magnetic field has a non-dipolar component, or that the magnetospheric sources are contributing to the electric fields or conductivity. From all listed factors in this work there is only the first factor taken into account, namely, that the dipole axis and the geographic axis do not coincide. The second and the third of the above factors are neglected here. Firstly, in the model GSM TIP the dipole approximation of the ambient geomagnetic field is used, therefore a non-dipol component naturally is neglected. Secondly, all calculations in this work were carried out taking into account the dynamo field generated by thermospheric winds only. Therefore the contribution of electric fields of magnetospheric convection is absent.

The inclusion of the new block of the electric field calculation in the model GSM TIP allows us to investigate the equatorial ionosphere. The calculations were carried out for quiet equinox conditions on 22.03.1987 during solar activity minimum ($F_{10.7} = 76$).

3 Calculation results and their discussion

Figure 1 shows the global distribution of the dynamo electric field potential in kV in a polar geomagnetic coordinate system (latitude-longitude), calculated on the basis of GSM TIP for 00:00 UT. The geomagnetic latitudes are shown by circles with steps of $30°$ from the geomagnetic pole to the geomagnetic equator for both hemispheres. Numbers on the equatorial circle denote the positions of longitudinal geomagnetic

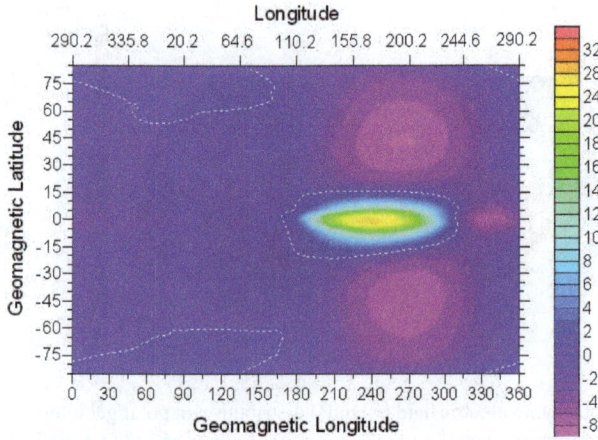

Fig. 2. Zonal current linear density obtained in the model GSM TIP in A/km for 23:00 UT.

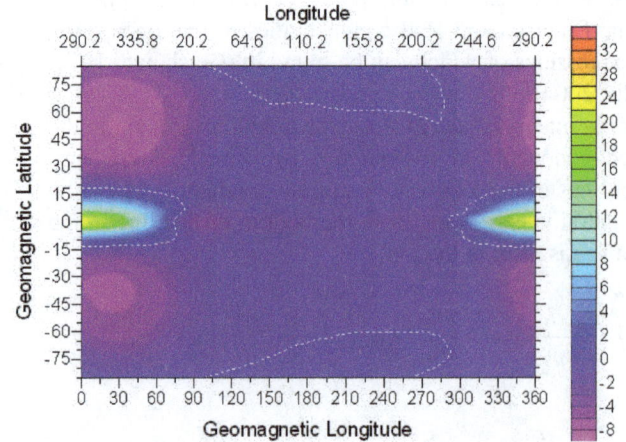

Fig. 4. Zonal current linear density obtained in the model GSM TIP in A/km for 15:00 UT.

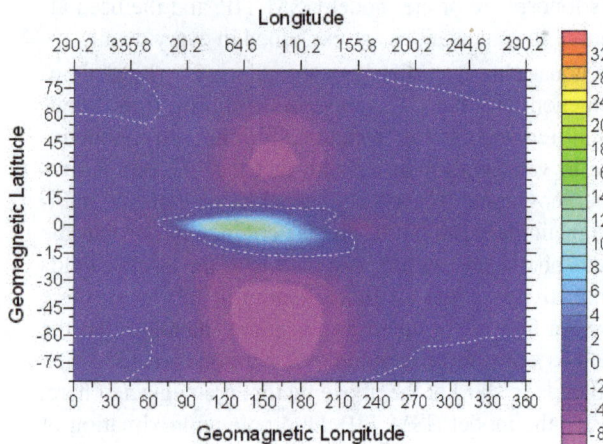

Fig. 3. Zonal current linear density obtained in the model GSM TIP in A/km for 06:00 UT.

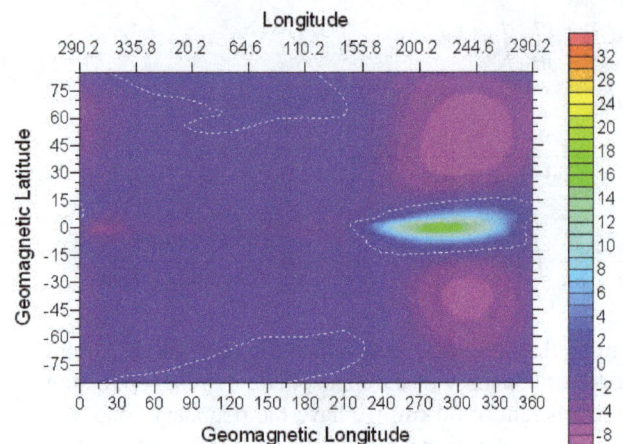

Fig. 5. Zonal current linear density obtained in the model GSM TIP in A/km for 20:00 UT.

meridians 0°, 90°, 180° and 270°. Also the time in a Solar-Magnetospheric coordinate system is shown.

Figures. 2–5 show the global distributions of zonal current linear density, obtained with the model GSM TIP for 23:00 UT, 06:00 UT, 15:00 UT and 20:00 UT in a Cartesian geomagnetic coordinate system (longitude-latitude). One can see that the maximal intensity of the zonal current linear density has maxima at 23:00 UT and 15:00 UT and minima at 06:00 UT and 20:00 UT. The greatest and smallest values of the equatorial electrojet maximal intensity during the day can differ by a factor of two. The longitudinal extent of the equatorial electrojet area does hardly reveal any UT-variation, although it is possible to note the greatest longitudinal extent at 06:00 UT. With amplification of the equatorial electrojet maximal intensity, its latitudinal extent slightly increases also (about 5–10°). At the same time, the equatorial electrojet in its maxima of intensity has approximately an

identical width, and in the minima of intensity the electrojet is narrow in the morning and wide in the afternoon.

Under symmetrical conditions in the Northern and Southern hemisphere the current at the upper boundary of the current-conducting layer should vanish. In reality, an asymmetry is always observed between the hemispheres, since the conductivities and thermospheric circulation are different on both sides of the geomagnetic equator. This asymmetry originates even during equinoctial conditions, because the axes of the Earth rotation and the geomagnetic dipole do not coincide. Therefore, the geomagnetic field lines remain on equipotential (due to the high field-aligned conductivity along these lines) only when currents are generated along the closed geomagnetic field lines. These field aligned currents almost instantaneously compensate the originating potential drop between the hemispheres. On open geomagnetic field lines, these currents are absent due to the absence of any relation between the hemispheres.

The majority of researchers speak now only about a longitudinal variation of various parameters of the near-Earth environment, including the equatorial electrojet, forgetting about UT-variation or identifying it with a longitudinal variation. We consider this as incorrect. Really, in an experiment it is very difficult, if at all possibly, to separate a longitudinal variation from a UT-variation. However it is very easy in numerical modeling.

Let us consider the sources of these variations. Originally it is supposed that a UT variation of the ionospheric parameters are connected with the mismatch of the axis of the geomagnetic dipole with the geographic axis of the Earth rotation. For any existence of longitudinal variations the presence of longitudinal sources is necessary. These sources must be located in well-defined places on the Earth's surface. For example, the South-Atlantic anomaly is a source of longitudinal variations. This source is connected with the presence of multipole components in the geomagnetic field. UT and longitude are independent variables. We can expand every function in series of these two independent variables. Herewith one term of the decomposition in the series will depend on longitude only, the other will depend on UT only, but the third (the cross terms of the decomposition) will depend on both variables. The UT variation will be described with the terms of the decomposition, depending on UT. The longitudinal variation will be described with the terms, depending on longitude. But cross terms will describe a dependency on both UT, and longitude. Most likely, the contribution of these terms will be small in comparison with the contribution of the main terms, describing longitudinal and UT variation.

In our model the geomagnetic field is approximated by a central dipole, the axis of which does not coincide with the geographical axis of Earth rotation. In this approximation of a geomagnetic field, there is no non-dipole component. Therefore the visible reasons for an existence of longitudinal variations are absent, while the mechanism of formation of a UT-variations is present. Hence, we shall speak about UT-variations, discussing results of the calculations. Also we shall speak about longitudinal variations, discussing experimental data in which UT and longitudinal variations are contained, since in this case we deal with real data.

The equatorial electrojet is known to peak around 11:00–12:00 LT. In our calculation results the maximum intensity of the equatorial electrojet appears around $\Lambda=240°$ (11:00 LT) for 23.00 UT, around $\Lambda=130°$ (10:40 LT) for 06:00 UT, around $\Lambda=0°$ (11:00 LT) for 15:00 UT, and around $\Lambda=280°$ (10:20 LT) for 20:00 UT. As for a displacement of the maximum of the electrojet into the morning sector of LT in comparison with real experimental data, this can be caused by neglecting tides on the bottom boundary of thermosphere and by neglecting the electric field of magnetospheric convection in our calculations.

In papers by Lühr et al. (2004), Manoj et al. (2006), Le Mouel et al. (2006) observational data of the equatorial elec-

trojet obtained with the CHAMP satellite were analyzed. The analysis of data has shown, that the equatorial electrojet represents a narrow formation on the geomagnetic equator. The width of the equatorial electrojet equals about 2000 km in the day-time ionosphere. At the same time the maximal intensity of the equatorial electrojet was estimated by Lühr et al. (2004) as 0.15 A/m on the average, whereas Manoj et al. (2006) obtained 0.04 A/m for the same value. The data of Manoj et al. (2006) correspond to a lower level of solar activity.

Comparing these data with results of our calculations, we can state their satisfactory consent in position and spatial size of the equatorial electrojet. In our calculations the width of the electrojet along latitude turns out a little greater, than in the experiment. This can be explained with the absence of the electric field of magnetospheric convection and the absence of thermospheric tides in our calculations which could lead to a modification of the spatial distribution of the zonal current in the Earth's ionosphere or with the rough spatial grid in our model. As for the maximal intensity of the equatorial electrojet in our calculations, its value of 35 A/km is very close to the measurements of Manoj et al. (2006).

A comparison of results of our model calculations with experimental data by Le Mouel et al. (2006) has shown, that:

1. Locations of calculated and observed electrojet at 06:00 UT, 08:00 UT and 22:00 UT practically coincide;

2. While in the experiment the longitudinal extent of the equatorial electrojet depends on UT and is minimal for the southern hemisphere, in the results of the model calculations the longitudinal extent of the equatorial electrojet is constant. This indicates that in the experiment there are sources of longitudinal and of UT variations of the equatorial electrojet (a real geomagnetic field with all features, including the South-Atlantic magnetic anomaly with which the minimal longitudinal extent of the equatorial electrojet in the southern hemisphere is possibly connected). In the model there are only sources of UT variation (dipole geomagnetic field). From this it is possible to draw the conclusion that the variability of the longitudinal extent of the equatorial electrojet is the consequence of a longitudinal variation;

3. In model calculations the counter electrojet is formed both in the morning and in the evening, but in the evening the counter electrojet is much stronger than in the morning. In the experiment the counter electrojet is formed only in the morning. This fact can be explained by the absence of thermospheric tides in the presented calculations.

Figure 6 shows the diurnal behavior of the maximal intensity of the equatorial electrojet. The plot in Fig. 6 has been obtained by a selection of the maximal values of the equatorial electrojet for each calculated UT time within the day,

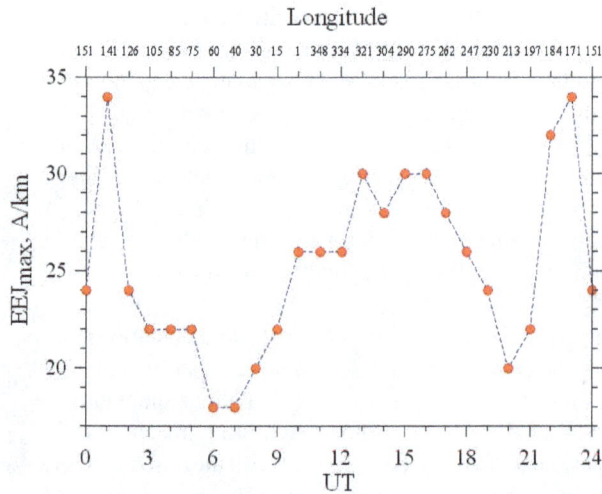

Fig. 6. Diurnal behavior of the maximal intensity of the equatorial electrojet obtained in the model GSM TIP in A/km.

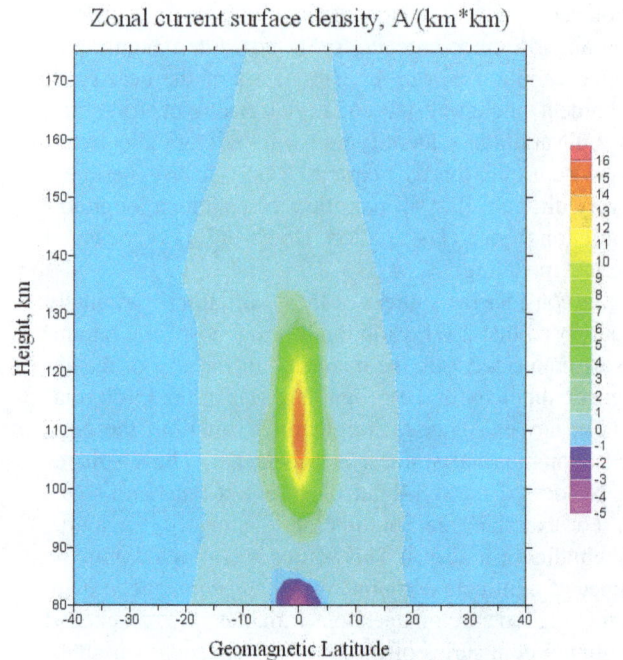

Fig. 7. Zonal current surface density obtained in the model GSM TIP in A/km^2 for 23:00 UT.

therefore each point on the abscissa of this plot corresponds to conditions close to noon for various geomagnetic longitudes at the geomagnetic equator. It is shown, that there is a precise semidiurnal harmonic in the diurnal behavior of the equatorial electrojet linear density with maxima at 23:00 UT and 15:00 UT and with minima at 06:00 UT and 20:00 UT.

In investigations by Ivers et al. (2003), Doumouya and Cohen (2004), and Le Mouel et al. (2006) obtained on the basis of observations of the equatorial electrojet with the CHAMP and Ørsted satellites, it was shown that the variability of the equatorial electrojet depends on time and longitude.

Ivers et al. (2003) and Le Mouel et al. (2006) have shown the existence of a quarter-diurnal (six-hour) harmonic in the equatorial electrojet, i.e. the presence of a UT or longitudinal behavior of the equatorial electrojet with four maxima and accordingly four minima. The maximal intensity of the equatorial electrojet in our calculations shows a semidiurnal harmonic behavior on UT. We used the geographical longitudes of the maxima from Ivers et al. (2003) in geomagnetic longitude at the geomagnetic equator. Then we recalculated these longitudes in UT which correspond to noon conditions in points: $\lambda = 0°–30°$ E corresponding to $\Lambda = 85°$ or 11:00 UT; $\lambda = 90°–120°$ E corresponding to $\Lambda = 175°$, or 05:00 UT; $\lambda = 180°–220°$ E corresponding to $\Lambda = 270°$, or 22:00 UT; $\lambda = 260°–290°$ E corresponding to $\Lambda = 345°$, or 17:00 UT. Thus we obtain a coincidence of maxima of the equatorial electrojet in the longitudinal ranges $\lambda = 180°–220°$ E (in our calculations 23:00 UT, and in experiment 22:00 UT) and $\lambda = 260°–290°$ E (in our calculations 15:00 UT, and in experiment 17:00'UT).

Le Mouel et al. (2006) have shown that the longitudinal behavior of the maximal intensity of the equatorial electrojet has four maxima and four minima. Three of the four maxima coincide with the maxima in the paper by Ivers et al. (2003).

From the papers by Doumouya and Cohen (2004) it is possible to pick out six extrema in the longitudinal behavior of the intensity of the equatorial electrojet. Three maxima and three minima from their work coincide with three maxima and three minima in the work by Le Mouel et al. (2006).

According to data from Doumouya and Cohen (2004) and Le Mouel et al. (2006) the longitudinal variation of the equatorial electrojet lays in the range 1.7–2.3 and 2.3–3.2, correspondingly. Figure 6 shows, that the magnitude of the UT variation of the equatorial electrojet equals approximately 2 in our calculations. We obtained this magnitude for quiet geomagnetic conditions during a spring equinox under a minimum of solar activity.

Figures 7–10 show the distributions of zonal current surface density, obtained with the model GSM TIP for 23:00 UT, 06:00 UT, 15:00 UT and 20:00 UT in a Cartesian geomagnetic coordinate system (longitude-altitude). One may notice that its UT-variation is much weaker and equals $\sim$1–3 A/km^2 and herewith the maximal intensity is $\sim$15–20 A/km^2. At the same time the latitudinal extent of the equatorial electrojet surface density is maximal at 23:00 UT and 15:00 UT and minimal at 06:00 UT and 20:00 UT.

A comparison of experimental data and results of model calculations by other authors has been made by Klimenko et al. (2006a). Comparison with simulation results of Richmond (1989) and Stening (1985) and with the experimental data from Stening (1985) has revealed a satisfactory consent for the vertical structure of the zonal current in the equatorial ionosphere.

Zonal current surface density, A/(km*km)

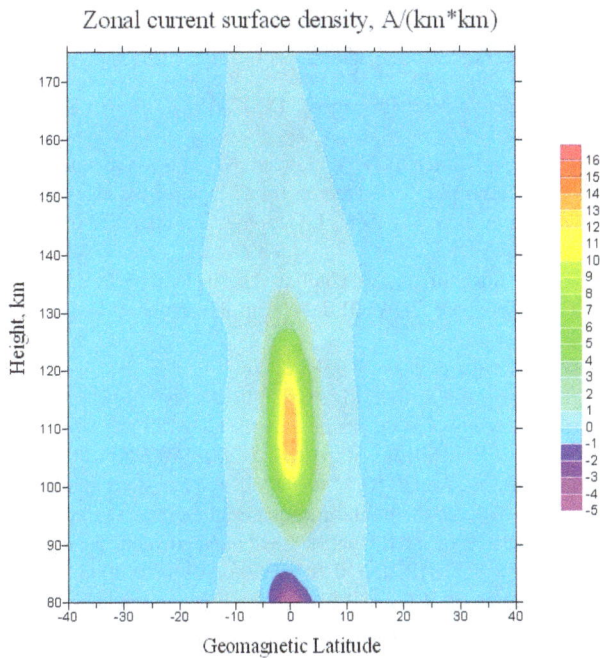

Fig. 8. Zonal current surface density obtained in the model GSM TIP in A/km^2 for 06:00 UT.

Zonal current surface density, A/(km*km)

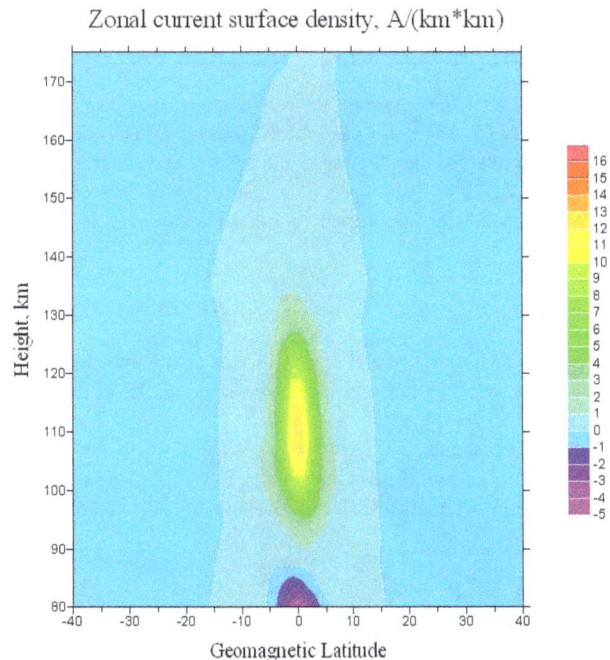

Fig. 9. Zonal current surface density obtained in the model GSM TIP in A/km^2 for 15:00 UT.

Zonal current surface density, A/(km*km)

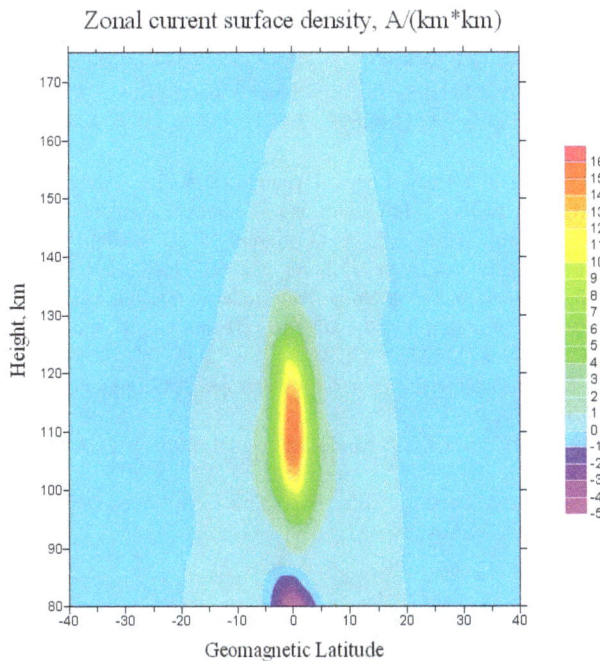

Fig. 10. Zonal current surface density obtained in the model GSM TIP in A/km^2 for 20:00 UT.

4 Summary

Numerical calculations have shown, that there is a semidiurnal harmonic in the diurnal behavior of the maximal intensity of the equatorial electrojet with maxima at 23:00 UT and 15:00 UT and with minima at 06:00 UT and 20:00 UT. The greatest and smallest values of its maximal intensity in the diurnal behavior can differ by a factor of two. The longitudinal extent of the area of the equatorial electrojet practically does not show any UT-variation. The variability of the longitudinal extent of the equatorial electrojet with respect to UT, which exists in experiment, is a longitudinal variation but not a UT variation. With amplification of the maximal intensity of the equatorial electrojet, its latitudinal extent slightly increases also. At the same time the equatorial electrojet in its maxima of intensity has approximately an identical width, and in the minima the electrojet is narrow in the morning and wide in afternoon. The comparison of experimental data with our calculation results has shown a satisfactory agreement in intensity, position and spatial size of the equatorial electrojet. A small displacement of the maximal intensity of the electrojet into the morning sector in comparison with experimental data, can be connected with the absence of thermospheric tides on the bottom boundary of the thermosphere and the absence of electric fields of magnetospheric convection in our calculations. The presence of the counter electrojet in the evening sector of local time in our calculations which is absent in experimental data, can possibly be explained by the absence of thermospheric tides in our calculations. The UT-variation of the surface density of the equatorial electrojet is much weaker and equals ∼1–3 A/km^2 and herewith the maximal intensity is ∼15–20 A/km^2. The latitudinal extent of the surface density of the equatorial electrojet is maximal at 23:00 UT and 15:00 UT and minimal at 06:00 UT and 20:00 UT.

Acknowledgements. The authors express sincere gratitude to M. Förster for useful discussions and for the presentation of our reports at the. Kleinheubacher Tagung of the URSI Landesausschuss Deutschland e.V., 2006, Miltenberg, Germany, 25–29 September 2006, thanks to which the opportunity of this publication was given.

References

Baker, W. G. and Martyn, D. F.: Electric Currents in the Ionosphere, I The Conductivity, Phil. Trans. Roy. Soc., London, A246, 281–294, 1953.

Chapman, S.: The equatorial electrojet as detected from the abnormal electric current distributions above Huancayo, Peru and elsewhere, Arch. Meteorol. Geophys. Bioklimatol, A4, 368–390, 1951.

Cowling, T. G.: The magnetic field of sunspots, Mon. Not. R. Astron. Soc., 94, 39–48, 1933.

Doumouya, V. and Cohen, V.: Improving and testing the empirical equatorial electrojet model with CHAMP satellite data, Ann. Geophys., 22. 3323–3333, 2004.

Egedal, J.: The magnetic diurnal variation of the horizontal force near the magnetic equator, Terr. Magn. Atmos. Electr., 52, 449–451, 1947.

Gershman, B. N.: Dynamics of ionospheric plasmas, Moscow, Nauka, 256p., 1974 (In Russian).

Gurevich, A. V., Krylov, A. L., and Tsedilina, E. E.: Electric Fields in the Earth's Magnetosphere and Ionosphere, Space Sci. Rev., 19, 59–160, 1976.

Heelis, R. A.: Electrodynamics in the low and middle latitude ionosphere: a tutorial, J. Atmos. Solar-Terr. Phys., 66, 825–838, 2004.

Ivers, D, Stening, R., Turner, J., et al.: Equatorial electrojet from Ørsted scalar magnetic field observations, J. Geophys. Res., 108, 1061, doi:10.1029/2002JA009310, 2003.

Klimenko, M. V., Klimenko, V. V., and Bruykhanov, V. V.: Comparison of two variants of model of the electric field in the ionosphere of the Earth, KSTU News, 8, 59–68, 2005 (In Russian).

Klimenko, M. V., Klimenko, V. V., and Bruykhanov, V. V.: Numerical Simulation of the Electric Field and Zonal Current in the Earth's Ionosphere: The Dynamo Field and Equatorial Electrojet, Geomagnetism and Aeronomy, 46, 457–466, 2006a.

Klimenko, V. V., Klimenko, M. V., and Bruykhanov, V. V.: Numerical modeling of electric field and zonal current in the Earth's ionosphere – Statement of the problem and test calculations, Matematicheskoye Modelirovaniye, 18, 77–92, 2006b (In Russian).

Korenkov, Yu. N., Klimenko, V. V., Forster, M., et al.: Calculated and observed ionospheric parameters for a Magion 2 passage and EISCAT data on June 31, 1990, J. Geophys. Res., 103, 14697–14710, 1998.

Le Mouel, J.-L., Shebalin, P., and Chuliat, A.: The field of the equatorial electrojet from CHAMP data, Ann. Geophys., 24, 515–527, 2006, http://www.ann-geophys.net/24/515/2006/.

Lühr, H., Maus, S., and Rother, M.: Noone-time equatorial electrojet: Its spatial features as determined by the CHAMP satellite, J. Geophys. Res., 109, A01306, doi:10.1029/2002JA009656, 2004.

Manoj, C., Lühr, H., Maus, S., et al.: Evidence for short spatial correlation lengths of the noontime equatorial electrojet inferred from a comparison of satellite and ground magnetic data, J. Geophys. Res., 111, A11312, doi:10.1029/2006JA011855, 2006.

Martyn, D. F., Cowling, T.G., and Borger, R.: Electric conductivity of the ionospheric D-region, Nature, Lond., 162, 142–143, 1948.

Namgaladze, A. A., Korenkov, Yu. N., Klimenko, V. V., et al.: Global Model of the Thermosphere-Ionosphere-Protonosphere System, Pure and Applied Geophysics (PAGEOPH), 127, 219–254, 1988.

Namgaladze, A. A., Korenkov, Yu. N., Klimenko, V. V., et al.: Numerical modelling of the thermosphere–ionosphere–protonosphere system, J. Atmos. Terr. Phys., 53, 1113–1124, 1991.

Richmond, A.D.: Modeling the ionosphere wind dynamo: A review, Pure and Appl. Geophys. (PAGEOPH), 131, 413–435, 1989.

Richmond, A. D.: Thermospheric Dynamics and Electrodynamics, in Sol.-Terr. Phys., Principles and Theoretical Foundations, edited by: Carovillano, R. L. and Forbes, J. M., D. Reidel Publishing Company, Dordrecht, Holland, 523–607, 1982.

Singh, A., and Cole, K.D.: A Numerical Model of the Ionospheric Dynamo I. Formulation and Numerical Technique, J. Atoms. Terr. Phys., 49, 521–527, 1987.

Stening, R.J.: Modeling the equatorial electrojet, J. Geopys. Res., 90, 1705-1719, 1985.

Sugiura, M. and Cain, J. G.: A Model Equatorial Electrojet, J. Geophys. Res., 71, 1869–1877, 1966.

Volland, H.: Atmospheric Electrodynamics, Springer-Verlag, Berlin, 1984.

9

LOFAR in Germany

W. Reich

Max-Planck-Institut für Radioastronomie, Auf dem Hügel 69, 53121 Bonn, Germany

Abstract. The LOw Frequency ARray – LOFAR – is a new fully digital radio telescope designed for frequencies between 30 MHz and 240 MHz centered in the Netherlands. In May 2006 ten German institutes formed the German LOng Wavelength consortium – GLOW – to coordinate its contributions and scientific interests to the LOFAR project. The first LOFAR station CS1 was installed in summer 2006 near Exloo/Netherlands. The second station IS-G1 is presently been placed in the immediate vicinity of the Effelsberg 100-m radio telescope near Bad Münstereifel/Germany. This contribution briefly describes the basic properties and aims of LOFAR, the aims of the GLOW consortium and the actual activities to install a LOFAR station at the Effelsberg site.

1 Introduction and background

Radio emission covers the longest wavelength range of the electromagnetic spectrum, where emission from space is been observed. This means low angular resolution even for large telescopes when compared with optical, infrared or X-ray telescopes. Classical single-dish radio telescopes like the Effelsberg 100-m dish have a huge collecting area and high sensitivity, however, at 3 cm (10 GHz) its angular resolution is about 1', which is not always sufficient for detailed studies. Higher angular resolution is achieved by synthesis telescopes like the Very Large Array (VLA) or the Westerbork Synthesis Radio Telescope (WSRT), where a number of small telescopes are distributed over distances of a few kilometres. The signals received by each element are combined into images of arcsecond angular resolution. Very Long Baseline Interferometry (VLBI) uses telescopes with distances up to the Earth diameter or even antennas in space, where the raw signals need to be stored on tapes or disks

and are being correlated off-line. That way radio astronomy becomes a record holder in angular resolution for any astronomical observation by producing images showing milli arcsecond structures. For all these observations classical radio telescopes as sketched in Fig. 1 are used, where a paraboloid dish is pointed towards the object under investigation and the receivers being installed in the telescopes focus to record its emission. At low frequencies, however, a different technology can be used (as sketched in Fig. 1), where the signals from numerous elements are electronically delayed in such a way that the waves from a certain direction are received in phase. This technology is presently limited to the long wavelength range. LOFAR is based on this concept of beam formation and utilizes actual digital components to allow observations of unprecedented flexibility over a wide continuous wavelength range.

2 The LOFAR concept and basic parameters

The ideas and the concept of LOFAR were developed at AS-TRON (Dwingeloo/Netherlands, see Bregmann, 2000). LOFAR is actually realized by ASTRON on behalf of the LOFAR Foundation, which includes a number of Universities and Research Institutes. However, LOFAR is more than a new radio telescope for observations in the long wavelength range with unprecedented sensitivity and angular resolution. Its basic concept is that of a sensor network with high-speed data links, which allows to connect any kind of sensors for non-astronomical applications as well. For more details see: www.lofar.org or www.lofar.de (in German). A brief description including the actual status of the LOFAR project was given by Falcke et al. (2006).

The antenna elements of LOFAR are inverted-V crossed dipole antennas for two orthogonal directions (NE-SW and SE-NW) allowing to simultaneously observe all Stokes parameters. The dipoles have a field of view close to the entire

Correspondence to: W. Reich
(wreich@mpifr-bonn.mpg.de)

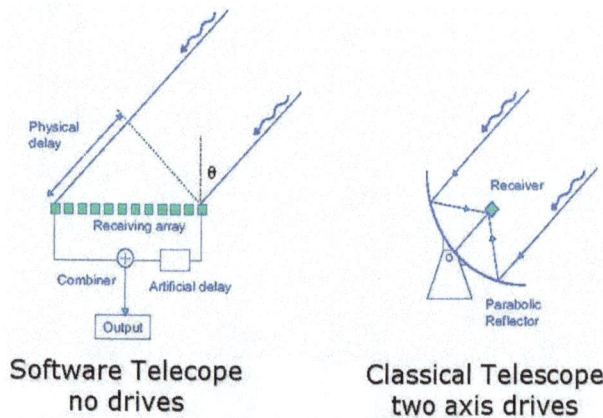

Software Telecope
no drives

Classical Telescope
two axis drives

Fig. 1. The pointing of a multi-element telescope like LOFAR in a certain direction is done by delaying the received signals from each element in such a way that the wavefront from a certain direction is received in phase. Classical telescopes are pointed mechanically.

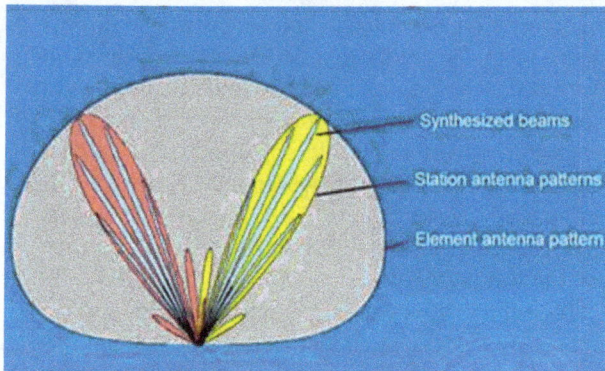

Fig. 2. LOFAR allows to form several beams. Within the beam of a single antenna element, which defines LOFAR's accessible sky coverage, a beam from a single station is formed. Its width is given by the size of the antenna field of a station (see Fig. 3). This means a field of view of about 7.5° at 50 MHz (LBA) and about 3° at 150 MHz (HBA). Combining many station beams to a "synthesized beam" refines the angular resolution according to the maximum distance of all stations. LOFAR allows to form several station and synthesized beams at the same time.

visible sky (Fig. 2). Two different types of dipoles are used: one for the frequency range from about 30 MHz to 80 MHz with a possibility to observe down to 10 MHz with reduced sensitivity. The other one is optimized for the frequency range from 110 MHz to 240 MHz. The frequency range between 80 MHz and 110 MHz is not accessible in Europe because of the allocation of strong transmitters for broadcasting services. 96 elements of each antenna type together form a "LOFAR station". The principle placement of the two antenna fields, the third field for other, non-astronomical sensors and the common station container hosting all electronics for local processing are shown in Fig. 3.

LOFAR: Station Layout

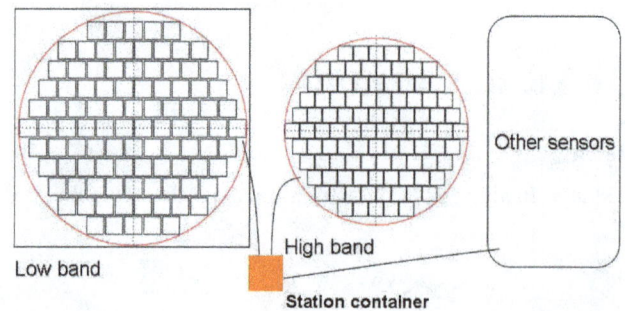

Fig. 3. Each LOFAR station has two antenna fields for the LBA (diameter about 65 m) and the HBA (diameter about 50 m) and a third field for non-astronomical sensors. The distribution of elements within each field was finally decided to be irregular (see Fig. 10) thus different from the distribution shown. The LOFAR cabinet hosting all the station electronics is placed between the two antenna fields.

Actually (end of 2006) the Low-Band Antennas (LBAs) are being available and ready for installation, while a final decision on the High-Band Antenna (HBA) design is to be made during 2007 and their mass production will start subsequently. Each LBA (for details see Fig. 6) has two integrated LNAs for the orthogonal dipoles on top of its mount. Two cables with a length of 110 m connect each antenna with the station container. This requires to put about 20 km cables into the ground for each antenna field. The signals are digitized and further processed by the station electronics in several ways depending on the requirements of the scheduled experiments. A total bandwidth of 32 MHz can be selected and split into a large number of narrow frequency channels within the available band pass of the low-band LNAs from 10 MHz to 80 MHz. The components of each station are remotely controlled by ASTRON using a commercial software package (PVSS II), which is also used to steer the station electronics according to the requirements of the scheduled observations. The data output produced by each LOFAR station is about 2 Gbit/s, which needs to be transferred in real-time to Groningen for correlation with the signals from other stations. Adding the requirements for the control of each station by the PVSS II software in total a data transfer of 3 Gbit/s must be provided by a fibre connection.

LOFAR is funded in the Netherlands for 77 stations. In addition a supercomputer *IBM Blue Gene/L* was installed in Groningen for the correlation of the data streams from all stations. The enormous raw data volume of LOFAR requires real time processing. No storage of the raw data is planned and even the reduced data volume is very large, which requires their distribution to the various data centres soon after processing.

32 of the 77 LOFAR stations are placed at its core close to

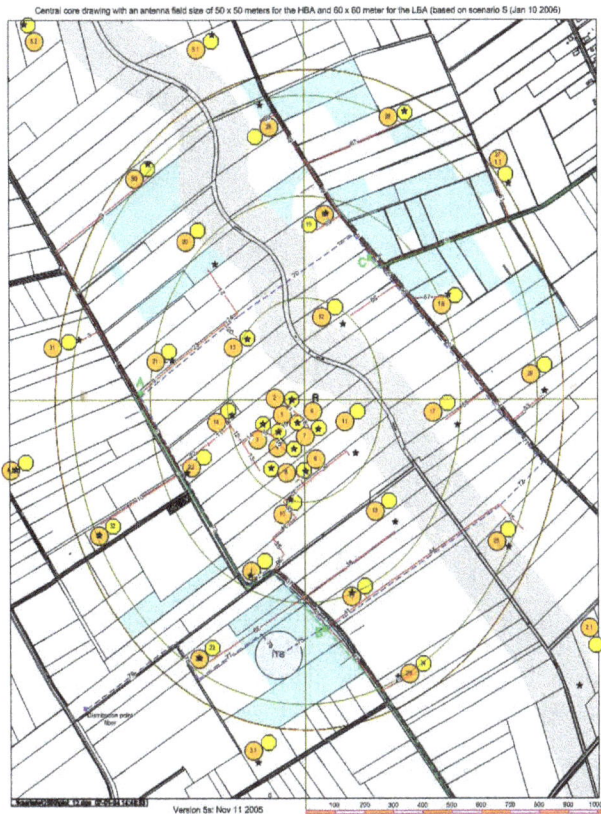

Central core drawing with an antenna field size of 50 x 50 meters for the HBA and 60 x 60 meter for the LBA (based on scenario 5 (Jan 10 2006)

Version 5a: Nov 11 2005

Fig. 4. Overview of the likely distribution of stations in the core area of LOFAR, which is located in a flat area with rural environment near Exloo/Netherlands. In the core area the density of LOFAR stations is highest. The two antenna fields (LBA and HBA) for each station are indicated.

Fig. 5. The planned distribution of LOFAR stations in the Netherlands.

Table 1. Nominal angular resolution of the LOFAR array as a function of maximum baseline at low-band (30 MHz/75 MHz or λ 10 m/4 m) and at high-band (120 MHz/240 MHz or λ 2.5 m/1.25 m).

Baseline	30 MHz	75 MHz	120 MHz	240 MHz
Core	1000"	400"	250"	125"
100 km	21"	8"	5"	3"
CS1 - IS-G1	8"	3"	2"	1"
1000 km	2"	0."8	0."5	0."3

the village of Exloo. Figure 4 shows the actual plan for the distribution of stations for the innermost area. The station density is highest at its core and decreases with increasing distance from Exloo. The planned distribution of stations in the Netherlands is shown in Fig. 5.

The first LOFAR station near Exloo Core Station 1 (CS1) (Fig. 6) was installed in summer 2006. CS1 differs from a standard LOFAR station as just 48 LBAs were so far installed. 48 more LBAs were distributed on nearby station locations for interferometric test purpose. Fibre connections to the *IBM Blue Gene/L* in Groningen were in place. The CS1 saw its "first light" in September 2006.

The sensitivity of the completed LOFAR exceeds that of any other long wavelength facility by up to about two order of magnitudes. According to Falcke et al. (2006) LOFAR's sensitivity after one hour of integration time, using a bandwidth of 4 MHz and dual polarization is 2, 1.3, 0.07 and 0.06 mJy for 30, 75, 120 and 200 MHz, respectively. The angular resolution being achieved with LOFAR depends on the wavelength of the observation and the maximum baselines available. Table 1 lists expected angular resolutions as

a function of wavelength for the LOFAR core, the Dutch LOFAR extending to about 100 km and the baseline CS1 - Effelsberg (IS-G1) of 260 km, which will be available soon.

The full digital concept of LOFAR allows to observe with eight independent station beams at the same time. This is realized by combining selected subbands in a station by a digital beamformer.

Some details of the LBA field can be seen from the CS1 image (Fig. 6). Each element is placed on a 3 m × 3 m ground plane. A plastic foil prevents plants to grow up. Two orthogonal dipoles are fixed by tent pegs and are connected to the LNAs, where two cables run down the central post and further underground to the LOFAR cabinet seen in the back of Fig. 6. This special air-conditioned outdoor cabinet is well shielded suppressing RFI transmission by about 40 dB.

3 GLOW – The German LOng Wavelength Consortium

ASTRON has started to realize LOFAR exclusively in the Netherlands, which is connected to funding conditions. The maximum LOFAR baselines in the Netherlands are of the order of about 100 km, which implies too low angular reso-

CS1 near Exloo

Fig. 6. The first LOFAR station CS1 (Core Station 1) was completed in summer 2006. This station has so far 48 LBAs. 48 more LBAs were distributed at some hundred metre distance for initial interferometric tests.

lutions (see Table 1) to realize many scientific aims. Already during early planning of the LOFAR project ASTRON initiated discussions with its neighbor countries seeking for their participation in LOFAR. A number of German astronomical and other research institutes quickly realized the scientific potential of LOFAR and expressed their interest to participate in an extended, which means a European LOFAR. An overview on LOFAR and the German engagement was given by Brüggen et al. (2006) and by Beck and Reich (2006).

The Max-Planck-Institut für Radioastronomie agreed to lead and to coordinate a German consortium participating in LOFAR. The GLOW consortium formed on a meeting at the Astrophysical Institute Potsdam on 5 May 2006. Ten institutes signed a Memorandum of Understanding (MoU), where the common interests of GLOW are formulated. The GLOW board consists of: M. Brüggen (IU Bremen, vice-chair), R. J. Dettmar (RUB Bochum), M. Steinmetz (AIP Potsdam), J. A. Zensus (MPIfR Bonn, chair). As the GLOW secretary R. Beck (MPIfR) was elected. A scientific working group (chair: B. Ciardi, MPA Garching) and a technical working group (chair: W. Reich, MPIfR) were also set on. Actual efforts by GLOW member institutes concentrate to secure funding of their LOFAR station and to prepare their installation. The Forschungszentrum Jülich prepares for large storage capacities (about 1 PByte) for LOFAR data. The working groups are busy to finalize proposals for "Key Science Projects", discuss fibre connection issues of common interest and how to make use of D-Grid facilities in future.

4   Scientific aims of GLOW

The scientific aims of the GLOW members were collected in a "White Paper", which was published on behalf of GLOW

GLOW Members:

- Astronomisches Institut der Ruhr-Universität Bochum
- Argelander-Institut für Astronomie Bonn
- Max-Planck-Institut für Radioastronomie Bonn
- International University Bremen
- Max-Planck-Institut für Astrophysik Garching
- Sternwarte Hamburg
- Forschungszentrum Jülich
- 1. Physikalisches Institut der Universität Köln
- Astrophysikalisches Institut Potsdam
- Thüringische Landessternwarte Tautenburg

by the MPIfR in 2005. A copy of the "White Paper" can be ordered from the MPIfR Bonn. There is a wide range of scientific interests from solar observations, Galactic and extragalactic research and in particular to study the Epoch of Reionization (EoR). A possible detection of signatures from the EoR was the main trigger to set up LOFAR. In a companion paper (Beck, this volume) describes the exciting aspects of LOFAR observations to investigate cosmic magnetic fields, which will complement high frequency results.

During the first phase of LOFAR operation most of the available observing time will be granted to so called "Key Science Projects" (KSPs). Dutch astronomers have already defined four KSPs: EoR, Surveys, Transients and Cosmic rays. International participation in these KSPs is presently discussed. GLOW expressed high interest in KSPs on solar research, cosmic magnetism and observations of Galactic and extragalactic jets.

5   Planned GLOW stations

It was estimated that about 10 to 12 stations distributed across Germany are needed to achieve an optimum high resolution image quality for an extended LOFAR. End of 2006 concrete plans exist for the installation of seven stations (Fig. 7). For some stations funding is not yet granted, but aside the Effelsberg station the Garching station (MPA), the Potsdam station (AIP) and the Tautenburg station (TLS) are funded and will be installed in due course. Actually intensive negotiations take place on the problem of the data transfer to the central processor in Groningen. The transfer of scientific data is not supported in Germany, while in some European countries it is actually free of charge. Renting fibre connections for the data link to Groningen will by far be the largest cost factor for the operation of the GLOW stations. Funds are required to allow a large number of institutes to participate in LOFAR.

Fig. 7. Location of the actually planned LOFAR station by the GLOW consortium. The LOFAR core in Exloo is also shown (AIP).

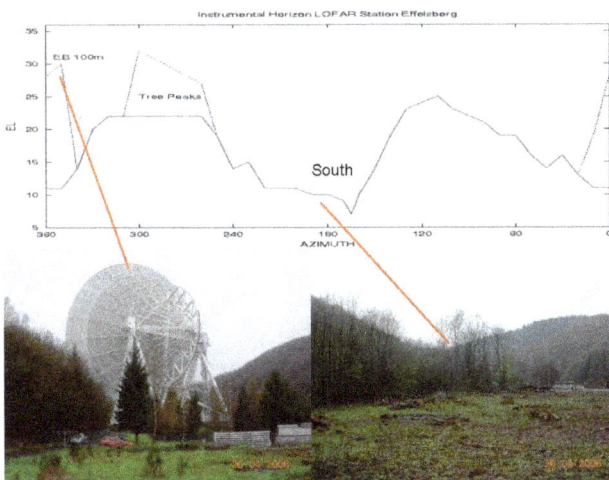

Fig. 9. ASTRON equipment to measure low frequency RFI is placed at the Effelsberg LOFAR site.

Fig. 8. Horizon for the LOFAR LBA field at Effelsberg. Optimum RFI shielding is expected towards east and west.

6 The Effelsberg LOFAR station

Following the CS1 the second LOFAR station "IS-G1" is actually placed about ∼200 m south of the Effelsberg 100 m telescope. The LOFAR site is a valley open to the south and the available area is just wide enough to place a complete

LOFAR field.

The advantage of a valley is a certain level of protection against terrestrial interference, which might become a limiting factor for sensitive observations in the long wavelength range of LOFAR. However, this means a somewhat restricted field of view. Fig. 8 shows the measured horizon at the approximate centre of the low-band field. Mobile RFI equipment provided by ASTRON (Fig. 9) was used for a successful two day RFI-measurement campaign in May 2006 by ASTRON personal, which proves the site to be well suited for LOFAR observations. However, another aspect must be considered at the Effelsberg site as the 100-m telescope with its outstanding sensitivity eventually will suffer from RFI generated by the LOFAR electronics, where the standard shielding might be not sufficient. It was therefore decided to place the standard LOFAR container in a much bigger (size: $9\,m\times5\,m\times2.5\,m$), isolated and completely welded container, where depending on the actual RFI situation a maximum additional shielding of about 60 dB can be obtained.

The distribution of the 96 elements within the LBA field is the same as for the CS1, where the main sub-field, however, has 48 elements. The positions of the IS-G1 LBAs is rotated by 65° (Fig. 10) relative to the CS1 positions. Each of the following stations will have a different rotation angle. This angle of 65° gives a minimum east-west extent, which fits in the local requirements due to a number of late environmental

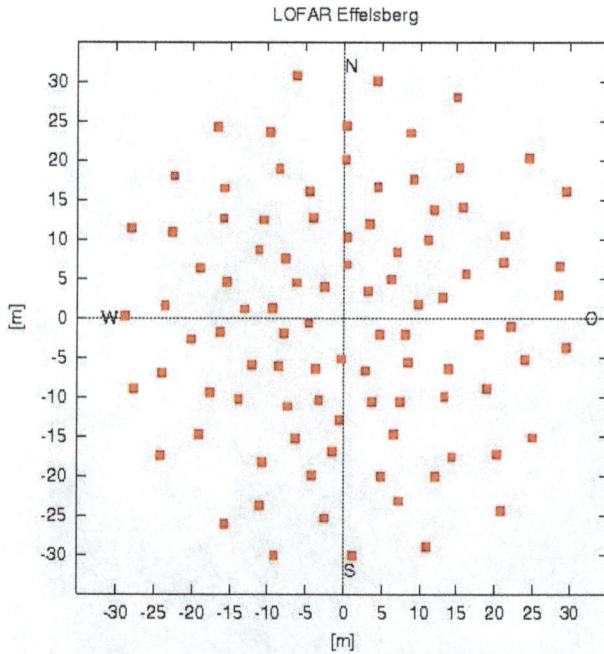

Fig. 10. The distribution of LBAs of the Effelsberg LOFAR station, which is rotated by 65° relative to that of the CS1 station near Exloo.

restrictions. The IS-G1 site is meanwhile (December 2006) prepared for the required flatness within a few cm, which was difficult to achieve for the dense, rather stony ground at the Effelsberg site.

Funded by the Max-Planck-Gesellschaft the required fibre connection from the LOFAR station to the MPIfR Bonn will be installed in 2007. Further data transfer from Bonn to Groningen will be realized by renting existing fibre links. The fibre connection Effelsberg - Bonn will be shared by using it also for eVLBI experiments where the 100-m telescope is involved in. These data are further transferred from Bonn to the JIVE correlator located at Dwingeloo/Netherlands.

In December 2006 MPIfR, ASTRON and the LOFAR Foundation signed a MoU on the cooperation in placing and operating the Effelsberg "IS-G1" station. Common interferometric observations between the CS1 and the IS-G1 are also agreed on as well as the common development of the so called "stand-alone" mode for a LOFAR station. In this mode a LOFAR station is operated as a quasi "single-dish" telescope, whose mode is of interest for most GLOW stations as well. A software interface is being developed aiming to analyze LOFAR station data with tools as they are available for observations with the Effelsberg 100 m telescope.

7   Conclusions

LOFAR is the first large digital radio telescope where many ambitious technical needs are going to be realized. Huge data streams have to be transferred across Europe and have to be processed by a dedicated supercomputer in real time. LOFAR promises revolutionary insights into the formation of the early Universe and will solve a number of astronomical key questions by its capability to observe at low radio frequency with unprecedented sensitivity.

The new technique realized with LOFAR and the operational experience will provide important constraints for the concept of the Square Kilometre Array (SKA), which is an international project expected to be realized in the next decade (see e.g. Beck, 2006). The SKA with one square kilometre collecting area will become the major international radio astronomical facility in the world.

Acknowledgements. I appreciate the close cooperation with ASTRON on all LOFAR related aspects and in particular Corina Vogt (LOFAR Science Office), who also provided Figs. 2 to 5 (copyright: ASTRON). I like to thank many colleagues at the MPIfR for their engagement in the installation of the LOFAR station at Effelsberg and preparing for its operation.

References

Beck, R.: Sterne und Weltraum, 9, 22–33, 2006.

Beck, R. and Reich, W.: Sterne und Weltraum, 9, 19–21, 2006.

Bregmann, J. D.: in Proc. SPIE Vol. 4015, Radio Telescopes, edited by: Butcher, H. R., 19–32, 2000.

Brüggen, M., Beck, R., and Falcke, H.: Reviews of Modern Astronomy, 19, 277–292, 2006.

Falcke, H., van Haarlem, M. P., de Bruyn, A. G. et al.: in: Highlights of Astronomy, 14, XXVth IAU General Assembly, edited by van der Hucht, K. A., 119–120, 2006.

10

The CPW-TEC project: Planetary waves in the middle atmosphere and ionosphere

C. Jacobi[1], N. Jakowski[2], A. Pogoreltsev[3], K. Fröhlich[1], P. Hoffmann[1], and C. Borries[2]

[1]Institute for Meteorology, University of Leipzig, Stephanstr. 3, 04103 Leipzig, Germany
[2]DLR, Institute of Communications and Navigation, Kalkhorstweg 53, 17235 Neustrelitz, Germany
[3]Russian State Hydrometeorological University, 98 Maloohtinsky St., Petersburg 195196, Russia

Abstract. Maps of vertically integrated electron density over the higher middle and polar latitudes that are regularly produced by DLR Neustrelitz are investigated with respect to planetary waves (PW) in the period range of several days. The results are compared with planetary wave analyses using stratospheric reanalyses. Case studies show that PW signatures in the ionosphere and neutral middle atmosphere waves have similar seasonal variations, indicating a possible coupling between these layers. Numerical modelling of the middle and upper atmosphere is performed to analyse the possible penetration of PW effects into the thermosphere. Numerical results show that direct propagation of PW to the thermosphere is not possible, and indirect ionospheric effects must be responsible for the wave coupling of the atmospheric layers. In the paper an overview of the data used and the methods applied in this project is given.

1 Introduction

Atmospheric planetary waves (PW) in the middle atmosphere are characterized by periods of 2 to 30 days (frequently called the "long-period range"). Usually, PW are not able to penetrate to altitudes above the lowermost thermosphere (Lastovicka, 2006). Nevertheless, oscillations with PW periods have been observed in the ionosphere, for example in ionospheric F-region maximum heights (Pancheva et al., 2002) or electron densities (Altadill et al., 2001, 2003). The simultaneous occurrence of PW in the ionosphere and stratosphere might be an indicator for vertical coupling between the middle atmosphere and the thermosphere/ionosphere system. However, the mechanisms that lead to such kind of coupling necessarily must be indirect.

Correspondence to: C. Jacobi
(jacobi@uni-leipzig.de)

There already exist some ideas about how PW may penetrate into the ionosphere. Pancheva et al. (2002) suggested a modulation of upward propagating tides through PW, or PW associated neutral wind variations inducing electric fields, which modulate the height of the maximum electron density hmF2 or the plasma density of the ionospheric F-region. Other possible mechanisms (see e.g. Altadill et al., 2001) that may lead to PW signatures in the ionosphere are the modulation of gravity waves (GW) by PW in the middle atmosphere. These GW may propagate to the ionosphere, and break there, thus transporting the PW signature to the upper atmosphere. Also possible is that winds in the mesopause region may cause a redistribution of major atmospheric components in the lower thermosphere and changes the mean velocity of mass transport. This affects the vertical transport of minor constituents and provides phase changes of the electron density in the E and F region (Mikhailov, 1983). Pancheva and Lysenko (1988) proposed that PW mesopause region winds may drive the atmospheric dynamo. Altadill et al. (2001) concluded from the result of case studies that the mechanisms which lead to PW like oscillations in the ionosphere are different from case to case.

To comprehensively investigate the possible coupling between the neutral middle atmosphere and the ionosphere at PW time scales, within the CPW-TEC (Climatology of Planetary Waves seen in TEC) project we make use of maps of Total Electron Content (TEC) in a vertical air column of the ionosphere. The analysis of these maps suggests signatures of PW in the ionosphere, so that the results obtained are compared with stratospheric wave analyses accompanied by numerical model experiments on PW propagation. The project is run in collaboration between DLR-IKN Neustrelitz, Leipzig University, and the Russian State Hydrometeorological University, St. Petersburg. In the following sections an overview of the data used and the methods applied in this project is given.

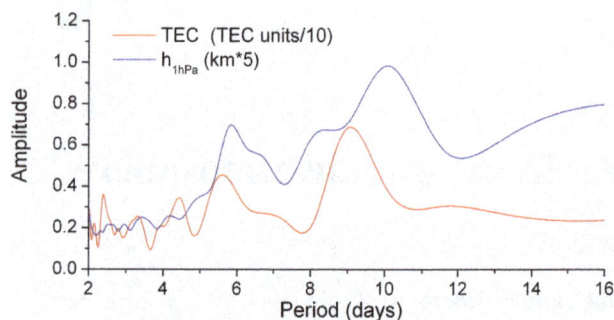

Fig. 1. Examples of amplitude spectra of GPS TEC and Met Office 1 hPa geopotential heights over Europe, 1 September – 20 November 2004.

2   Data base and analysis

TEC maps are regularly produced at DLR Neustrelitz using ground based GPS measurements available from the International GPS Service (e.g. Jakowski, 1996; Jakowski et al., 2002; w3swaci.dlr.de/index.htm). Maps covering the European sector are available since 1995, while similar maps from 50° N to the pole are produced since 2002. The TEC values are mapped and blended into a TEC model, which was established by the DLR (Jakowski et al., 1996) especially for TEC map construction. Here we use the results of this data assimilation from the pole to 50° N. These TEC maps are processed by DLR Neustrelitz with a time resolution of ten minutes and an accuracy of 1–2 TEC units (1 TEC unit = 10^{16} electrons/m^2).

For the analysis of stratospheric PW we use UK Met Office assimilated data (Swinbank and O'Neill, 1994) at 1 hPa. To avoid disturbances through auroral processes in this analysis we restrict ourselves to the analysis of data at 52.5° N. An example of amplitude spectra of GPS TEC and stratospheric 1 hPa geopotential heights is shown in Fig. 1. The time window analysed here (1 September – 20 November 2004) has been chosen because a quasi 6-day oscillation has been detected in mesopause region radar winds (not shown here), so that autumn 2004 was chosen as a sort of first test whether long-period oscillations may be visible in both middle atmosphere and ionosphere. Oscillations at periods of 5–8 days are frequently found in the middle atmosphere under equinox conditions (e.g. Jacobi et al., 1998), so from this point of view the selected time interval can be considered as typical. Similarly, in Fig. 1 there are peaks at a period close to 6 days in both the stratosphere and ionosphere spectra. A TEC oscillation with a 9-day period is accompanied by a broad spectral maximum in the stratospheric geopotential height amplitudes, but it remains questionable whether this can be considered as the signature of middle atmosphere-ionosphere coupling. Periods of 6 and 9 days may be owing to the Rossby quasi 5 and 10-day waves. However, from simple spectra alone the global structure of the underlying waves

cannot be revealed.

To investigate and interpret the planetary scale disturbances, we have to separate them into the travelling and standing PW components (Hayashi, 1971; Schäfer, 1979). At first, the amplitudes and phases for the zonal harmonics $m=1,2,3$ at 52.5° N are estimated by using the singular value decomposition (SVD) algorithm. The hourly results for TEC are daily averaged. A pair of time series is constructed representing two eastward and westward moving waves, those signals are registered at two zonal grid points with a distance equal to the phase shift between the sine and cosine functions divided by the zonal wavenumber. The coefficients for the eastward and westward travelling waves are calculated by using the phase-difference method (Pogoreltsev et al., 2002) for both waves. The Fourier amplitudes and phases of the two components are calculated for a time segment of 48-days. After this, the procedure is repeated while shifting the window by one day to study the time dependency of wave activities during 2004. To filter waves the spectral information at specific period bands are combined and transformed back into the time domain. For this resulting time series a wavelet amplitude spectrum is calculated.

When analysing wavelike structures in the ionosphere by TEC or electron density data one has to keep in mind that the thermosphere is only ionized to a minor degree even at F2 layer heights. Thus, wavelike variations in TEC or electron density usually trace processes propagating in the thermosphere via interaction of neutral gas and charged particles. Consequently, PW like periods should be related either to intrinsic PW type waves in the thermosphere or corresponding modulations of other wave types such as atmospheric GW. A direct modulation of the plasma density could be induced by upward propagating electric field variations excited/modulated by planetary waves which still exist in lower thermospheric heights. Therefore, the term "ionospheric PW" is, strictly speaking, not correct. Nevertheless, for convenience we shall subsequently use this term, but have to keep in mind that it does not describe an intrinsic PW in the ionospheric plasma.

3   Some results

Figure 1 indicates that long-period oscillations at times may be visible in both the middle atmosphere and the ionosphere, although corresponding peaks in the spectra cannot be considered as a proof of coupling processes and may be accidental. To analyse possible corresponding wave activity in the quasi 5- and 10-day range in more detail, in Fig. 2 time series of the mean amplitudes of waves with wavenumber 1–3 (eastward and westward) in the period range 3–7 days are shown, as a representation of the quasi-5-day oscillation from Fig. 1. The stratospheric waves show a clear prevalence of westward travelling waves with wavenumber m=1 and, at times, $m=2$. This is also visible for the TEC data, however,

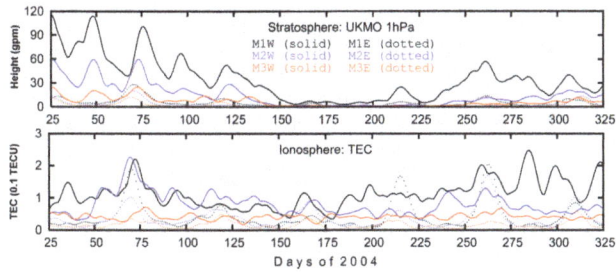

Fig. 2. Time series of wavenumber 1–3 westward (M1W, M2W, M3W) and eastward (M1E, M2E, M3E) travelling wave amplitudes for 1 hPa geopotential height (in gpm) and TEC (in 0.1 TEC units) in the period window 3–7 days.

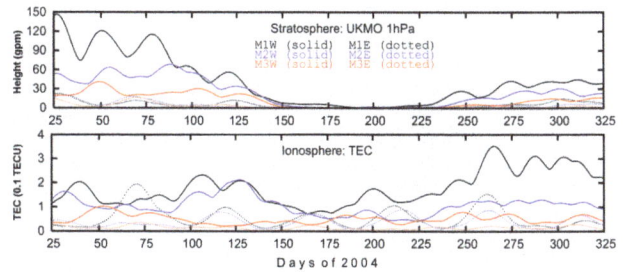

Fig. 3. As in Fig. 2, but for the 7–12 day period range.

correlation between the stratospheric and ionospheric westward waves is weak in most part of the year, except few time intervals, e.g. around day #75, and in autumn. Occasionally a 5-day eastward wave is visible in TEC, which in some cases is connected with a weak eastward wave in the stratosphere also.

The corresponding signature of a 10-day wave is shown in Fig. 3. As is the case with the 3–7 day oscillations in Fig. 2, the well-known seasonal cycle of stratospheric PW with maximum activity in winter and weak activity in summer is, to a certain degree, also visible in TEC. The correspondence between stratospheric and ionospheric wave activity is not strong, but still some corresponding peaks of activity can be seen, e.g. in spring for westward travelling $m=1$ waves.

Note that the large summer values of TEC amplitudes around day 200 are accompanied by a geomagnetic storm which may disturb PW signatures in TEC. Here we did not analyse a possible influence of geomagnetic variations on the ionosphere which may give rise to long-period ionospheric variations also at time scales of PW. On the other hand, PW induced oscillations of winds in the dynamo region may lead to geomagnetic oscillations at periods of PW (e.g. Kohsiek et al., 1995; Jarvis, 2006), so that in turn the external forcing of the geomagnetic field carefully has to be distinguished from PW forcing. Detailed analyses of external influence is a topic for further research.

Apart from the seasonal cycle at times there is a correspondence between the variation of wave activity both in the stratosphere and ionosphere, so, for example, in the westward propagating wave 1 around days 100 and 125 in Fig. 3 and after day 250 in Fig. 2. This correspondence leads to the conclusion that there is a possible coupling between the middle atmosphere and ionosphere. However, this correspondence is not continuously apparent during 2004, so that coupling processes between the atmospheric layers, if they exist, are variable, and intermittent.

4 Numerical modelling

The Middle and Upper Atmosphere Model (MUAM) is a 3D mechanistic model of the atmospheric circulation extended from the 1000 hPa surface up to the heights of the ionospheric F2-layer. It is based on the Cologne Model of the Middle Atmosphere-Leipzig Institute for Meteorology (COMMA-LIM, Fröhlich et al., 2003). The MUAM is a grid-point model with horizontal (latitude/longitude) resolution of $5 \times 5.625°$, and with up to 60 levels spaced evenly in the non-dimensional log-pressure height (scale height $x=-ln(p/p_0)$, $p_0=1000$ hPa] with a constant step size of about 0.4. The upper boundary is placed at $x=24$ which corresponds to the geopotential height of 300–400 km depending on thermospheric temperature. The PW to be analysed can be explicitly forced in the model.

Additionally to the standard radiative scheme used in the COMMA-LIM version (Fröhlich et al., 2003) the EUV heating in the thermosphere has been included. Solar fluxes and absorption coefficients for each EUV spectral interval and each constituent were calculated using the model proposed by Richards et al. (1994). The constant value of 0.366 for the EUV heating efficiency has been used as recommended by Roble (1995).

To integrate the prognostic equations, the initial Cauchy problem was splitted (Marchuk, 1967; Strang, 1968) into a set of simpler problems according to the physical processes considered, which in our case are vertical diffusion of momentum and heat on one side, and all other processes on another side. Finally, we use the Matsuno (1966) time-integration scheme with a time step of 100 s. This was necessary because the leap-frog scheme used in COMMA-LIM becomes unstable in the upper thermosphere where viscous and thermal conduction terms are dominant.

First results of numerical modelling of planetary waves are shown in Fig. 4. We consider here the stationary wave, and the westward propagating 10-day wave as examples. Other waves will be analysed in future investigations. In Fig. 4, zonal wind amplitudes of the stationary PW and the westward propagating 10-day wave are presented. As background conditions those of January had been chosen. Note that the waves do not propagate to the thermosphere, which shows

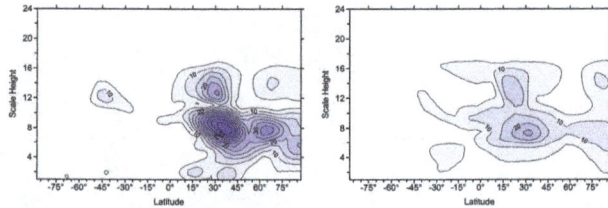

Fig. 4. Zonal wind amplitudes of the stationary PW (left panel) and the westward propagating wavenumber 1 Rossby 10-day wave (right panel), as modelled with the MUAM model for January conditions.

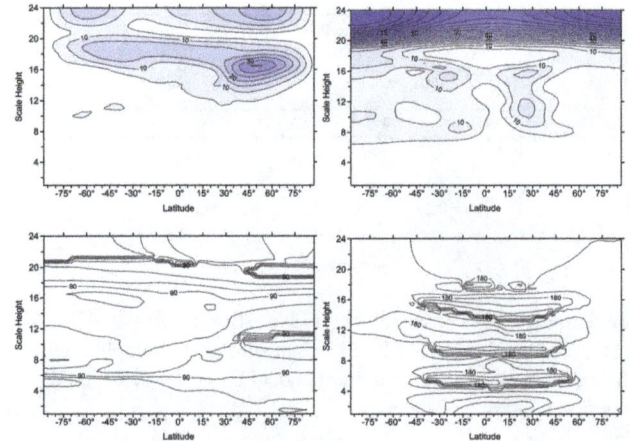

Fig. 5. Amplitudes (upper panels) and phases (lower panels, 360° equal 24 h) of the semidiurnal (left panels) and diurnal tide (right panels), as modelled with the MUAM model for January conditions.

that a direct propagation of waves to the ionosphere is not possible as expected. Similarly, the solar tides only propagate to the lowermost thermosphere (upper panels of Fig. 5). The strong amplitudes in the thermosphere are not owing to propagating tides, but to in situ forcing of the thermosphere in the course of the day. This can be deduced from the phase behaviour of the diurnal tide, which clearly changes above a height of about 18 scale heights. For the semidiurnal tide, the change in phase behaviour is not so clearly expressed, but there is still a tendency visible that for heights above about 21 scale heights no further phase change with height occurs, and thus the semidiurnal oscillations there are due to in situ heating of the thermosphere.

Acknowledgement. This study was supported by Deutsche Forschungsgemeinschaft under JA 640/3-1 and JA 836/19-1 within the special priority program SPP 1176 "CAWSES" and by RBFR under grant RFBR-DFG 05-05-04001. Stratospheric Reanalyses have been provided by UK Met Office through BADC.

5 Conclusions

We have shown, using Northern Hemisphere TEC data as well as stratospheric reanalyses that there probably exist coupling processes that lead to corresponding bursts of PW activity in the middle atmosphere and ionosphere, especially at periods and wavenumbers known as typical for the quasi-5 and 10-day PW. This coupling is obviously intermittent, and the underlying processes are not yet known. A numerical modelling experiment using an essentially neutral middle atmosphere model shows that both PW and tides do not propagate to the ionospheric F-region, so that direct PW propagation as well as tidal modulation through PW, according to these first results, may be ruled out as candidates for the coupling process. However, the numerical model at this stage only includes prescribed ionospheric electron densities and fixed magnetic fields, so that electrodynamic processes are not included in detail. In addition, the MUAM model still includes an improved Lindzen-type GW parameterisation (see Jacobi et al., 2006, and references therein) that is not specifically designed to describe those GW that are able to propagate to and force corresponding travelling ionospheric disturbance propagation in the thermosphere/ionosphere. Especially the question of the wavelength and frequency spectrum of those waves, and how this can be incorporated in a parameterisation, still has to be considered. Therefore, at this stage the question of coupling processes still remains open.

References

Altadill, D., Apostolov, E. M., Sole, J. G., and Jacobi, C.: Origin and development of vertical propagating oscillations with periods of planetary waves in the ionospheric F region, Phys. Chem. Earth (C), 26, 387–393, 2001.

Altadill, D., Apostolov, E. M., Jacobi, Ch., and Mitchell, N. J.: 6-day westward propagating wave in the maximum electron density of the ionosphere, Ann. Geophysicae, 21, 1577–1588, 2003.

Fröhlich, K., Pogoreltsev, A., and Jacobi, Ch.: Numerical simulation of tides, Rossby and Kelvin waves with the COMMA-LIM model, Adv. Space Res., 32, 863–868, doi:10.1016/S0273-1177(03)00416-2, 2003.

Hayashi, Y.: A general method of resolving disturbances into progressive and retrogressive waves by space Fourier and time cross-spectral analyses, J. Meteor. Soc. Jpn., 49, 125–128, 1971.

Jacobi, Ch., Schminder, R., and Kürschner, D.: Planetary wave activity obtained from long-term (2–18 days) variations of mesopause region winds over Central Europe (52° N, 15° E), J. Atmos. Solar-Terr. Phys., 60, 81–93, 1998.

Jacobi, Ch., Fröhlich, K., and Pogoreltsev, A.: Quasi two-day-wave modulation of gravity wave flux and consequences for the planetary wave propagation in a simple circulation model, J. Atmos. Solar-Terr. Phys., 68, 283–292, 2006.

Jakowski, N.: TEC monitoring by using satellite positioning systems, in: Modern Ionospheric Science, edited by: Kohl, H., Rüster, R., and Schlegel, K., EGS, Katlenburg-Lindau, ProduServ GmbH Verlagsservice, Berlin, 371–390, 1996.

Jakowski, N., Heise, S., Wehrenpfennig, A., Schlüter, S., and Reimer, R.: GPS/GLONASS-based TEC measurements as a contributor for space weather forecast, J. Atmos. Solar-Terr. Phys., 64, 729–735, 2002.

Jarvis, M. J.: Planetary wave trends in the lower thermosphere – Evidence for 22-year solar modulation of the quasi 5-day wave, J. Atmos. Solar-Terr. Phys., 68, 1902–1912, 2006.

Kohsiehk, A., Glassmeier, K., and Hirooka, T.: Periods of planetary waves in geomagnetic variations, Ann. Geophysicae, 13, 168–176, 1995.

Lastovicka, J.: Forcing of the ionosphere by waves from below, J. Atmos. Solar-Terr. Phys., 68, 479–497, 2006.

Marchuk, G. I.: Numerical Methods in Weather Prediction (in Russian), Gidrometeoizdat, Leningrad, 1967 (English edn., Academic Press, New York, 1974).

Matsuno, T.: A finite difference scheme for time integrations of oscillatory equations with second order accuracy and sharp cutoff for high frequencies, J. Meteorol. Soc. Jpn., 44, 76–84, 1966.

Mikhailov, A. V.: Mechanism of in phase variations of electron concentration between E and F2 ionospheric layers (in Russian), Geomagn. Aeronomy, 23, 557–561, 1983.

Pancheva, D. and Lysenko, I.: Quasi-two-day fluctuations observed in the summer F-region electron maximum, Bulg. Geophys. J., 14, 41–51, 1988.

Pancheva, D., Mitchell, N., Clark, R. R., Drobjeva, J., and Lastovicka, J.: Variability in the maximum height of the ionospheric F2-Layer over Millstone Hill (September 1998 – March 2000); influence from below and above, Ann. Geophysicae, 20, 1807–1819, 2002.

Pogoreltsev, A. I., Fedulina, I. N., Mitchell, N. J., Muller, H. G., Luo, Y., Meek, C. E., and Manson, A. H.: Global free oscillations of the atmosphere and secondary planetary waves in the MLT region during August/September time conditions, J. Geophys. Res., 107, 4799, doi:10.1029/2001JD001535, 2002.

Richards, P. G., Fennelly, J. A., and Torr, D. G.: EUVAC: A solar EUV flux model for aeronomic calculations, J. Geophys. Res., 99, 8981–8992, 1994. (correction, J. Geophys. Res., 99, 13 283, 1994).

Roble, R. G.: Energetics of the mesosphere and thermosphere, in: The Upper Mesosphere and Lower Thermosphere: A Review of Experiment and Theory, Geophys. Monogr. Ser., Vol. 87, edited by: Johnson, R. M. and Killeen, T. K., 1–21, AGU, Washington, D. C., 1995.

Schäfer, J.: A space-time analysis of tropospheric planetary waves in the northern hemisphere, J. Atmos. Sci., 36, 1117–1123, 1979.

Strang, G.: On the construction and comparison of difference schemes, SIAM J. Numer. Anal., 5, 516–517, 1968.

Swinbank, R. and O'Neill, A.: A stratosphere-troposphere data assimilation system, Mon. Weather Rev., 122, 686–702, 1994.

11

Long-term Measurements of Nighttime LF Radio Wave Reflection Heights over Central Europe

C. Jacobi[1] and D. Kürschner [2]

[1]Institute for Meteorology, University of Leipzig, Stephanstr. 3, 04103 Leipzig, Germany
[2]Institute for Geophysics and Geology, University of Leipzig, Collm Observatory, 04779 Wermsdorf, Germany

Abstract. The nighttime ionospheric absolute reflection height of low-frequency (LF) radio waves at oblique incidence has been measured continuously since late 1982 using 1.8 kHz sideband phase comparisons between the sky wave and the ground wave of a commercial 177 kHz LF transmitter. The dataset allows the analysis of long-term trends and other regular variations of the reflection height. Beside the clear signal of the 11-year solar cycle a quasi-biennial oscillation is visible in LF reflection heights, which is correlated to the equatorial stratospheric wind field. A long-term decreasing reflection height trend is found, confirming results from other measurements and theoretical estimations. The results can be interpreted as a long-term decrease of the height levels of fixed electron density in the lower E region, reflecting a long-term cooling trend of the middle atmosphere.

1 Introduction

It has been known for a long time that the middle and upper atmosphere may serve as an indicator for climate variability. For instance, it has been shown that in general the increase of greenhouse gases results in a cooling of the stratosphere (Ramaswamy et al., 2001) and mesosphere (Beig et al., 2003). Considering the upper mesosphere/lower thermosphere region, this leads to a descent of layers of constant pressure, which may be monitored using radio wave reflection heights, which are connected with altitudes of constant electron density that in turn are controlled by the pressure profile (von Cossart and Taubenheim, 1987; Bremer and Berger, 2002).

At shorter time scales, one of the most prominent neutral middle atmospheric circulation patterns is the equatorial quasi-biennial oscillation (QBO, e.g. Naujokat, 1986). It influences the mid- and high-latitude middle atmosphere as well (Holton and Tan, 1980; Labitzke, 2004), so that, for

example, during QBO west phases the winter stratospheric vortex is deeper and colder than during QBO east phases. The influence of the QBO on the upper atmosphere, however, is less clear. Jarvis (1997) found a QBO modulation of the semidiurnal tide expressed in ground geomagnetic variations. Jacobi et al. (1996), analysing Collm Central Europe lower thermosphere winds found that only the winter zonal prevailing winds are stronger during QBO west phase. This is in accordance with the stratospheric behaviour described by Holton and Tan (1980), but upper mesosphere wind measurements over Saskatoon, Canada, although revealing some biennial or quasi-biennial periodicity (Namboothiri et al., 1994), did not show a clear correspondence with the equatorial QBO. To conclude, the results on the QBO effect on the MLT region still appears to be inconclusive to a certain degree.

To investigate the long-term variability of the upper middle atmosphere, we analysed nighttime low-frequency (LF) radio wave absolute reflection heights. When interpreting these data, however, it has to be taken into account that these height variations show the signal of a mixture of different processes, namely changes of neutral atmosphere pressure level heights and changes in lower E-region ionisation. Therefore, the results have to be interpreted with care when conclusions about atmospheric variability should be drawn.

2 Description of the measurements

Low frequency 177 kHz radio waves from a commercial radio transmitter are registered at Collm Observatory, Germany (distance to transmitter 170 km). The virtual reflection heights h', referring to the reflection point at $52.1°$ N, $13.2°$ E, are estimated using measured travel time differences between the ground wave and the reflected sky wave through phase comparisons on sporadic oscillation bursts of the amplitude modulated LF radio wave in a small modulation frequency range around 1.8 kHz (Kürschner et al., 1987). Because here we are primarily interested in a qualitative

Correspondence to: C. Jacobi
(jacobi@uni-leipzig.de)

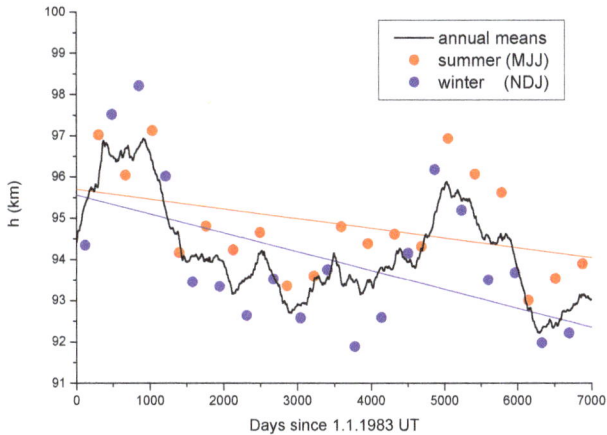

Fig. 1. Time series of annual mean virtual reflection heights (black line). Also given are seasonal (90 day) means for summer (May–July, red dots) and winter (November–January, blue dots).

analysis of heights, the original virtual heights are used here. For most part of the measurements, the retardation effect is small, as has been found from wave field calculations. Virtual reflection heights range between roughly 80 km during day and more than 100 km during nighttime for the ordinary component.

The reflection height measurements have started in September 1982 and are carried out continuously since then. During daylight hours, particularly in the summer months height measurements are not possible due to the strong D-region absorption then. Therefore, and because the radio wave reflection height exhibits a strong diurnal variation, it is not possible to average over all measured height values of one day to derive daily mean heights. Therefore, a least squares fit was applied to detect the nighttime mean height and a semidiurnal variation:

$$h'(t) = h_0 + a \cdot \sin \frac{2\pi}{12h}t + b \cdot \cos \frac{2\pi}{12h}t + \varepsilon, \quad (1)$$

with h_0 as the daily mean , $\Delta h = (a^2 + b^2)^{1/2}$ as the 12-h oscillation amplitude and ε as indefinite height variations. Each regression analysis was applied to 15 days of half-hourly mean h' values, and the resulting mean nighttime height was attributed to the centre of the respective time window. The window then was shifted by one day, and the procedure was repeated.

The h_0 time series presented here differs somewhat from the phase-height measurements presented by von Cossart and Taubenheim (1987) or Bremer and Berger (2002) in such a way, that we measure continuously during the night, while phase-height measurements are taken once a day at daytime and refer to a constant zenith angle. In addition the transmitter distance for the phase-height measurements was much longer. Therefore our average height is found at the lower nighttime E-region well above 90 km, while the phase-height measurements show reflection heights at about 82 km in the D-region.

3 Long-term trends

The time series of reflection heights is shown in Fig. 1. The data are 1-year mean values, shifted by 1 day. Also shown are seasonal (90 day) means for summer and winter. Besides a decadal variation, a clear long-term decrease is visible. Linear fits are added. It can be seen that the trends are negative. The decrease amounts to -90 ± 40 m/yr in summer, and -170 ± 60 m/yr in winter. During equinoxes (not shown here) the respective values are -110 ± 40 m/yr (spring) and -100 ± 45 m/yr (autumn). Hence, the decrease takes place in each season, with maximum values in winter. This is qualitatively in agreement with the results of Bremer and Berger (2002), who found stronger decrease in D-region reflection heights in December and January.

Note also that the estimated trend may slightly overestimate the real trend, since the begin and the end of the time series are not exactly in the same phase of the solar cycle. To remove this effect from the trend analyse we calculated a multiple linear regression analysis, including simultaneously the 11-year oscillation and a possible linear trend. This analysis revealed nearly the same values of annual height decrease.

If we assume that both composition and effective recombination coefficient remain constant during the time interval under consideration, we may assume that the reflection heights represent heights of constant pressure, so that, following Bremer and Berger (2002) we may calculate the mean temperature change through integrating the hydrostatic equation from a lower reference level to our reflection height for the past (i.e. 1983) and present (i.e. 2001) conditions:

$$\int_{48km}^{z_2} \frac{1}{T} dz = \int_{48km-\Delta z_1}^{z_1-\Delta z_2} \frac{1}{T(z) - \Delta T} dz. \quad (2)$$

We use the CIRA (Fleming et al., 1990) temperatures, upper reference levels of $z_2 = 95.5$ km for summer and winter, 96 km for spring, and 94 km for autumn as "past" conditions, and the Δz_2 values given above. The change of reference height near 48 km was taken as $\Delta z_1 = 10$ m/yr after Taubenheim (1994). The resulting temperature trends are 0.93 K/yr for winter, 0.57 K/yr for spring, 0.47 K/yr for summer, and 0.61 K/yr for autumn. Comparing these data with those presented by Beig et al. (2003), we find that our results are not outside the variety of measurements, but the trends are stronger than the average of other methods.

Of course, our assumption of constant air composition and recombination coefficient are not realistic. Therefore, our derived trends are not more than a first rough estimate of temperature trends. While the assumption of constant composition may lead to an underestimation of trends (see Bremer and Berger, 2002), the influence of the solar cycle will lead to a potential overestimation of trends, because the beginning of the time series falls into solar minimum, while the end is near solar maximum.

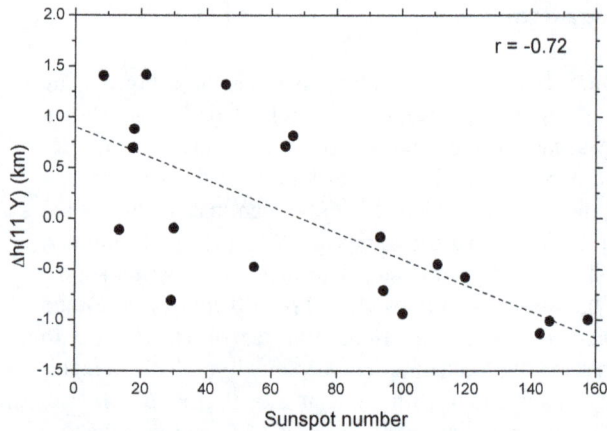

Fig. 2. Annual mean differences of the low-pass filtered winds (cut-off 4 yrs) vs. the annual mean sunspot number.

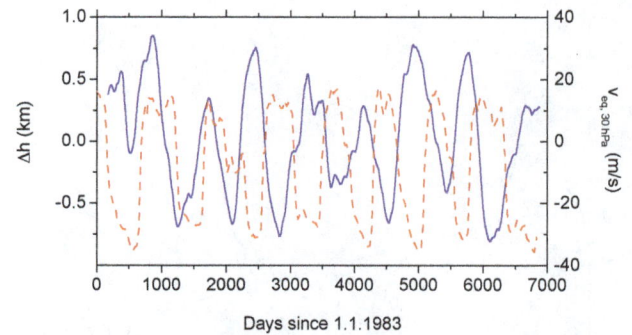

Fig. 3. Time series of band pass (2.2 years) filtered reflection heights (blue line, left axis), and equatorial winds at 30 hPa (red dashed line, right axis).

4 Solar cycle

To show variations in the period range of months to years, we applied transverse band-pass filters to the time series. Apart from the annual and semiannual cycle, an 11-year variation and the QBO effect appears to be the prominent features. To highlight the solar cycle dependence, in Fig. 2 the low-pass filtered h_0 data are plotted vs. the annual mean sunspot numbers. The reflection heights are lower during solar maximum (see also Fig. 1) than during solar minimum, which is easily understandable from the stronger ionisation during times of high solar activity.

The amplitude of the 11-year oscillation amounts to about 1 km, i.e. the difference between solar maximum and minimum is 2 km. This is more than that has been given in literature (Entzian, 1967). However, we refer to a different height region than these results do. In addition, it is known that the neutral stratosphere and mesosphere exhibit a signal of the 11-year solar cycle as well (e.g., van Loon and Labitzke, 1990; Keckhut et al., 1995), which may contribute to the effect shown in Fig. 2. This means, on time scales of solar variability we cannot clearly distinguish between the solar effect through E-region ionisation and a possible atmospheric response.

5 QBO

In Fig. 3 the band-pass filtered reflection heights, with a resonant frequency of 2.2 years are shown together with the equatorial winds at 30-hPa heights. The lengths of the cycles are varying from 22 to 32 months, with a mean length of 27 months. This is well within the ranges of the stratospheric QBO (Naujokat, 1986). It can be seen that for QBO West years in most cases the reflection heights are lower than those during QBO East years. This would fit to the concept of a cooler middle atmosphere during QBO West years, which is in correspondence with the stratospheric behaviour during winter (Holton and Tan, 1980).

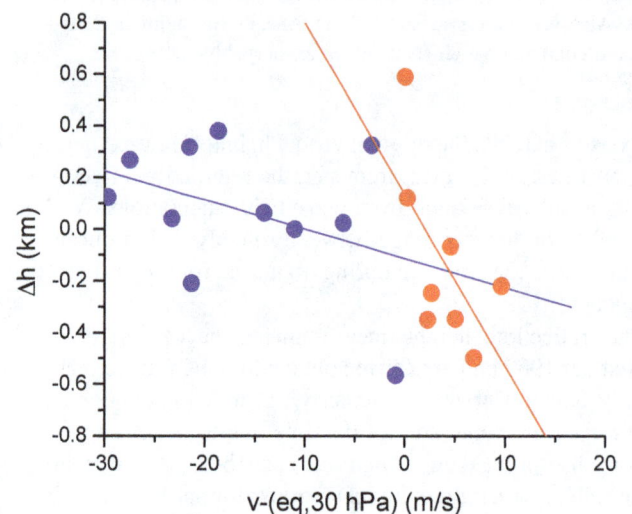

Fig. 4. Annual mean band pass (2.2 years) filtered reflection heights vs. equatorial winds at 30 hPa.

Reflection height differences vs. equatorial QBO are presented in Fig. 4. It can be seen from separate linear fits for QBO East and QBO West years, however, that for equatorial westerlies (and weak easterlies) there is a possible effect on the midlatitude lower ionosphere, but for QBO East years this effect is weak.

6 Conclusions

We have shown nighttime LF reflection heights measured at Collm, Germany. The long-term time series reveals a solar cycle, a long-term decreasing trend, and a QBO. While the influence of the 11-year solar cycle at least partly is obviously due to increasing radiation during solar maximum, the other effects are probably of atmospheric origin, connected with a long-term cooling of the mesosphere and the equatorial stratospheric circulation.

There has been a long debate about trends in ionospheric parameters as possible indicators for atmospheric variations (Bencze and Poor, 1996; Ulich and Turunen, 1997; Lastovicka and Bremer, 2004; Bremer, 2004). Our LF measurements may provide another piece of evidence that long-term cooling of the atmosphere, as well as interannual atmospheric variability, really changes the lower ionospheric regions. However, when interpreting ionospheric results in terms of atmospheric variability, care must be taken to distinguish between in situ ionospheric effects and influences from below.

Acknowledgements. This study has been supported by BMBF under grant 07ATF10 (MEDEC).

References

Beig, G., Keckhut, P., Lowe, R. P., Roble, R. G., Mlynczak, M. G., Scheer, J., Fomichev, V. I., Offermann, D., French, W. J. R., Shepherd, M. G., Semenov, A. I., Remsberg, E. E., She, C. Y., Lübken, F. J., Bremer, J., Clemesha, B. R., Stegman, J., Sigernes, F., and Fadnavis, S.: Review of mesospheric temperature trends, Rev. Geophys. 41, 1015, doi:10.1029/2002RG000121, 2003.

Bencze, P. and Poor, A.: On the effect of greenhouse gases in the upper atmosphere, Acta Geod. Geoph. Hung., 31, 439–444, 1996.

Bremer, J.: Investigations of long-term trends in the ionosphere with world-wide ionosonde observations, Adv. Radio Sci., 2, 253–258, 2004.

Bremer, J. and Berger, U.: Mesospheric temperature trends derived from ground-based LF phase-height observations at midlatitudes: comparison with model simulations, J. Atmos. Solar-Terr. Phys., 64, 805–816, 2002.

Entzian, G.: Der Sonnenfleckenzyklus in der Elektronenkonzentration der D-Region, Kleinheubacher Berichte, 12, 309–313, 1967.

Fleming, E. L., Chandra, S., Barnett, J. J., and Corney, M.: Zonal mean temperature, pressure, zonal wind, and geopotential height as function of latitude, Adv. Space Res. 10 (12), 11–59, 1990.

Holton, J. R. and Tan, H.-C.: The influence of the equatorial quasi-biennial oscillation on the global circulation at 50 mb, J. Atmos. Sci., 37, 2200–2208, 1980.

Jacobi, Ch., Schminder, R., and Kürschner, D.: On the influence of the stratospheric quasi-biennial oscillation on the mesopause zonal wind over Central Europe, Meteorol. Zeitschrift, N. F., 5, 318–323, 1996.

Jarvis, M. J.: Latitudinal variation of quasi-biennial oscillation modulation of the semidiurnal tide in the lower thermosphere, J. Geophys. Res., 102, 27 177–27 187, 1997.

Keckhut, P., Hauchecorne, A., and Chanin, M. L.: Midlatitude long-term variability of the middle atmosphere: Trends and cyclic and episodic changes, J. Geophys. Res., 100, 18 887–18 897, 1995.

Kürschner, D., Schminder, R., Singer, W., and Bremer, J.: Ein neues Verfahren zur Realisierung absoluter Reflexionshöhenmessungen an Raumwellen amplitudenmodulierter Rundfunksender bei Schrägeinfall im Langwellenbereich als Hilfsmittel zur Ableitung von Windprofilen in der oberen Mesopausenregion, Z. Meteorol., 37, 322–332, 1987.

Labitzke, K.: On the signal of the 11-year sunspot cycle in the stratosphere and its modulation by the quasi-biennial oscillation, J. Atmos. Solar-Terr. Phys., 66, 1151–1157, 2004.

Lastovicka, J. and Bremer, J.: An overview of long-term trends in the lower ionosphere below 120 km, Surveys in Geophysics 25, 66–99, 2004.

Namboothiri, S. P., Meek, C. E., and Manson, A. H.: Variations of mean winds and solar tides in the mesosphere and lower thermosphere over scales ranging from 6 months to 11a. Saskatoon 52° N, 107° W, J. Atmos. Terr. Phys., 56, 1313–1325, 1994.

Naujokat, B.: An update of the observed quasi-biennial oscillation of the stratospheric winds over the tropics, J. Atmos. Sci., 43, 1873–1877, 1986.

Ramaswamy, V., Chanin, M.-L., Angell, J., Barnett, J., Gaffen, D., Gelman, M., Keckhut, P., Koshelkov, Y., Labitzke, K., Lin, J.-J. R., O'Neill, A., Nash, J., Randel, W., Rood, R., Shine, K., Shiotani, M., and Swinbank, R.: Stratospheric temperature trends: Observations and model simulations, Rev. Geophys., 39, 71–122, 2001.

Taubenheim, J., Berendorf, K., Krüger, W., and Entzian, G.: Height dependence of long-term trends in the middle atmosphere, EGS XIX General Assembly, Session ST3, Grenoble, 1994.

Ulich, T. and Turunen, E.: Evidence for long-term cooling of the upper atmosphere in ionosonde data, Geophys. Res. Lett., 24, 1103–1106, 1997.

Van Loon, H. and Labitzke, K.: Association between the 11-year solar cycle and the atmosphere, Part IV: The stratosphere, not grouped by the phase of the QBO, J. Clim., 8, 827–836, 1990.

Von Cossart, G. and Taubenheim, J.: Solar cycle and long-period variations of mesospheric temperatures, J. Atmos. Terr. Phys., 49, 303–307, 1987.

12

The 8-h tide in the mesosphere and lower thermosphere over Collm (51.3° N; 13.0° E), 2004–2011

Ch. Jacobi and T. Fytterer

Institute for Meteorology, University of Leipzig, Stephanstr. 3, 04103 Leipzig, Germany

Correspondence to: Ch. Jacobi (jacobi@uni-leipzig.de)

Abstract. The horizontal winds in the mesosphere and lower thermosphere (MLT) at heights of about 80–100 km have been measured continuously since summer 2004 using an all-sky 36.2 MHz VHF meteor radar at Collm, Germany (51.3° N, 13° E). A climatology of the 8-h solar tide has been constructed from these data. The amplitude shows a seasonal behaviour with maximum values during the equinoxes, and it is generally increasing with altitude. The largest amplitudes are measured in autumn, partly reaching values up to $15\,\mathrm{m\,s^{-1}}$. The phase, defined as the time of maximum eastward or northward wind, respectively, has earlier values in winter and later ones in summer. Except for summer, the phase difference between the zonal and meridional components is close to +2 h, indicating circular polarization of the tidal components. The vertical wavelengths are short in summer (~20 km) but significantly longer during the rest of the year. The terdiurnal tide is generally assumed to originate from either a terdiurnal component of solar heating or nonlinear interaction between the diurnal and semidiurnal tide. Analysing monthly means reveals positive correlation during the spring maximum, but negative correlation in autumn.

1 Introduction

The dynamics of the mesosphere and lower thermosphere (MLT) are strongly influenced by atmospheric waves, including the solar tides with periods of a solar day and its harmonics. Their wind amplitudes usually maximise around 100–120 km (e.g. Hagan et al., 1995). In these regions, their amplitudes are of the order of magnitude of the mean wind. As a result, the solar tides drive the global circulation and more accurate knowledge leads to a better understanding of the wind fields in the MLT. Shorter period waves often have smaller amplitudes, so that in the past the diurnal tide (DT)

and the semidiurnal tide (SDT) have attracted more attention. But recently, also the terdiurnal tide (TDT) has been considered to play an important role as well, because occasionally their amplitudes are as large as the ones of DT and SDT.

DT and SDT are essentially forced by absorption of solar radiation through tropospheric water vapour and stratospheric ozone, respectively. But the forcing mechanism of the TDT is still under debate, and both direct solar heating in the lower and middle atmosphere and nonlinear interaction between the DT and SDT (Teitelbaum et al., 1989) or gravity waves are also thought to excite the TDT. Observations in the Arctic mesosphere showed a relationship between the vertical wavelengths of the TDT, SDT and DT when the TDT had large amplitudes. This indicates, at least to a certain degree, the existence on nonlinear coupling (Younger et al., 2002). Model analyses gave partly inconclusive results. Akmaev (2001) concluded from circulation model calculations that nonlinear interaction makes a contribution to the excitation of the TDT. Smith and Ortland (2001) performed calculations that indicate that the direct solar forcing of the TDT is the dominant mechanism at middle and high latitudes, while nonlinear interactions contribute to the TDT at low latitudes. Huang et al. (2007) used a nonlinear tidal model and concluded that the migrating TDT can be significantly excited by the nonlinear interaction between the diurnal and semidiurnal tides in the MLT region. These models have used prescribed lower boundary TDT fields. Du and Ward (2010) analysed CMAM Global Circulation Model results with respect to correlations between TDT and DT/SDT on the short-term or seasonal time scale. Since the correlation was significant essentially on the seasonal and not on the short-term scale, they concluded that possible correlation is due to corresponding source or propagation condition variability rather than nonlinear interaction. To summarize, model results still show somewhat inconclusive results, and the dominant forcing mechanism of the TDT is still not clear.

Observations of the TDT are relatively rare. A global characteristics at 95 km, derived from Upper Atmosphere Research Satellite/High Resolution Doppler Imager (UARS/HRDI) measurements, have been presented by Smith (2000). There, a winter amplitude maximum was visible. The characteristics of the TDT observed by radar have been described on few occasions (e.g. Beldon et al., 2006; Jacobi, 2011). Often, the zonal amplitude has been found to be larger than the meridional one. A clear seasonal cycle is apparent, with smaller amplitudes in summer and two maxima in spring and autumn, while the latter one is dominating. The amplitudes range from 1–$10\,\mathrm{m\,s^{-1}}$ at altitudes at $\sim$90 km depending on season and latitude. Observations at high latitudes have shown no spring maximum below 95 km (Younger et al., 2002), while it was clearly visible at mid-latitudes (Beldon et al., 2006; Jacobi, 2011). The phase varies with latitude and season. Often the phase difference between the zonal and meridional components is close to $+2\,$h, indicating a circular polarized wave. Noticeable differences only occur during summer (Beldon et al., 2006). The reported values of the vertical wavelength strongly vary, and are sometimes contradicting. Observations made at Arctic latitudes show wavelengths of about 25–90 km (Younger et al., 2002). Considering the results at mid-latitudes (ranging from $\sim$22–>1000 km in the course of a year) by Namboothiri et al. (2004) or Thayaparan (1997) and the observations at lower latitudes ($\sim$12–32 km) reported by Tokumoto et al. (2007), some differences are seen.

In this paper some features of the TDT measured at Collm, Germany (51.3° N; 13.0° E) are presented, using the dataset from August 2004 to June 2011. This data represents an update of the one presented by Fytterer and Jacobi (2011) and Jacobi (2011).

2 Collm meteor radar wind measurements and tidal analysis

The data used in this study have been measured by VHF meteor radar at Collm Observatory, Germany (51.3° N, 13° E). It is operating since July 2004, and the 7-yr dataset from August 2004–July 2011 is used here to investigate the peculiarities of the TDT. The meteor radar operates at 36.2 MHz with a pulse repetition frequency of 2144 Hz by using a transmitter of 6 kW peak power with a pulse width of 13 μs. Meteor reflections are detected by a five element antenna array forming an asymmetric cross, acting as an interferometer. This allows the calculation of azimuth and zenith angle. In combination with range measurements, the meteor trail position is determined. Radial wind velocity along the line of sight is obtained from the Doppler phase progressions with time at each receiver, which are averaged to form the mean Doppler frequency. The number of meteors varies between approximately 3500 and 7000 per day, depending on season (e.g. Arras et al., 2009). Meteors are mainly detected between 80 and

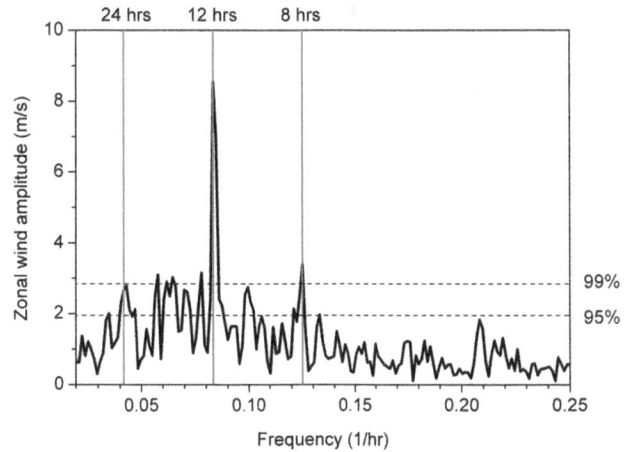

Fig. 1. Example of a zonal mean wind spectrum over Collm. The data used are March 2005 hourly means over the 85–95 km altitude region. The dashed lines indicate the 95 % and 99 % significance levels according to a χ^2-test.

100 km altitude, with a maximum around 90 km. Individual radial winds calculated from the meteors are summarized to form hourly or half-hourly mean values through projection of the horizontal wind on the individual radial winds, using least-squares fitting and assuming that vertical winds are negligible (Hocking et al., 2001; Jacobi, 2011). In Fig. 1, a sample FFT spectrum derived from hourly mean winds that have been calculated using meteors between 85 and 95 km is presented. While, as expected, the major signal is owing to the SDT, a peak at 8 h is visible as well, which is, at least in this case, even larger than the well-known DT peak at 24 h.

For characterizing the horizontal wind field in the MLT, in the following the observed height interval is divided in six not overlapping height gates centred at 82, 85, 88, 91, 94 and 98 km. Owing to the meteor detection rates strongly decreasing with altitude above 95 km, the real mean altitude of the uppermost height gate is only 97 km (Jacobi, 2011). The amplitudes and phases of the TDT within each height interval in a given 15-day time interval are calculated by a multiple regression analysis of 15 days of half-hourly zonal or meridional wind components, which includes the mean wind, as well as the 8-, 12-, and 24 h oscillations. Thus, the horizontal winds are modelled as:

$$v_\mathrm{m}(t) = v_0 + \sum_{i=1}^{3} a_i \sin\left(\frac{2\pi}{P_i}\right)t + b_i \cos\left(\frac{2\pi}{P_i}\right)t, \qquad (1)$$

with v_m as the measured half-hourly (either zonal or meridional) wind, v_0 as the prevailing (mean) wind and using the periods $P_1 = 24\,$h, $P_2 = 12\,$h, $P_3 = 8\,$h. The amplitudes A_3 and phases T_3 of the TDT are calculated from the regression coefficients a_3 and b_3 through:

$$A_3 = \sqrt{a_3^2 + b_3^2},\ T_3 = \frac{P_3}{2\pi}\mathrm{atan}\left\{\frac{a_3}{b_3}\right\}. \qquad (2)$$

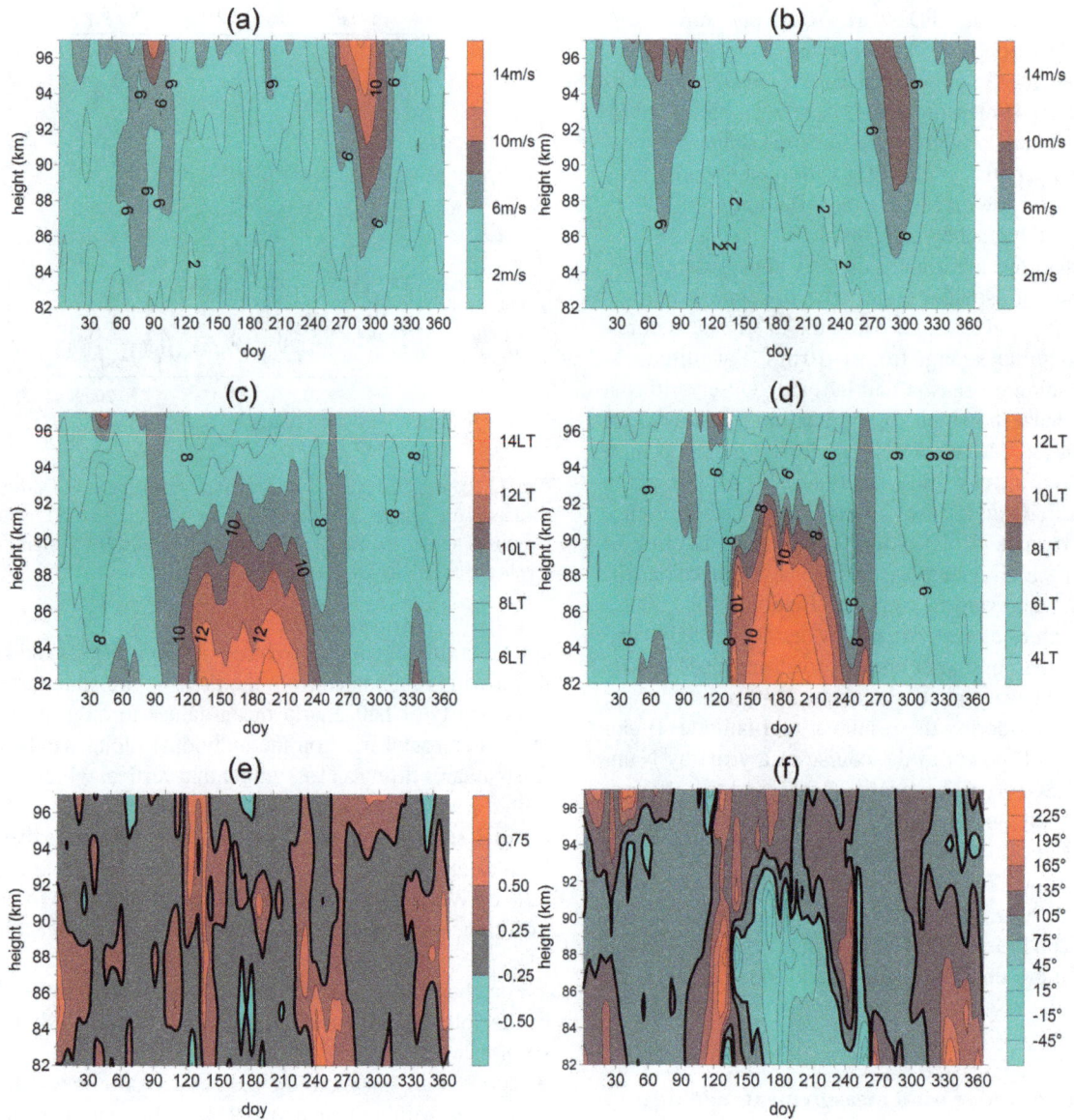

Fig. 2. Zonal amplitudes (**a**) and phases (**b**), meridional amplitudes (**c**) and phases (**d**) of the terdiurnal tide over Collm. Relative amplitude differences after Eq. (3) are shown in panel (**e**); the regions of amplitude differences smaller than 25 % are indicated by heavy isolines. Phase differences $\Delta\varphi$ in degrees are given in panel (**f**). Positive values of the phase difference indicate later westerly than northerly wind maxima. The regions of $\Delta\varphi = 90 \pm 15°$ are indicated by heavy isolines. Data are 7-yr means constructed from regression analyses including 15 days of data each, updated from Jacobi (2011).

3   7-yr mean tidal amplitudes and phases

Figure 2 presents TDT zonal (a) and meridional amplitudes (b), and zonal (c) and meridional phases (d). Note that the scaling of the meridional phase is shifted by 2 h with respect to the zonal one to more clearly indicate seasons and regions where the TDT components are in quadrature. The average standard error has been taken from daily analyses, i.e. applying Eqs. (1) and (2) but including only one day of data each. For the amplitudes below 91 km, the error ranges between 0.50–0.53 m s^{-1}, and increases to 0.62/0.66 m s^{-1}

for the zonal/meridional component at 97 km. The average phase standard error is 0.19 h for both components, largely independent of height. Standard errors of individual 15-day mean amplitudes approximately range between 1 and 2 m s^{-1}. Zonal and meridional amplitudes both maximise in autumn, while a second maximum is found in spring. Generally, amplitudes tend to increase with altitude during the entire year. An exception is the region below $\sim$90 km in midsummer, where a slight tendency for constant or decreasing amplitudes with height is visible. On an average, zonal and meridional amplitudes are of similar order of magnitude.

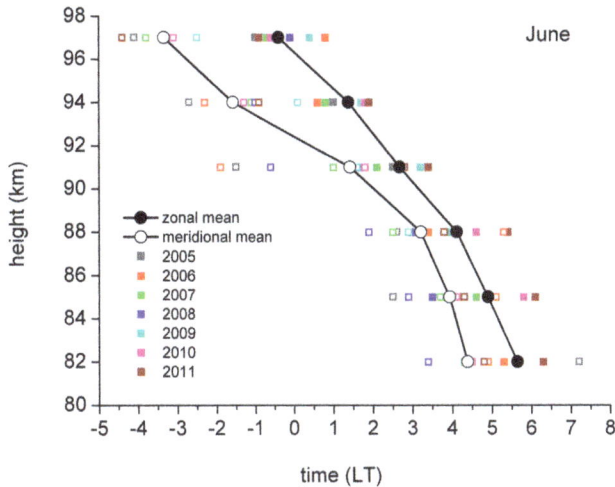

Fig. 3. June TDT phases for individual years (small squares) and for 7-yr means (large circles).

Figure 2e presents relative amplitude differences ΔA_3:

$$\Delta A_3 = 2 \frac{A_{3,z} - A_{3,m}}{A_{3,z} + A_{3,m}}, \qquad (3)$$

with $A_{3,z}$ and $A_{3,m}$ as the zonal and meridional TDT amplitudes. Zonal amplitudes are slightly smaller than meridional ones during the March maximum at the upper height gates, while the contrary is the case during the autumn maximum. During early and late summer, there is a tendency for slightly larger zonal than meridional amplitudes. The summer meridional amplitudes below 90 km appear to be smaller than the zonal ones, but generally the amplitudes in summer are very small, and differences are insignificant.

TDT phases (Fig. 2c and d) clearly change between winter and summer. Phase change with height is very small during winter, and therefore the vertical wavelength is very long then, generally exceeding 200 km. Since the wavelengths are calculated from the phase change with altitude, such large values cannot be determined with accuracy. However, the results are in agreement with the ones by Namboothiri et al. (2004). On the contrary, there is a strong phase change with altitude in summer. This results in a short wavelength of about 20 km, which is even shorter than those reported by Namboothiri et al. (2004) and Thayaparan (1997). The zonal and meridional phases are close to being in quadrature when the amplitude is strong, i.e. in autumn and late winter/early spring. In summer, zonal and meridional phase difference changes with altitude, with a rather sharp change around 90 km (Fig. 3), and connected with a change in amplitude gradient (see Fig. 2a, b). This indicates that at least two TDT modes are present in the summer mid-latitude MLT.

4   Interannual variability

Interannual variability of the TDT amplitudes and phases at 91 km is shown in Fig. 4. Here the 7-yr mean, calculated from individual data each based on 15 days of measurements, is represented as the solid line, while the dots show the respective amplitudes (upper panels) and phases (lower panels) of single years. In most months, the amplitude variability is very strong, and during some years amplitudes of less than 1 m s^{-1} are occasionally found. An exception is represented by the autumn maximum (a multi-year average of $\sim 9 \text{ m s}^{-1}$), which has considerably large amplitudes in every year. On the contrary, the spring maximum ($\sim 7 \text{ m s}^{-1}$ on an average) is less stable. Beldon et al. (2006) showed that for the case of the UK radar the spring amplitudes are not larger than the winter ones. The phases generally range within about 3 h from year to year, but the interannual variability is smaller in March and October, when the amplitudes are larger. During summer, the meridional phase is extremely variable, which is partly due to the choice of the height gate presented here. At 91 km, the transition between the two summer modes is seen (Fig. 3), and while the zonal component shows the same vertical phases gradient at all heights considered, the meridional one changes with altitude, and the 91 km value obviously belongs to different modes in different years.

Since the phase variability is weakest during spring and autumn, the March and October amplitudes are well suited to investigate interannual variability. Du and Ward (2010) have used correlation analysis between the TDT and the DT or SDT, respectively, from CMAM model results in order to explain possible coupling processed between tides. They found significant correlation between TDT and SDT amplitudes, which, however, change with height and latitude and show a rather complicated picture in the MLT near 50° N. In Fig. 5 time series of monthly mean amplitudes at 88 km are shown for these months. For comparison, and to indicate possible corresponding variability with the SDT, the SDT amplitudes are also shown. During March there is a positive correlation between TDT and SDT amplitudes (correlation coefficient $r = 0.56$, however, r increases for lower height gates, reaching $r = 0.72$ at 82 km), while in October the correlation is negative ($r = -0.59$, and its magnitude also increasing at lower heights). However, the time series used is still too short to draw substantial conclusions, and the correlations are not significant at the 95 % level. Correlation between TDT and mean wind (not shown here) is weak and inconclusive.

5   Conclusions

The Collm meteor radar has been operated nearly continuously since summer 2004, measuring daily mean winds and tidal amplitudes and phases. Here we focus on the TDT, and a 7-yr climatology is presented. Generally, from a point measurement as is used here, one cannot prove that the analysed

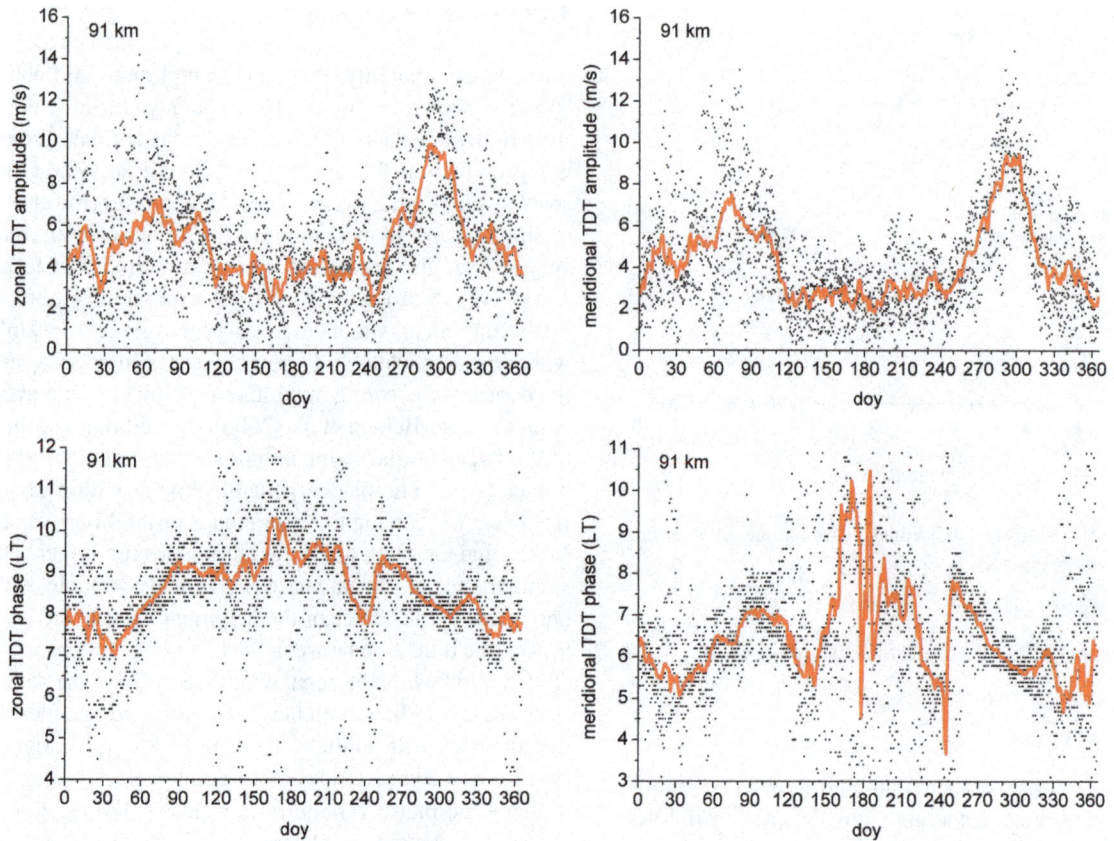

Fig. 4. Zonal (left panels) and meridional (right panels) amplitudes (upper panels) and phases (lower panels) of the TDT over Collm at 91 km. The solid line denotes the 7-yr mean while the dots represent data from individual years. Data are calculated from regression analyses based on 15-days of half-hourly means each.

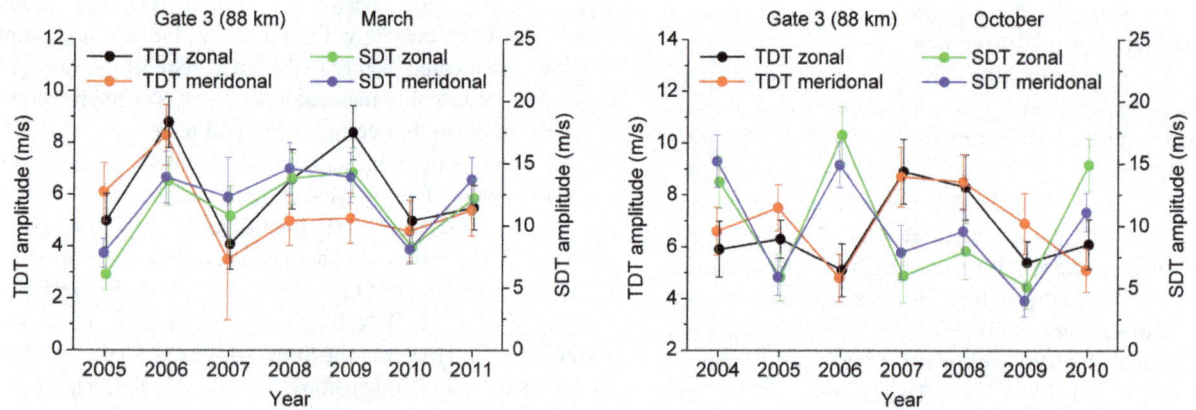

Fig. 5. Time series of monthly mean TDT and SDT amplitudes at 88 km in March (left panel) and October (right panel).

oscillations really belong to waves rather than other kind of fluctuations. However, we have seen from comparison with literature that some essential features of the 8-h oscillation at Collm – e.g. the seasonal cycle of the vertical wavelength – are likewise found at other stations, which indicates that the 8-h feature is a large-scale one and not a local pattern only. Furthermore, circular polarization of the horizontal compo-

nents at those times when the amplitude is large is consistent with an upward propagating inertio-gravity wave. Therefore we may consider the 8-h oscillation at Collm as the signature of the TDT, although a final proof is not given.

Maximum amplitudes of the TDT are found in spring and autumn. The autumn maximum is quite stable, while the spring maximum is broader and less stable; sometimes small amplitudes are found. Vertical wavelengths are short (~20 km) in summer, but very long in winter.

Interannual variability of amplitudes has been analysed for March and October, when the phases are only weakly varying from year to year. TDT amplitudes are positively correlated with SDT ones in spring, but negatively in autumn. This indicates a possible interaction between TDT and SDT, but the underlying process is not clear and probably different in different seasons.

Acknowledgements. The authors thank Peter Hoffmann, Leipzig, for useful discussion, and Falk Kaiser, Leipzig, for maintaining the radar measurements. Topical Editor Matthias Förster thanks Edward Kazimirovsky and an anonymous reviewer for their help in evaluating this paper.

References

Akmaev, R. A.: Seasonal variations of the terdiurnal tide in the mesosphere and lower thermosphere: a model study, Geophys. Res. Lett., 28, 3817–3820, 2001.

Arras, C., Jacobi, Ch., and Wickert, J.: Semidiurnal tidal signature in sporadic E occurrence rates derived from GPS radio occultation measurements at midlatitudes, Ann. Geophys., 27, 2555–2563, 2009,
http://www.ann-geophys.net/27/2555/2009/.

Beldon, C. L., Muller, H. G., and Mitchell, N. J.: The 8-hour tide in the mesosphere and lower thermosphere over the UK, 1988–2004, J. Atmos. Sol.-Terr. Phys., 68, 655–668, 2006.

Du, J. and Ward, W. E.: Terdiurnal tide in the extended Canadian Middle Atmospheric Model (CMAM), J. Geophys. Res., 115, D24106, doi:10.1029/2010JD014479, 2010.

Fytterer, T. and Jacobi, Ch.: Climatology of the 8-hour tide over Collm (51.3° N, 13° E), Rep. Inst. Meteorol. Univ. Leipzig, 48, 23–32, 2011.

Hagan, M. E., Forbes, J. M., and Vial, F.: On modeling migrating solar tides, Geophys. Res. Lett., 22, 893–896, 1995.

Hocking, W. K., Fuller, B., and Vandepeer, B.: Real-time determination of meteor-related parameters utilizing modern digital technology, J. Atmos. Sol.-Terr. Phys., 63, 155–169, 2001.

Huang, C. M., Zhang, S. D., and Yi, F.: A numerical study on amplitude characteristics of the terdiurnal tide excited by nonlinear interaction between the diurnal and semidiurnal tides, Earth Planets Space, 59, 183–191, 2007.

Jacobi, Ch.: 6 year mean prevailing winds and tides measured by VHF meteor radar over Collm (51.3° N, 13.0° E), J. Atmos. Sol.-Terr. Phys., 75–76, 81–91, doi:10.1016/j.jastp.2011.04.010, 2011.

Namboothiri, S. P., Kishore, P., Murayama, Y., and Igarashi, K.: MF radar observations of terdiurnal tide in the mesosphere and lower thermosphere at Wakkanai (45.4° N, 141.7° E), Japan, J. Atmos. Sol.-Terr. Phys., 66, 241–250, 2004.

Smith, A. K.: Structure of the terdiurnal tide at 95 km, Geophys. Res. Lett., 27, 177–180, 2000.

Smith, A. K. and Ortland, D. A.: Modeling and analysis of the structure and generation of the terdiurnal tide, J. Atmos. Sci., 58, 3116–3134, 2001.

Teitelbaum, H., Vial, F., Manson, A. H., Giraldez, R., and Massebeuf, M.: Non-linear interactions between the diurnal and semidiurnal tides: terdiurnal and diurnal secondary waves, J. Atmos. Sol.-Terr. Phys., 51, 627–634, 1989.

Thayaparan, T.: The terdiurnal tide in the mesosphere and lower thermosphere over London, Canada (43° N, 81° W), J. Geophys. Res., 102, 21695–21708, 1997.

Tokumoto, A. S., Batista, P. P., and Clemesha, B. R.: Terdiurnal tides in the MLT region over Cachoeira Paulista (22.7° S, 45° W), Rev. Bras. Geofís., 25, 69–78, 2007.

Younger, P. T., Pancheva, D., Middleton H. R., and Mitchell, N. J.: The 8-h tide in the Arctic mesosphere and lower thermosphere, J. Geophys. Res., 107, 1420, doi:10.1029/2001JA005086, 2002.

13

EUV-TEC proxy to describe ionospheric variability using satellite-borne solar EUV measurements

C. Unglaub[1]**, Ch. Jacobi**[1]**, G. Schmidtke**[2]**, B. Nikutowski**[1,2]**, and R. Brunner**[2]

[1]University of Leipzig, Institute for Meteorology, Stephanstr. 3, 04103 Leipzig, Germany
[2]Fraunhofer IPM, Heidenhofstraße 8, 79110 Freiburg, Germany

Correspondence to: C. Unglaub (unglaub@uni-leipzig.de)

Abstract. An updated version of a proxy, termed EUV-TEC, describing the global total primary photoionisation is calculated from satellite-borne EUV measurements assuming a model atmosphere consisting of four major atmospheric constituents. Regional number densities of the background atmosphere are taken from the NRLMSISE-00 climatology. For calculation the Lambert-Beer law is used to describe the decrease of the radiation along their way through the atmosphere. The EUV-TEC proxy thus describes the ionospheric response to solar EUV radiation and its variability. EUV-TEC is compared against the global mean total electron content (TEC), a fundamental ionospheric parameter created from vertical TEC maps derived from GPS data. Strong correlation between these indices is found on different time scales. Results show that the EUV-TEC proxy represents the ionsopheric variability better than the conventional solar index F10.7 does, especially during high and moderate solar activity.

1 Introduction

The EUV (Extreme Ultraviolet) radiation is defined as the wavelength range between 10 nm and 121 nm (ISO, 2007). It is completely absorbed in the terrestrial atmosphere at altitudes above 50 km. The absorption occurs mainly in the upper atmosphere, i.e. the thermosphere/ionosphere system, and therefore solar EUV radiation is the most important energy source at altitudes above 100 km. It interacts with the atoms and molecules in this region through photodissociation and, at wavelengths up to 102 nm, through photoionisation, thereby leading to the development of the planetary ionosphere. However, independent from the respective mechanism EUV radiation absorption finally causes heating of the thermosphere.

The total electron content (TEC) of the atmosphere is a fundamental ionospheric parameter defined as the electron density integrated along a path under consideration. To determine TEC, the ionospheric influence of radio wave propagation paths may be used, because the ionospheric effect on the propagation velocity depends on the radio wave frequency and the ionospheric electron density integrated along the radio wave propagation path. Because GPS-satellites emit two coherent frequencies the total electron content along the line of sight between the GPS satellite and a ground-based receiver can be deduced, and subsequently may be converted into vertical TEC. Thus measured TEC here is defined as the height integrated electron density between ground and the satellite orbit (Aggarwal, 2011).

Solar EUV radiation varies on different time-scales where the 11-yr Schwabe sunspot cycle causes the primary decadal-scale irradiance variability and the Carrington rotation with an average period of 27 days causes the primary short-term variability. Consequences are strong changes of temperature, composition, density, electron density and ion content of the upper atmosphere. This can affect Low Earth Orbiting (LEO) satellites through variable atmospheric drag, and disturb communication and navigation signals (Woods, 2008).

The solar activity is often described by simple solar indices like the solar radio flux F10.7, which is defined as the solar radio emission at a wavelength of 10.7 cm. However, the primary factor that controls TEC variations and the variability of thermospheric density and temperature is the solar EUV radiation (Emmert and Picone, 2010; Maruyama, 2010) and a nonlinear relationship between F10.7 and EUV fluxes has been found (Liu et al., 2011). Especially during the extended last solar minimum from 2007 to 2009, F10.7 is not an ideal proxy for solar EUV irradiance (Lühr and Xiong, 2010; Chen et al., 2011). Thus, there is a need for updated EUV indices to describe the ionospheric variability.

In this paper we describe progress in constructing a new ionospheric proxy, EUV-TEC, which is intended to explain solar induced ionospheric variability, because the ionospheric electron content is primarily determined by the direct photoionisation induced by the incident solar EUV radiation (Lean et al., 2011). The proxy thus describes ionospheric, not solar variability, in response to the changing sun. The proxy may be used for space weather monitoring and ionospheric research. EUV-TEC is calculated from satellite-borne instruments measuring the EUV radiation considering the modified composition of the atmosphere which is caused by the EUV radiation. This proxy will be compared with F10.7 and the global mean TEC to demonstrate that the ionospheric variability is described better by EUV-TEC than conventional indices like F10.7.

2 EUV-TEC calculation

Solar EUV radiation is nearly completely absorbed in the upper atmosphere. It interacts with the atmospheric gas in this region and thus the EUV radiation will be attenuated. The decrease of the radiation along its propagation path is described by Lambert-Beer's law:

$$dI(\lambda, z) = I_0(\lambda) \cdot \sum_i \sigma_i(\lambda) \cdot n_i(z) \cdot ds, \qquad (1)$$

where dI is the absorbed radiation along the radiation path ds through the atmosphere dependent on the local radiation flux I_0, the absorption cross section σ_i, both dependent on the wavelength λ and the number densities n_i of the respective gas.

EUV-spectra with a resolution of 1 nm are available from the Solar EUV Experiment (SEE) on board the TIMED satellite (Woods et al., 2000, 2005) since February 2002 to date. We use version 10 level 3 products available at LASP, University of Colorado, through http://lasp.colorado.edu/see/see_data.html. Additional EUV data are available from the SOLar Auto-Calibrating EUV/UV Spectrometers (SolACES) Experiment on the ISS since 2008 up to the present (Schmidtke et al., 2006a,b; Nikutowski et al., 2011) with a resolution of 1 nm, too. SolACES has in-flight absolute calibration capability and therefore can be used for validation of other EUV data. However because of the orbit of the ISS a continous measurement is not possible so that only few EUV-spectra from SolACES are available.

For the calculation of EUV absorption and ionisation a spherical model atmosphere is assumed around a spherical model earth surface. The model atmosphere consists of the four major constituents O, N, O_2 and N_2. It reaches from the ground to an altitude of 1000 km with a resolution of 1 km. The absorption and ionisation cross sections are taken from Metzger and Cook (1964) and Fennelly and Torr (1992) and were averaged to get a 1 nm resolution like the EUV-spectra have (Unglaub et al., 2011). For the calculation we assume

that only photons contribute to the ionisation. Secondary ionisation processes are neglected.

An earlier version of EUV-TEC has been determined using globally averaged number densities (Unglaub et al., 2011). Now, regional thermospheric composition profiles from NRLMISE-00 (Picone et al., 2002) are used. To obtain the proxy, first a sphere with 6370 km radius is assumed surrounded by 1000 spherical shells with 1 km distance representing the atmospheric layers. Then, the points of intersection between the radiation paths of the incoming EUV-radiation and the spherical shells are determined. Subsequently these intersection points are converted into geographical coordinates considering the declination. Thus, the regional thermospheric densities can be calculated by the NRLMSISE-00 model for every particular atmospheric layer.

To calculate the primary ionisation rates, Eq. (1) is numerically integrated along each radiation path through the layers of the atmosphere. The path lengths ds through each particular layer is deduced from the intersection points between the radiation paths and the spherical shells. The primary ionisation is calculated for each layer along the radiation paths, integrated over one day and multiplied by the area where the radiation impacts. This results in the total ion production rate per day in the atmosphere. By dividing the production rate by the surface of the earth, the EUV-TEC proxy is obtained representing the global mean ion production per day and m^2 in the atmosphere.

The regional thermospheric densities and the primary ionisation were calculated with a horizontal resolution of 220 km and a temporal resolution of 4 h. This modified calculation and a refinement of the abort criterion causes a slightly larger daily ion production rate in the atmosphere than has been obtained by Unglaub et al. (2011).

3 EUV-TEC proxy: Results

To check how good EUV-TEC mirrors the ionospheric variability, the proxy has been compared against a global daily mean TEC created from gridded vertical TEC maps recorded with the IGS tracking network (Hernandez-Pajares et al., 2009). The datasets are available every 2 h for different longitudes and latitudes with a horizontal resolution of 2.5°. They were weighted with the cosine of their geographic latitude und thereafter a global diurnal mean was calculated. To compare the indices, EUV-TEC, TEC, and F10.7 data were each normalized by subtracting their mean value between July 2002 and June 2007 and dividing through their respective standard deviation. We chose this time period because EUV datasets from TIMED/SEE are available from February 2002 so a complete solar cycle is not available with a comparatively small data amount during solar maximum conditions that can be used for the normalization. This procedure also ensures that the peculiarities of the recent solar minimum are highlighted, since this time interval is not

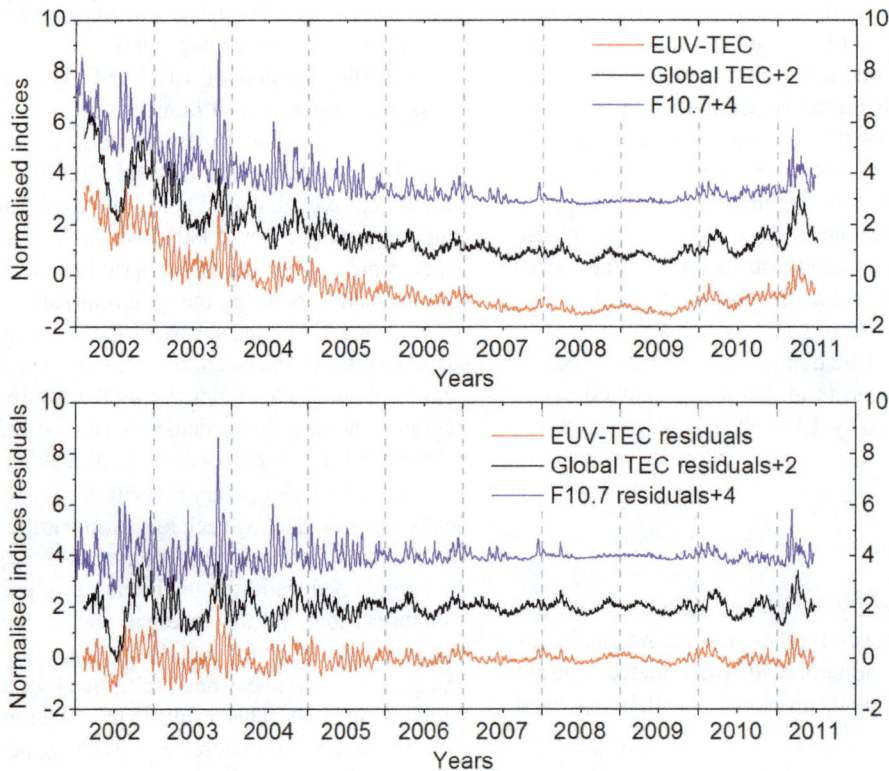

Fig. 1. February 2002 to June 2011 time series of the normalized EUV-TEC, global TEC and F10.7 (upper panel) and of the residuals after subtracting annual mean data (lower panel).

included in the data used for normalization. As a side effect, however, the average of the subsequently presented proxy data is not zero.

The time series of normalized EUV-TEC, normalized F10.7 and normalized global TEC are shown in Fig. 1 (upper panel). All data are uncorrected for earth orbit effect to study the solar influence on the atmosphere. Comparing the tree time series at low solar activity the seasonal pattern is well visible in both global TEC and EUV-TEC. In contrast during solar minimum F10.7 attains nearly consistent values showing a less marked seasonal pattern with a smaller amplitude than global TEC and EUV-TEC have. The seasonal pattern of EUV-TEC and F10.7 is mainly caused by the earth orbit effect where the earth is in perihelion at the beginning of January and in aphelion at the beginning of July. Thus, the indices attain the largest values at the turn of the year and the smallest values in the midyear. In addition to the earth orbit effect global TEC shows a half-year oscillation with two maxima in spring and autumn which can describe neither EUV-TEC nor F10.7, because it is dynamically induced.

Unglaub et al. (2011) have shown that a stronger correlation exists between the global mean TEC and the EUV-TEC proxy than between global TEC and F10.7 during 2002–2009. The updated EUV-TEC proxy shown in Fig. 1 is strongly correlated with global TEC, too, with a correlation coefficient of $r = 0.95$, whereas the correlation coefficient

between global TEC and F10.7 is $r = 0.89$ for data from February 2002 to June 2011. Thus EUV-TEC describes the ionospheric variability, including long-term and short-term variability, better than F10.7 during 2002–2011. The strong correlations essentially result from the 11-yr solar cycle, because all indices attain smaller values with decreasing solar activity. To subtract this trend, the data were smoothed by adjacent averaging over 365 days and the smoothed values were subtracted from the normalized indices to get the residuals that describe only short-term variability and the seasonal cycle. These residuals are shown in Fig. 1 (lower panel).

In the left panel of Fig. 2 the normalized residuals of EUV-TEC are shown vs. the normalized residuals of global TEC. A significant correlation between these indices with a correlation coefficient of $r = 0.68$ is obtained. This correlation is slightly stronger than one between the normalized residuals of the EUV-radiation in the wavelenght range from 5 nm to 102 nm and the normalized residuals of global TEC ($r = 0.682$ vs. $r = 0.679$). The updated EUV-TEC proxy describes the ionospheric variability better than the earlier version of the EUV-TEC proxy ($r = 0.68$ vs. $r = 0.67$ from 9 February 2002 to 31 December 2010) (Unglaub et al., 2011). The normalized residuals of F10.7 vs. the normalized residuals of global TEC are shown in the right panel of Fig. 2. There is a substantially weaker correlation than between the residuals of EUV-TEC and global TEC, with a correlation

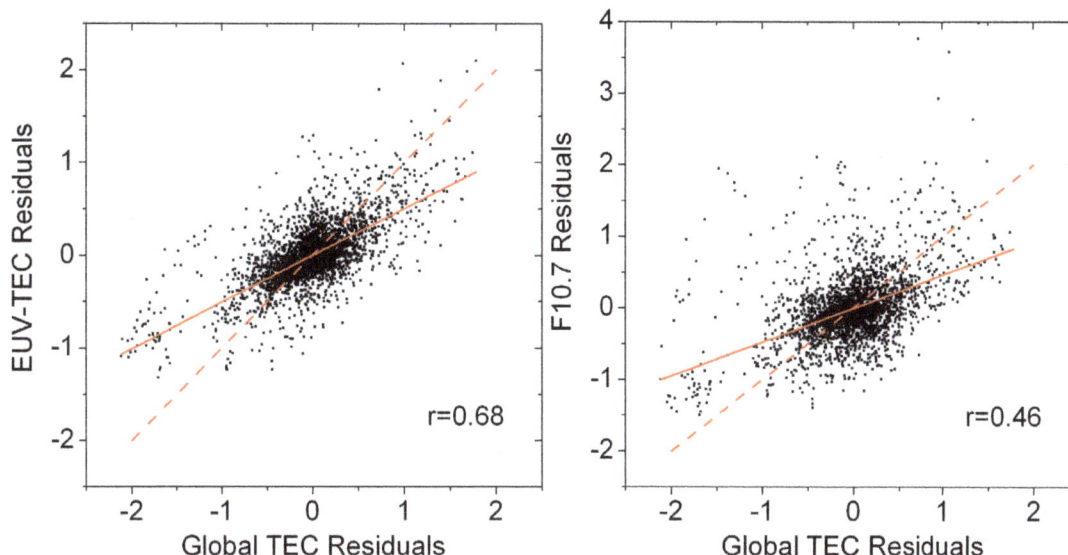

Fig. 2. Normalized EUV-TEC residuals (left panel) and normalized F10.7 residuals (right panel) vs. normalized global TEC residuals.

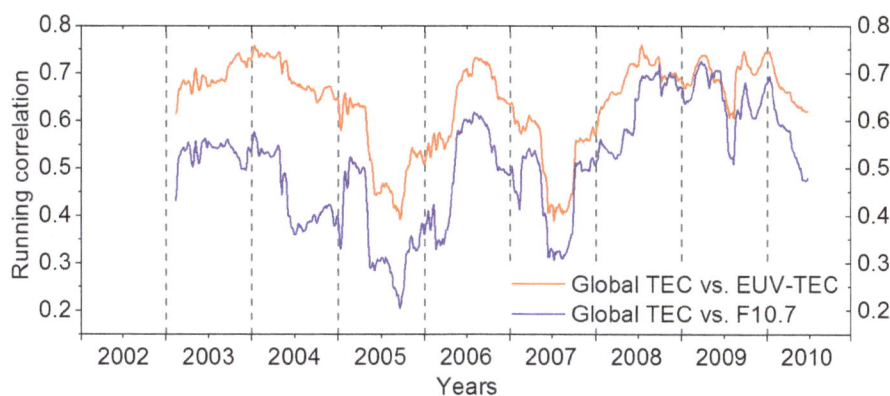

Fig. 3. Running correlations (365 days window) of the normalized global TEC residuals with the normalized EUV-TEC residuals (red line) and the normalized F10.7 residuals.

coefficient of $r = 0.46$. Thus, during the time interval 2002 to 2011 the ionospheric short-term TEC variability and its seasonal pattern are clearly better described by EUV-TEC than by F10.7. Because of the dynamical influence the seasonal pattern of global TEC has larger amplitudes than the seasonal pattern of EUV-TEC and F10.7 has. This can be seen as well in Fig. 2.

Figure 1 also reveals that on the one hand during solar minimum both the curves of EUV-TEC and global TEC are smooth, although they show partly different seasonal cycles, such that the semiannual global mean TEC signature, which is a dynamical feature, is not represented in EUV-TEC. On the other hand, EUV-TEC and global TEC short-term variability connected with the Carrington rotation, partly differ during high and moderate solar activity. To investigate possible changes of correlation in the course of the seasonal cycle, in Fig. 3 the running correlations, each based on 365 days of

data, between the normalized global TEC residuals and the normalized EUV-TEC residuals (red curve) or the normalized F10.7 residuals (blue curve), respectively, are shown. A continuously stronger correlation between the normalized global TEC residuals and the normalized EUV-TEC residuals than between the normalized global TEC residuals and the normalized F10.7 residuals exists during high and moderate solar activity. The correlation between global TEC and F10.7 partly becomes as small as $r = 0.2$ while the correlation between global TEC and EUV-TEC is stronger, with $r > 0.35$. However, for solar minimum conditions the correlations are similar. As a consequence, EUV-TEC describes ionospheric short-term variability better than F10.7 during high and moderate solar activity.

For solar minimum conditions the seasonal pattern is the dominant source of variability, because the sun is very quiet resulting in weak fluctuations. The similar correlations indicate that on an annual time scale F10.7 represents the seasonal pattern of global TEC as well as EUV-TEC does, merely with a different amplitude.

4 Conclusions

From satellite borne solar EUV measurements a new version of the EUV-TEC proxy, representing global mean photoionisation rates, have been calculated. The regional number densities of the background atmosphere are taken from the NRLMSISE-00 climatology. The EUV-TEC proxy describes the influence of solar variability on the ionosphere, and therefore can be considered as an ionospheric proxy, which may be used for the analysis of space weather effects on the upper atmosphere. It was compared with a global mean TEC created from vertical TEC maps. EUV-TEC shows a strong correlation with global TEC. Stronger correlations between EUV-TEC and global TEC than between the conventional solar index F10.7 and global TEC are found on different time scales. During 2002–2011 EUV-TEC describes the ionospheric variability, including both short-term and long-term variability, slightly better than F10.7 does. On an annual time scale EUV-TEC represents the short-term variability of the ionosphere distinctly better than F10.7 does, especially during high and moderate solar activity. On the whole, the EUV-TEC proxy performs better than F10.7 to describe ionospheric variability.

Acknowledgements. TIMED-SEE data has been provided by LASP, University of Colorado, through http://lasp.colorado.edu/see/see_data.html. TEC data has been provided by NASA through ftp://cddis.gsfc.nasa.gov/gps/products/ionex/. F10.7 indices have been provided by NGDC through ftp://ftp.ngdc.noaa.gov/STP/SOLAR_DATA/.

Topical Editor Matthias Förster thanks Norbert Jakowski and an anonymous reviewer for their help in evaluating this paper.

References

Aggarwal, M.: TEC variability near northern EIA crest and comparison with IRI model, Adv. Space Res., 48, 1221–1231, 2011.

Chen, Y., Liu, L., and Wan, W.: Does the F10.7 index correctly describe solar EUV flux during the deep solar minimum of 2007–2009?, J. Geophys. Res., 116, A04304, doi:10.1029/2010JA016301, 2011.

Emmert, J. T. and Picone, J. M.: Climatology of globally averaged thermospheric mass density, J. Geophys. Res., 115, A09326, doi:10.1029/2010JA015298, 2010.

Fennelly, J. A. and Torr, D. G.: Photoionization and photoabsorption cross sections of O, N_2, O_2 and N for aeronomic calculations, Atom. Data Nucl. Data Tables, 51, 321–363, 1992.

Hernandez-Pajares, M., Juan, J. M., Sanz, J., Orus, R., Garcia-Rigo, A., Feltens, J., Komjathy, A., Schaer, S. C., and Krankowski, A.: The IGS VTEC maps: a reliable source of ionospheric information since 1998, J. Geod., 83, 263–275, 2009.

ISO: ISO 21348:2007(E), Space environment (natural and artificial) – Process for determining solar irradiances, ISO, p. 12, 2007.

Lean, J. L., Meier, R. R., Picone, J. M., and Emmert J. T.: Ionospheric total electron content: Global and hemispheric climatology, J. Geophys. Res., 116, A10318, doi:10.1029/2011JA016567, 2011.

Liu, L., Chen, Y., Le, H., Kurkin, V. I., Polekh, N. M., and Lee, C.-C.: The ionosphere under extremely prolonged low solar activity, J. Geophys. Res., 116, A04320, doi:10.1029/2010JA016296, 2011.

Lühr, H. and Xiong, C.: IRI-2007 model overestimates electron density during the 23/24 solar minimum, Geophys. Res. Lett., 37, L23101, doi:10.1029/2010GL045430, 2010.

Maruyama, T.: Solar proxies pertaining to empirical ionospheric total electron content models, J. Geophys. Res., 115, A04306, doi:10.1029/2009JA014890, 2010.

Metzger, P. H. and Cook, G. R.: A reinvestigation of the absorption cross sections of molecular oxygen in the 1050–1800 Å region, J. Quant. Spectrosc. Radiat. Transfer, 4, 107–116, 1964.

Nikutowski, B., Brunner, R., Erhardt, Ch., Knecht, St., and Schmidtke, G.: Distinct EUV minimum of the solar irradiance (16–40 nm) observed by SolACES spectrometers onboard the International Space Station (ISS) in August/September 2009, Adv. Space Res., 48, 899–903, 2011.

Picone, J. M., Hedin, A. E., and Drob, D. P.: NRLMSISE-00 empirical model of the atmosphere: statistical comparisons and scientic issues, J. Geophys. Res., 107, 1468, doi:10.1029/2002JA009430, 2002.

Schmidtke, G., Fröhlich, C., and Thuillier, G.: ISS-SOLAR: Total (TSI) and spectral (SSI) irradiance measurements, Adv. Space Res., 37, 255–264, 2006a.

Schmidtke, G., Brunner, R., Eberhard, D., Halford, B., Klocke, U., Knothe, M., Konz, W., Riedel, W.-J., and Wolf, H.: SOL-ACES: Auto-calibrating EUV/UV spectrometers for measurements onboard the International Space Station, Adv. Space Res., 37, 273–282, 2006b.

Unglaub, C., Jacobi, Ch., Schmidtke, G., Nikutowski, B., and Brunner, R.: EUV-TEC proxy to describe ionospheric variability using satellite-borne solar EUV measurements: first results, Adv. Space Res., 47, 1578–1584, 2011.

Woods, T. N.: Recent advances in observations and modeling of the solar ultraviolet and X-ray spectral irradiance, Adv. Space Res., 42, 895–902, 2008.

Woods, T. N., Bailey, S., Eparvier, F., Lawrence, G., Lean, J., McClintock, B., Roble, R., Rottmann, G. J., Solomon, S. C., Tobiska, W. K., and White, O. R.: TIMED Solar EUV Experiment, Phys. Chem. Earth (C), 25, 393–396, 2000.

Woods, T. N., Eparvier, F., Bailey, S., Chamberlin, P., Lean, J., Rottmann, G. J., Solomon, S. C., Tobiska, W. K., and Woodraska, D. L.: Solar EUV Experiment (SEE): Mission overview and first results, J. Geophys. Res., 110, A01312, doi:10.1029/2004JA010765, 2005.

14

Multi beam observations of cosmic radio noise using a VHF radar with beam forming by a Butler matrix

T. Renkwitz, W. Singer, R. Latteck, and M. Rapp

Leibniz Institute of Atmospheric Physics at the Rostock University, Schloss-Str. 6, 18225 Kühlungsborn, Germany

Abstract. The Leibniz-Institute of Atmospheric Physics (IAP) in Kühlungsborn started to install a new MST radar on the North-Norwegian island Andøya (69.30° N, 16.04° E) in 2009. The new Middle Atmosphere Alomar Radar System (MAARSY) replaces the previous ALWIN radar which has been successfully operated for more than 10 years. The MAARSY radar provides increased temporal and spatial resolution combined with a flexible sequential point-to-point steering of the radar beam. To increase the spatiotemporal resolution of the observations a 16-port Butler matrix has been built and implemented to the radar. In conjunction with 64 Yagi antennas of the former ALWIN antenna array the Butler matrix simultaneously provides 16 individual beams. The beam forming capability of the Butler matrix arrangement has been verified observing the galactic cosmic radio noise of the supernova remnant Cassiopeia A. Furthermore, this multi beam configuration has been used in passive experiments to estimate the cosmic noise absorption at 53.5 MHz during events of enhanced solar and geomagnetic activity as indicators for enhanced ionization at altitudes below 90 km. These observations are well correlated with simultaneous observations of corresponding beams of the co-located imaging riometer AIRIS (69.14° N, 16.02° E) at 38.2 MHz. In addition, enhanced cosmic noise absorption goes along with enhanced electron densities at altitudes below about 90 km as observed with the co-located Saura MF radar using differential absorption and differential phase measurements.

1 Introduction

The Leibniz-Institute of Atmospheric Physics (IAP) in Kühlungsborn has been studying the dynamics and structure of the lower and middle atmosphere at polar latitudes for more than 15 years. For this purpose IAP operated for

Correspondence to: T. Renkwitz
(renkwitz@iap-kborn.de)

about 10 years the ALWIN MST radar system at 53.5 MHz (Latteck et al., 1999) on the North-Norwegian island Andøya (69.30° N, 16.04° E). A phased antenna array consisting of 144 Yagi antennas was used to form 6 degree wide beams on transmission and reception. Especially, the characteristics of Polar Mesospheric Summer Echoes (PMSE) have been studied with high time resolution. In 2009 IAP started to build the more flexible and powerful successor system MAARSY (Middle Atmosphere Alomar Radar System, Latteck et al., 2010). The new radar consists of a phased array of 433 individual 3-element Yagi antennas arranged in an equilateral grid structure and the same amount of transceiver modules. Since the installation of this versatile radar system observations in the troposphere and mesosphere with, until now, sequential beam steering have been carried out.

In the current manuscript, we present an approach of simultaneous multi beam capability as extension to the MAARSY system. For this purpose a 16-port Butler matrix has been built, which allows to form simultaneously 16 individual beams in combination with 64 Yagi antennas of the former ALWIN antenna array.

In the following chapters we describe the architecture of the beam forming matrix and the system used for the observations presented in this study. Afterwards we describe two methods to verify the beam forming capability of the Butler matrix arrangement by the observation of cosmic radio noise originating from a radio source. Furthermore we will compare events of local ionospheric absorption monitored with this multi beam architecture and the co-located imaging riometer AIRIS. Finally we present a summary of the results and conclusions.

2 System description

The ALWIN64 antenna array consists of 64 4-element Yagi antennas composed of 16 individual antenna groups with a regular antenna spacing of 3.97 m. These 16 antenna groups were either directly connected to the receiver to e.g. generate a narrow beam or were switched to a 16-port Butler matrix.

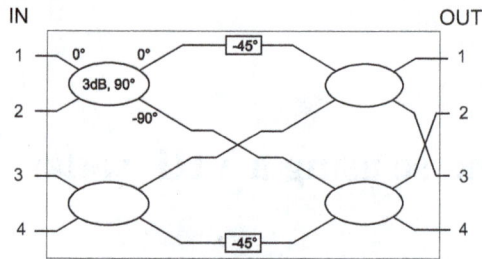

Fig. 1. Layout of a 4-port Butler matrix, where the ellipses depict 90° half-power combiners and the rectangles marked with −45° denote phase shifters.

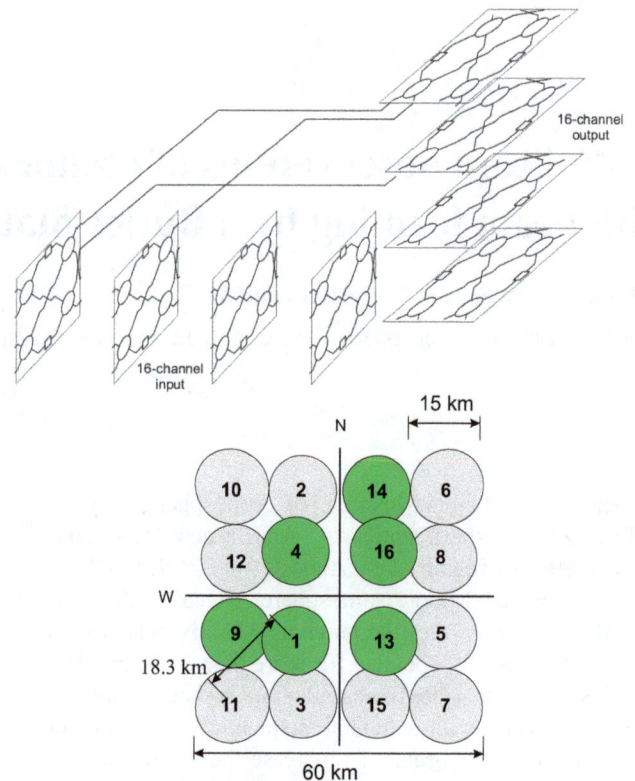

Fig. 2. Top: Layout of the 16-channel Butler matrix used in this study composed of a concatenation of eight 4-port Butler matrices. Bottom: The idealized multi beam radiation pattern (side lobes removed) in a projection to 85 km height generated with the combination of the 16-port Butler matrix and ALWIN64 antenna array. The numbers denoted in the ellipses refer to the corresponding port of the Butler matrix, where the beams marked in green are used within this study.

It is generally known that a radar beam can be steered off the broadside by implying a progressive time- or phase-offset to the individual array antennas. The idea of a Butler-matrix is to generate several simultaneous beams by the superposition of the corresponding phases needed to excite every beam individually. A Butler matrix built in hardware is generally composed of half-power 90° combiners and phase shifters, first described by Butler and Lowe (1961). Each half-power 90° combiner merges signals from two sources to one output adding a 90° phase shift, while both inputs are isolated to each other. The same combiner may also be used to split one input into two outputs including the 90° phase shift for one port. Additionally, these combiners provide a fourth port which is normally used to dump the energy that got reflected at the output due to impedance mismatch. In the case of adequate impedance match this port is isolated from the others. Due to this property of a half-power 90° combiner the individual in- and outputs of a Butler matrix are isolated to each other. To arrange an appropriate phase distribution to the output ports, in half of the paths through the Butler matrix, phase shifters are included. In such a structure the total number of available beams is determined by the amount of independent receivers and antenna feeds. A 4-port Butler matrix simultaneously generates four individual in- and outputs and is depicted in Fig. 1. In combination of four adjacent groups of the ALWIN64 antenna array with a 4-port Butler matrix four individual beams are formed simultaneously at zenith angles of approximately ±5° and ±14°. For the current 16 channel radar receiver of the MAARSY system a 16-port Butler matrix was built by the concatenation of eight 4-port Butler matrices. Using this 16-port Butler matrix with the ALWIN64 array 16 individual beams with linear polarization and a beam width of approximately 9° are generated. For the height of PMSE-layers, this beam width results in a target area of roughly 14 km diameter for each single beam. This configuration of beam forming matrix and the ALWIN64 antenna array have been described in Renkwitz et al. (2010). The layout of the 16-channel Butler matrix and the idealized beam pattern (side lobes removed) are shown in Fig. 2. After the integration of the Butler-matrix into the ALWIN-system receiver, measurements of the losses in mag-

nitude and phases of the Butler-matrix and the antenna array have been carried out to ensure the theoretical performance of the setup. The magnitude and phase measurements of the total 16-port Butler matrix underlined the appropriateness of the selected components used to built the matrix. Furthermore, the results of this examination have been used to simulate the radiation pattern of the system with NEC (Numerical Electromagnetics Code) including mutual coupling of the individual antennas and the influence of lossy ground. The computed radiation pattern for all 16 beams are presented in Fig. 3. The deviation to the initially assumed radiation pattern was found to be negligible. Within these 16 individual beams, four beams with an off-zenith direction of 6.8°, eight beams with an off-zenith direction of 15° and four beams with an off-zenith direction of 20° are generated. Generally, with the increase of off-zenith beam pointing angle the radiation pattern deteriorates as the attenuation of side lobes is decreased. The reason for this degradation in the radiation pattern is the arrangement of four antennas to a group with a common feed point. Furthermore this antenna array

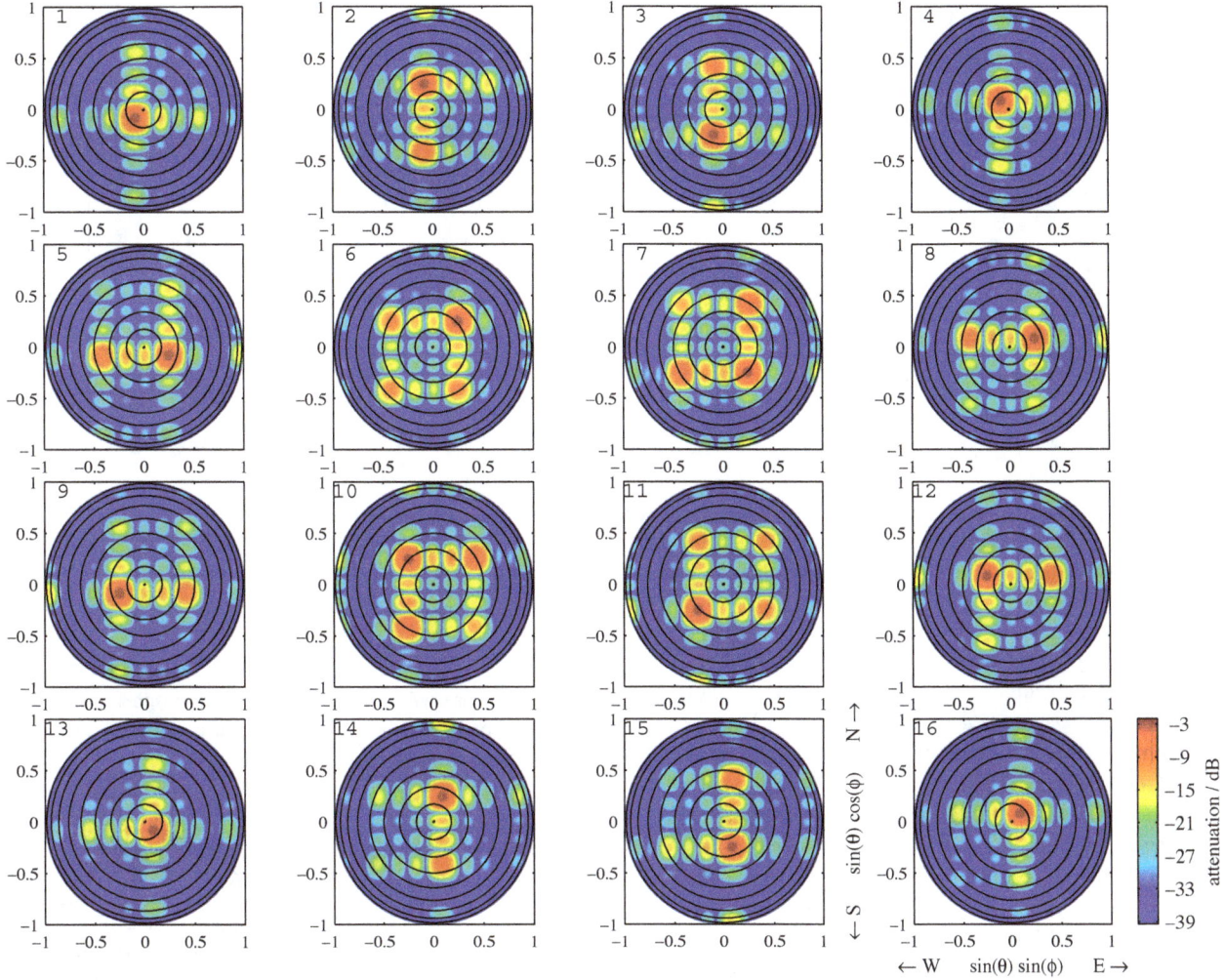

Fig. 3. Computed radiation pattern of the multi beam configuration using a 16-port Butler matrix and the 64 Yagi antenna array ALWIN64. Depicted are the radiation pattern of 16 simultaneously generated beams relative to the maximum gain with an overlay of equidistant rings of each 10° zenith angle. For this simulation the Numerical Electromagnetics Code (NEC) with the Norton-Sommerfeld approximation was used implicating the mutual coupling of antennas and the influence of imperfect ground. The corresponding beam number is depicted in the upper left corner of each individual pattern.

architecture, where the individual antennas are placed on a regular squared grid structure, induces the generation of grating lobes, which can be seen for 15 and 20° off-zenith beam pointing. Consequentially we generally focused on using the innermost beams with an off-zenith angle of 6.8°. However, for monitoring localized effects in the atmosphere also the remaining beams can be used with adequate cautiousness. To prohibit potential errors due to the misalignment of the individual receiver gain settings, we adjusted the observed noise power to a common reference. Therefore all channels of the ALWIN receiver have been terminated with a 50Ω load while the power of all channels have been obtained with the same receiver and experiment settings employed afterwards, especially for the gain and bandwidth. Thus, for each individual receiver we derived an averaged noise power and we defined

the receiver 1 as common reference. Subsequently the incident noise of the following observations have been adjusted in relation to this common reference receiver noise, which allows to compare the incident noise power in subsequent experiments.

3 Observations and discussion

Since the installation of the 16-port Butler matrix in November 2009 the ALWIN64 antenna array has been used in passive mode to sample cosmic noise. Six out of 16 beams have been selected to verify the functionality of the Butler matrix e.g. by monitoring the supernova remnant Cassiopeia A and were therefore connected to the ALWIN 6-channel receiver.

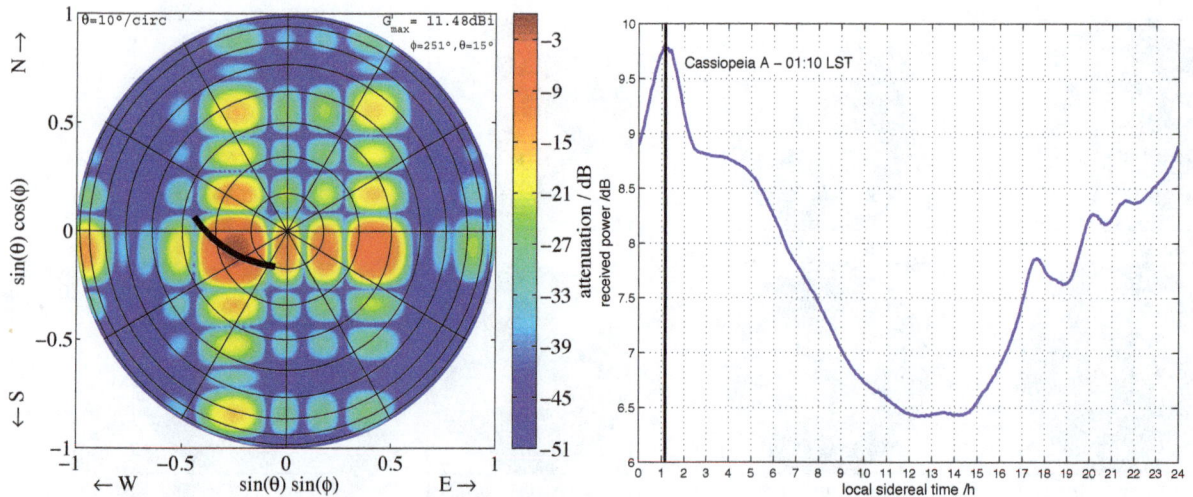

Fig. 4. Left: radiation pattern of the ALWIN-Butler beam 9 overlayed with the trajectory projection of Cassiopeia A.
Right: Quiet Day Curve for the ALWIN-Butler beam 9 monitoring the passage of Cassiopeia A. The passage of Cassiopeia A through the antenna beam appears on the local sidereal time 01:10 with an average of received noise power of 9.8 dB over receiver noise.

The corresponding beams are marked in Fig. 2 and a summary of the selected beams is presented in Table 1. For this study interference-free data from 15 days of observations in November 2009 have been used to estimate a Quiet Day Curve (QDC) for each individual beam. With the estimated QDC it was possible to derive the exact time of the apparent crossing of Cassiopeia A through the corresponding antenna beam. For the observation of Cassiopeia A the beam direction of the ALWIN-Butler beam 9 was found to give the best overlap with the path of the radio source. The radiation pattern of this beam including a projection of the trajectory of Cassiopeia A and the received noise power detected at ALWIN-Butler beam 9, relative to the local sidereal time, are illustrated in Fig. 4. The maximum power for the crossing of Cassiopeia A was determined to be 9.8 dB above receiver noise at the sidereal time 01 h 10 m 36 s. At this specific local sidereal time, the position of Cassiopeia A in the sky has been calculated as follows: azimuth 240.9° and zenith 15.5° relative to the position of the radar. However, the direction of maximum gain of the corresponding ALWIN-Butler beam has been calculated with NEC to be generated at the azimuth angle 251.2° and zenith angle of 15.2° with a half-power beam width of 9.3°. The overlay of the simulated radiation pattern and the trajectory of Cassiopeia A revealed the best fit for the direction of azimuth 246.5° and zenith 16.6°. This leads to the conclusion of a likely offset in the pointing direction of about 6° for ALWIN-Butler beam 9. A possible reason for this beam pointing error might be an impedance mismatch between the antennas and the Butler matrix. In this case the earlier described isolation between the in- and output ports might be deteriorated resulting in a modification of the generated beam steering phases. However, during the installation of the Butler matrix no conspic-

Table 1. Identifiers and directions of beams of the ALWIN-Butler and AIRIS riometer used for the comparison of ionospheric absorption. Azimuth and zenith angles are denoted by ϕ and θ respectively. A projection of these beams to 85 km height is shown in Fig. 5.

location	ALWIN-Butler 69.30° N, 16.04° E	AIRIS riometer 69.15° N, 16.03° E
beam	1: $\phi = 225°$, $\theta = 6.8°$	25: $\phi = 0°$, $\theta = 0°$
no.	9: $\phi = 251°$, $\theta = 15.2°$	17: $\phi = 315°$, $\theta = 20.4°$
&	13: $\phi = 135°$, $\theta = 6.8°$	18: $\phi = 0°$, $\theta = 14.2°$
directions	16: $\phi = 45°$, $\theta = 6.8°$	19: $\phi = 45°$, $\theta = 20.4°$

uous impedances have been found. The return loss due to impedance mismatch of the individual antenna groups have been measured to be about -20 dB in worst case, which describes a still well matched antenna.

Another valuable experiment is the comparison of derived cosmic noise absorption to data of the co-located riometer (relative ionospheric opacity meter) AIRIS (69.14° N, 16.02° E, Andoya Rocket Range, and Lancaster University, SPEARS Group). The AIRIS riometer is almost identical to the IRIS riometer built earlier in Kilpisjärvi, Finland, which have been described and used by e.g. del Pozo et al. (2002) and Kero et al. (2007). A riometer measures the incident noise power received from galactic radio sources and its damping caused by its propagation through the D-region of the ionosphere. It is well known from the magneto-ionic theory that this damping occurs as a consequence of collisions between free electrons and neutrals and hence maximises at altitudes between about 80–90 km. The current state of this absorption layer may vary in height and intensity depending

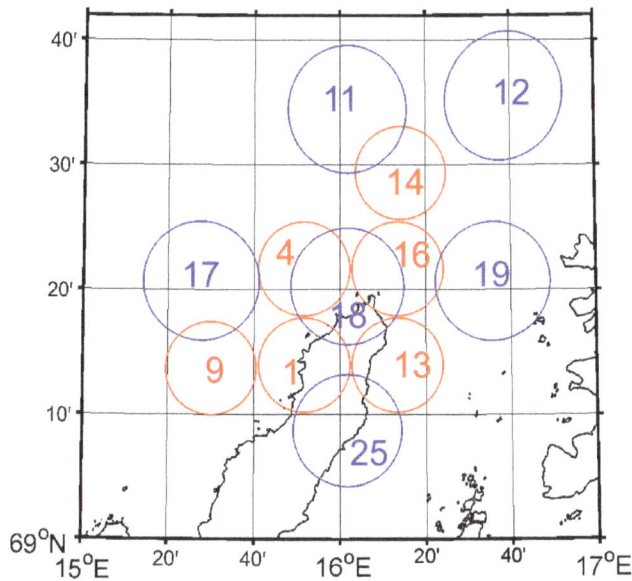

Fig. 5. Projection of the co-located ALWIN-Butler beams (red) and AIRIS beams (blue): The corresponding ellipses depict the target area for the beams at 85 km according to the assumed half-power beam width.

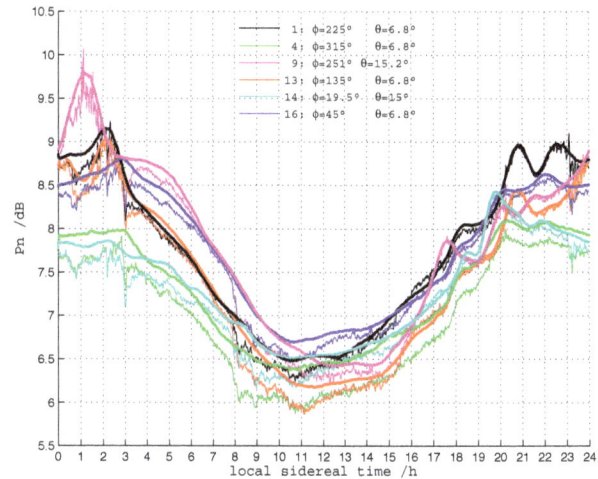

Fig. 6. Quiet Day Curve from interference-free data of 15 days (bold) and the received noise power for 6 ALWIN-Butler beams obtained for the time period of 24 November, 12:00 UT to 25 November, 12:00 UT.

on solar and geomagnetic activity. Additionally, the precipitation of highly energetic particles may lead to localized enhanced ionization and electron densities of significant order. The imaging riometer AIRIS provides absorption data for 49 individual spatially spread beams of circular polarization with a minimum beam width of roughly 11° to 14°. As some beam directions of AIRIS and ALWIN64 using the Butler matrix are co-located, a comparison of absorption data derived with both systems appears to be valuable to verify the theoretically simulated beam pattern. To compare absorption events with both AIRIS riometer and the multi beam configuration of ALWIN64 the system has been subsequently used to estimate the cosmic noise absorption at 53.5 MHz during events of enhanced solar and geomagnetic activity as an indicator for enhanced ionization at altitudes below 90 km.

The dataset of the AIRIS riometer used in this study for the subsequent comparison consists of already estimated ionospheric absorption with a coverage of 7 days (18 November 12:00 UT to 25 November 12:00 UT). Within this dataset the observations from 24 to 25 November 2009 represent the most significant signatures of ionospheric absorption. Furthermore at the same time we have seen no interferences in the ALWIN-Butler data which qualifies this specific time slot for the following analysis. In a first step corresponding beams of ALWIN-Butler and AIRIS have been selected which should show highly correlated observations. A projection of the chosen beams of ALWIN-Butler and AIRIS for a height of 85 km is depicted in Fig. 5. In this study we focus on four beam combinations (summarized in Table 1), which are co-located for an assumed observation height of 85 km.

The AIRIS system is located about 20 km south and therefore generally greater zenith beam pointing angles are necessary to overlap with beams of the ALWIN system. As to be seen below, these beams show overlapping regions with well correlated signals. The dataset of ALWIN-Butler and AIRIS has been averaged to 30 s samples and interpolated to the sidereal time of the beforehand estimated QDC for the ALWIN data. The time series of cosmic noise absorption for the chosen combinations of ALWIN-Butler and AIRIS beams are depicted over local sidereal time in Fig. 7, converted from the examined time period of 24 to 25 November, 12:00–12:00 UT. Generally, enhanced cosmic noise absorption goes along with enhanced electron densities at altitudes below about 90 km. To verify the observations of AIRIS and ALWIN-Butler and the estimated absorption we have analyzed data obtained by the MF radar Saura (Singer et al., 2008, 2010), co-located to the AIRIS riometer. The median electron density profiles derived by the MF radar using differential absorption and differential phase measurements for the two time slots of 00:08–00:38 UT and 02:59–03:29 UT are depicted in Fig. 8. At the time of low absorption at 00:08–00:38 UT consistently low electron densities have been observed, while for the time of significantly higher absorption also increased electron densities for the entire height coverage have been found. However, these profiles are derived for heights lower than 80 km, we nevertheless assume a reliable detection of atmospheric absorption derived by AIRIS and the ALWIN-Butler system at the time period used in this study. The time series in general show a good agreement of the selected beam combinations of ALWIN-Butler and AIRIS. However, at times around the local noon (LST: 14:30–17:30) the estimated absorption of AIRIS and ALWIN-Butler differ significantly. During that time on

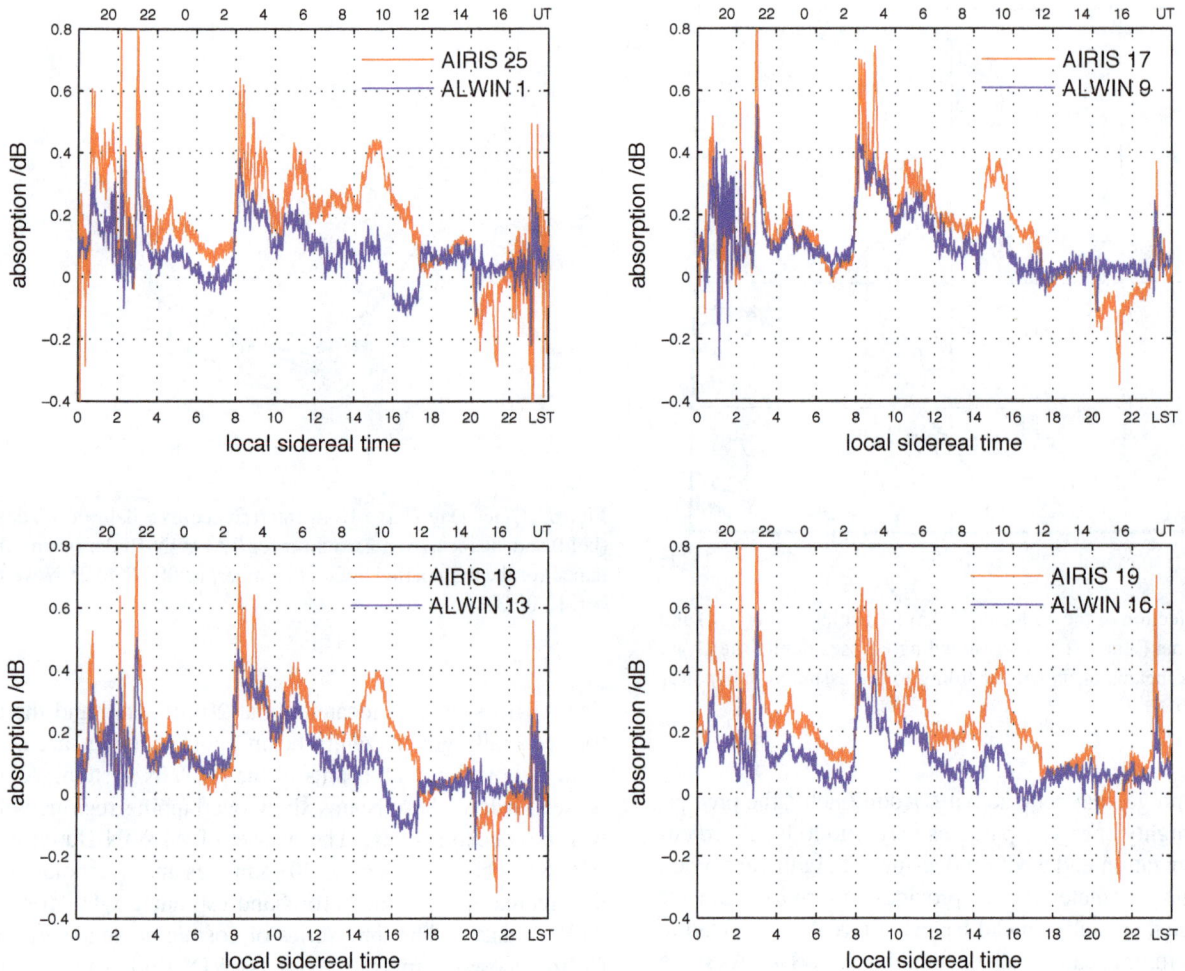

Fig. 7. Time series of absorption observed with ALWIN-Butler and AIRIS riometer for co-located beams.

25 November, the noise power detected with ALWIN-Butler has been enhanced, contrary to the observations of AIRIS. Additionally, for the complete dataset of seven days of the AIRIS riometer we found a daily increase of estimated negative absorption at times between LST 20:00–22:00. This regular diurnal repetition suggests to be a local interference at the AIRIS site. Hence, those questionable times have been discarded for subsequent analysis. The remaining data of 19 h observation time have been used to derive correlograms (Fig. 9), the estimation of regression and correlation coefficients, which are presented in corresponding scatter plots of the absorptions of both systems. In general we see a very good agreement for all beam combinations with high correlation coefficients between 0.89 and 0.97. Interestingly, we found specific signatures of absorption events with differing intensity and exact time in the individual beams (e.g. at local sidereal time around 02:00 and 04:00) implying rather localized sources of these events. The sun instead would lead to a homogenous ionization, which should be seen in all beams in equal strength and therefore can be excluded as a source of

the observed absorption events. Additionally, the influence of the sun to the observations have to be rather minimal as the observations have been performed in polar winter time. The systems AIRIS and ALWIN-Butler operate at significant different frequencies and the observed cosmic noise absorption data should follow a specific relation. Similar observations at different VHF frequencies have been performed by Campistron et al. (2001) that confirm the following reasoning. The cosmic noise power penetrating into the ionosphere is attenuated in dependence on frequency on the travel to the receiver. The ratio of the radio wave absorption of the ALWIN-Butler system and AIRIS system can be approximated by Eq. (1) following the magneto-ionic theory for a lossless medium.

$$\frac{L_{ALWIN}}{L_{AIRIS}} = \left(\frac{f_{AIRIS}}{f_{ALWIN}} \right)^2 \tag{1}$$

In this equation L_{ALWIN}, L_{AIRIS} and f_{ALWIN}, f_{AIRIS} denote the derived absorption values and the operating frequencies of the systems. We note that the cosmic noise signals incident on the ionosphere are also frequency dependent in themselves and obey a similar relation as given in Eq. (1) but with

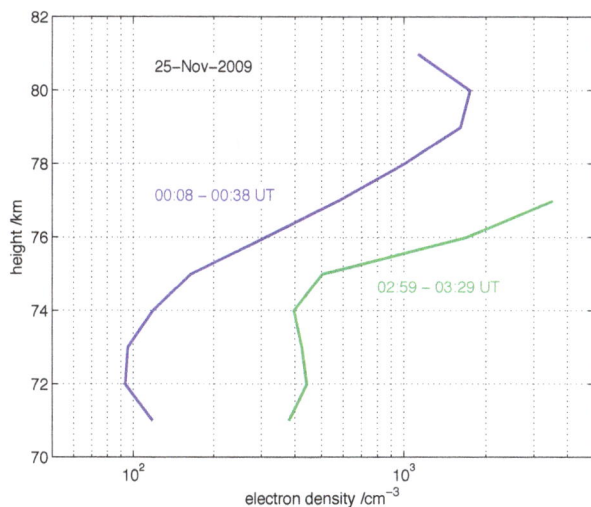

Fig. 8. Median profiles of electron densities derived with the MF radar Saura for 25 November 2009.

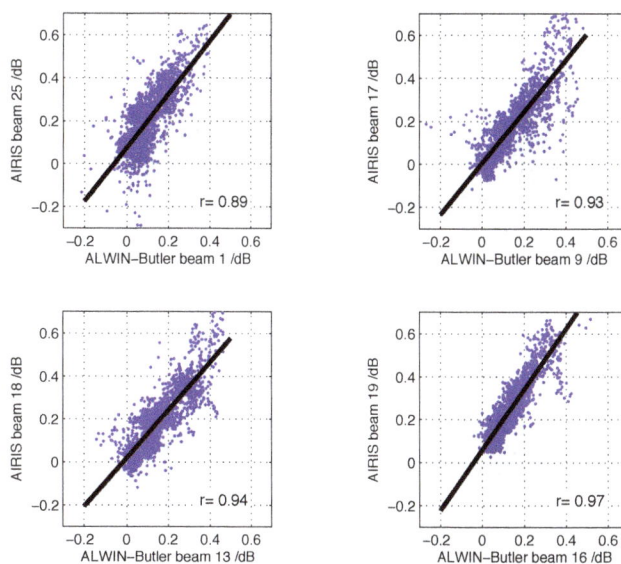

Fig. 9. Correlogram of absorptions estimated for the co-located ALWIN-Butler and AIRIS beams.

exponents between 2.0 and 2.5 (Campistron et al., 2001, and references therein). However, since only relative changes of observed radio noise are considered, the latter frequency dependence cancels out and does not need to be considered further. In addition, we note that Eq. (1) is only valid in the case of weak D-region ionization. This requirement is, however, fully met with cosmic noise absorption values of 0.4 dB only (see Fig. 9). Using the operating frequencies of 53.5 MHz and 38.2 MHz for the ALWIN-Butler und AIRIS system respectively, absorption ratios between 0.43 and 0.51 are estimated for the frequency dependence mentioned above.

Histograms of the estimated absorption ratios for the compared beams of ALWIN-Butler and AIRIS are depicted in Fig. 10 with 75 bins each histogram. To start with, two out of four histograms reveal mean ratios close to the expected ratio of about 0.5. The greatest similarity can be seen for the comparison of ALWIN64 Butler beam 16 and AIRIS beam 19, which initially have not been anticipated as those beams are theoretically separated by more than the estimated beam width (Fig. 5). The comparison of this beam combination resulted in a very good correlation factor of 0.97. Furthermore the correlogram (Fig. 9) and absorption ratio of both ALWIN-Butler and AIRIS for the corresponding beams (Fig. 10) underline the similarity. Nevertheless, the other beam combinations in fact still show high correlations of 0.89 to 0.94. However the cross sections of the considered beams likely still have no perfect overlap and therefore partially differing observation volumes are compared. The histograms for the beam comparisons ALWIN-Butler 16 to AIRIS 19 and ALWIN-Butler 1 to AIRIS 25 show a narrow spread of the absorption ratios and likewise holds for the standard deviation. The mean and median values obtained for both comparisons are close to 0.51 respectively, which is in very good agreement to the theoretical

expected value of 0.51 (Eq. 1). The comparisons of the remaining beams (ALWIN-Butler 9 to AIRIS 17 and ALWIN-Butler 13 to AIRIS 18) have shown reliable distribution of absorption ratios with a concentration of the mean and median values around 0.75. However, the correlation factors of at least 0.93 for the compared beams exhibit a very high similarity of the data. This deviation may result from the following potential aspects, (a) lack of perfect overlap of the corresponding beams, (b) different beam pointing zenith angles of the individual beams and therefore different path lengths including a 2–3° greater beam width for AIRIS results in a significant difference in the size of target areas, (c) imperfect estimation of the QDC, which may be significant as for the main part of the day absorptions of even less than 0.2 dB have been estimated.

Additionally, from the earlier described experiment, monitoring the noise power received from Cassiopeia A, we concluded a likely offset of beam pointing by about 6° in azimuth. For the comparison of ALWIN-Butler 9 and AIRIS 17 this 6° offset in beam pointing leads to an increased spatially displacement of the compared beams as it has been assumed in advance. Furthermore, the slightly greater half-power beam widths published for the AIRIS beams than the beams simulated for ALWIN-Butler should result in a decrease of the diurnal QDC variation and an increase of the detected absorption as it has been described by Friedrich et al. (2002) and Harrich et al. (2003). An increase of absorption for the AIRIS system due to a broader beam is contrary to the findings in the beam comparisons with absorption ratios of about 0.75 and is therefore not a valid explanation. Especially, as in the earlier comparisons with similar differences in the beam widths have nevertheless resulted in good agreements to the theoretical frequency dependence. Therefore a

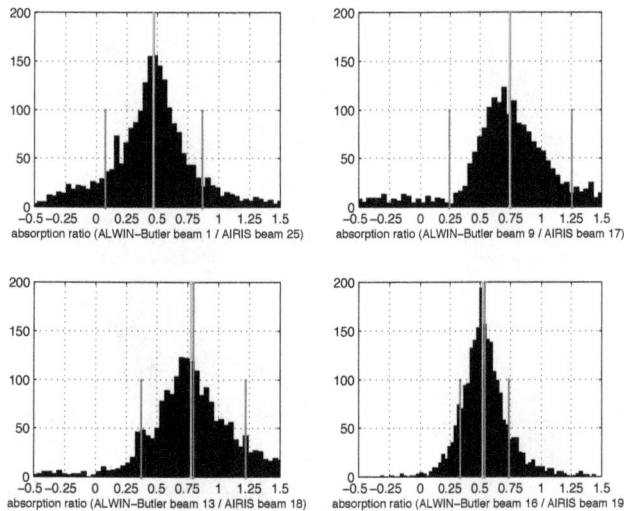

Fig. 10. Histograms of the absorptions ratio between the co-located beams of ALWIN-Butler and AIRIS. Mean and standard deviation are marked with red vertical lines while the median values are marked in green.

flaw in the estimated QDCs and/or likely events of localized effects due to precipitating particles that have been seen with the imperfect matched beam positions and sizes of the two systems may explain the observed deviations. Again, these two beam comparisons have shown very high correlation factors and thus suggesting the existence of co-aligned or at least partially overlapped beams.

4　Summary and conclusions

In 2009 the Leibniz-Institute of Atmospheric Physics in Kühlungsborn started to built the new powerful, and more flexible MST radar MAARSY as the successor of the AL-WIN radar. With both systems studies of the dynamics and the structure of the polar middle atmosphere have been carried out. The combination of a 16-port Butler matrix with 64 antennas of the former ALWIN array allow observations with simultaneously 16 individual beams. To verify the capability of this beam forming matrix the cosmic radio noise of the supernova remnant Cassiopeia A has been observed. We found a diurnal Quiet Day Curve with reasonable dynamic and a characteristic peak originating from Cassiopeia A. With an overlay of the simulated radiation pattern of one beam with the trajectory of Cassiopeia A we derived the supposedly effective direction observing the radio source. Additionally, from the observations we obtained the associated position of Cassiopeia A. We found a discrepancy between the simulation and the observation, which leads to a likely offset in the beam pointing of about $6°$. The multi beam configuration of ALWIN-Butler has also been used in experiments to estimate the cosmic noise absorption at 53.5 MHz. During the period of 18 to 25 November we have observed cosmic noise

absorption events with different beams of the ALWIN-Butler configuration and subsequently compared the observations to the co-located imaging riometer AIRIS. To support the observations of ALWIN-Butler and AIRIS, we derived electron density profiles using differential absorption and differential phase measurements from Saura MF radar. We found an increase of electron densities between 70 and 80 km height at the same times when ALWIN-Butler and AIRIS have estimated an increase of absorption. The simultaneous absorption observations with co-located beams of ALWIN-Butler and AIRIS were highly correlated, up to 97%, leading to the assumption that the beams examined in this study form very similar or at least nearby target areas. Additionally, in two examples we have seen a good agreement to the frequency dependency related to the magneto-ionic theory of Appleton-Hartree.

Generally, the capability of the beam forming matrix implemented to the MST radar have been verified. Furthermore, we found a good agreement in the observations of ALWIN-Butler and AIRIS implying co-located beam positions and therefore in general beam pointing directions in accordance with the assumed theoretical radiation pattern.

Acknowledgements. The authors would like to thank M. Gausa from the Andøya Rocketrange and the Space Plasma Environment and Radio Science (SPEARS) group, Department of Physics, Lancaster University (UK), permitting access to the data of the AIRIS riometer. Topical Editor Matthias Förster thanks Ernst-Dieter Schmitter and Ernst Fürst for their help in evaluating this paper.

References

Andoya Rocket Range: AIRIS riometer, http://alomar.rocketrange. no/iris-and.html, download in January 2011.

Butler, J. and Lowe, R.: Beam Forming Matrix Simplifies Design of Electronically Scanned Antennas, Electronic Design, 9, 170–173, 1961.

Campistron, B., Despaux, G., Lothon, M., Klaus, V., Pointin, Y., and Mauprivez, M.: A partial 45 MHz sky temperature map obtained from the observations of five ST radars, Ann. Geophys., 19, 863–871, doi:10.5194/angeo-19-863-2001, 2001.

del Pozo, C. F., Honary, F., Stamatiou, N., and Kosch, M. J.: Study of auroral forms and electron precipitation with the IRIS, DASI and EISCAT systems, Ann. Geophys., 20, 1361–1375, doi:10.5194/angeo-20-1361-2002, 2002.

Friedrich, M., Harrich, M., Torkar, K., and Stauning, P.: Quantitative measurements with wide-beam riometers, J. Atmos. Solar-Terr. Phys., 64, 359–365, 2002.

Harrich, M., Friedrich, M., Marple, S. R., and Torkar, K. M.: The background absorption at high latitudes, Adv. Radio Sci., 1, 325–327, doi:10.5194/ars-1-325-2003, 2003.

Kero, A., Enell, C.-F., Ulich, Th., Turunen, E., Rietveld, M. T., and Honary, F. H.: Statistical signature of active D-region HF heating in IRIS riometer data from 1994–2004, Ann. Geophys., 25, 407–415, doi:10.5194/angeo-25-407-2007, 2007.

Lancaster University, SPEARS Group: AIRIS riometer, http://www.dcs.lancs.ac.uk/iono/cgi-bin/riometers?orderdir= asc;id=64, download in January 2011.

Latteck, R., Singer, W., and Bardey, H.: The ALWIN MST radar – Technical design and performances, Proceedings of the 14th ESA Symposium on European Rocket and Balloon Programmes and Related Research, 1999.

Latteck, R., Singer, W., Rapp, M., and Renkwitz, T.: MAARSY – the new MST radar on Andøya/Norway, Adv. Radio Sci., 8, 219–224, doi:10.5194/ars-8-219-2010, 2010.

Renkwitz, T., Singer, W., and Latteck, R.: Study of multibeam ability for the VHF MST ALWIN radar system, Proceedings of the 12th International Workshop on technical and Scientific Aspects of MST radars, pp. 127–130, edited by: Swarnalingam, N. and Hocking, W. K., Canadian Association of Physicists, Ottawa, Ontario, Canada, ISBN 978-0-9867285-0-1, 2010.

Singer, W., Latteck, R., and Holdsworth, D.: A new narrow beam Doppler radar at 3 MHz for studies of the high-latitude middle atmosphere, Adv. Space Res., 41, 1488–1494, 2008.

Singer, W., Latteck, R., Friedrich, M., Wakabayashi, M., and Rapp, M.: Seasonal and solar activity variability of D-region electron density at 69° N, J. Atmos. Solar-Terr. Phys., doi:10.1016/j.jastp. 2010.09.012, 2010.

15

Visibility of Type III burst source location as inferred from stereoscopic space observations

M. Y. Boudjada[1], P. H. M. Galopeau[2], M. Maksimovic[3], and H. O. Rucker[1]

[1]Space Research Institute, Austrian Academy of Sciences, Graz, Austria
[2]Université Versailles St-Quentin, CNRS/INSU, LATMOS-IPSL, Guyancourt, France
[3]LESIA – Observatoire de Paris-Meudon, Meudon, France

Correspondence to: M. Y. Boudjada (mohammed.boudjada@oeaw.ac.at)

Abstract. We study solar Type III radio bursts simultaneously observed by RPWS/Cassini, URAP/Ulysses and WAVES/Wind experiments. The observations allows us to cover a large frequency bandwidth from 16 MHz down to a few kHz. We consider the onset time of each burst, and estimate the corresponding intensity level. Also we measure the Langmuir frequency as observed on the dynamic spectra recorded by the Ulysses spacecraft. The distances of Wind, Ulysses and Cassini spacecraft, with regard to the Sun, were in the order of 1 AU, 2.4 AU and 4.5 AU, respectively. The spacecraft trajectories were localized in the ecliptic plane in the case of Wind and Cassini, and for Ulysses in the southern hemisphere (i.e. heliocentric latitude higher than $-50°$). Despite the different locations, the spectral patterns of the selected solar bursts are found to be similar between 10 MHz and 2 MHz but unlike at lower frequency. We discuss the variation of the intensity level as recorded by the three spacecraft. We show that the reception system of each experiment affected the way the Type III burst intensity is measured. Also we attempt to estimate the electron beam along the interplanetary magnetic field where the trajectory is an Archimedean spiral. This leads us to infer on the visibility of the source location with regard to the spacecraft position.

1   Introduction

Non-thermal radio bursts are caused by energetic electrons from the Sun, which convert part of their energy into electromagnetic radiation via an emission mechanism (Ginzburg and Zheleznyakov, 1958). This mechanism involves the conversion of electron beam excited by Langmuir waves into escaping radiation at the fundamental (F) and second harmonic (H) of the electron plasma frequency. Direct observation of non-thermal electrons and plasma waves in space, in association with Type III bursts, provided an evidence for the plasma emission mechanism (Lin, 1973). The distribution function of these non-thermal electrons indeed demonstrated the generation of Langmuir waves which derive their energy from the non-thermal electrons; the intensity of the radio bursts depends on the non-thermal electron density and energy (Dulk et al., 1998).

In a dynamic spectrum, the Type III burst appears as an intense band of emission drifting rapidly with $df/dt \sim 100\,\mathrm{MHz\,s^{-1}}$ from high to low frequencies. Since the corona and the interplanetary (IP) medium are magnetized plasmas, propagation of electrons occurs along open magnetic field lines. Investigations led to a number of advances in studying electron beams as well as inferring the Archimedean spiral structure of the IP magnetic field, (Fainberg and Stone, 1970). Knowledge of the emission directivity is very important when studying the emission mechanism and propagation effects in the medium, particularly at low frequency. At lower frequencies, namely at hectometer and kilometer wavelengths, the widespread visibility of the Type III emission was reported by MacDowall et al. (1982). The visibility is associated to the beaming of the solar radio burst which consists of at least two components: a Gaussian core of half width about 60°, and a very broad halo that extends to 180° (Dulk et al., 1985). Later the first stereoscopic directivity measurements were reported by Poquerusse et al. (1996) and Hoang et al. (1997), using comparisons of flux densities between the radio receiver on the Ulysses spacecraft and the radio spectrographs ARTEMIS at the highest

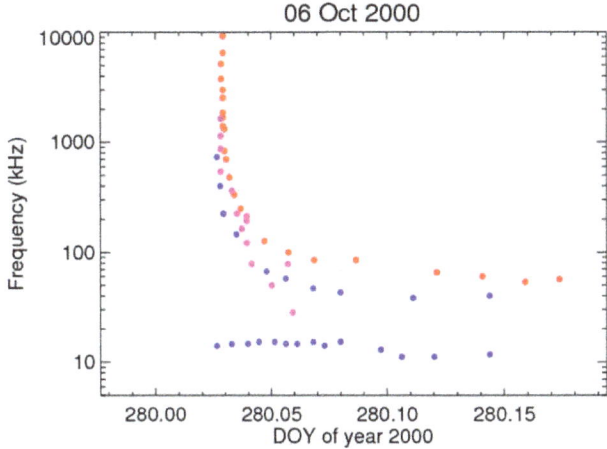

Figure 1. Spectral shape of the Type III solar bursts recorded on 6 October 2000 by Wind/WAVES (red points), Cassini/RPWS (violet points) Ulysses/URAP (blue points) experiments

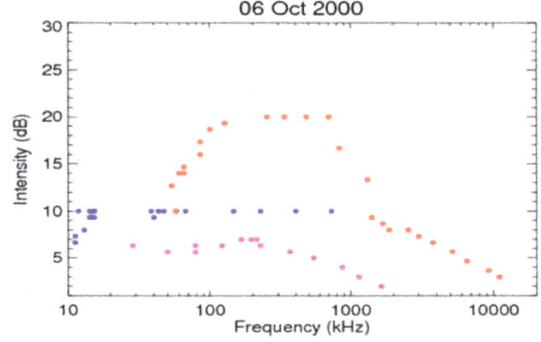

Figure 2. Variation of the intensity level (expressed in dB) versus the frequency range (in kHz) for the solar type III burst recorded on 6 October 2000

frequencies, between 100 and 500 MHz. More recently, Bonnin et al. (2008) investigated the directivity of solar type III radio bursts at hectometer and kilometer wavelengths using radio data recorded simultaneously by the Wind and Ulysses spacecraft. They derived for the first time the average directivity diagram of Type III bursts in two dimensions, longitude and latitude relative to the emission source. The diagram in longitude was found to shift east of the local magnetic field direction at the source. The angular width and eastward shift of the diagram are observed to increase with decreasing frequencies.

In this paper we attempt to analyze the visibility of solar Type III bursts observed simultaneously by Wind, Ulysses and Cassini spacecraft. In Sect. 2, we describe the radio space experiment and their mutual positions in the ecliptic plane. The method of data investigation is given in Sect. 3 where we consider that the electrons at the origin of the Solar Type III burst follows the interplanetary magnetic field. The main outcomes are discussed in the Sect. 4.

2 Radio space experiments

The data used in this investigation were acquired by URAP, WAVES and RPWS experiments on board Ulysses, Wind and Cassini spacecraft, respectively. The URAP instrument (Stone et al., 1992) measures the wave electric fields in two bands. The high-band receivers between 52 and 940 kHz, and the low-band receivers cover the range from 1.25 to 48.50 kHz. On the Wind satellite we investigate the data recorded by the super heterodyne receivers RAD1 and RAD2, respectively, in the frequency range 20–1024 kHz and 1.075–13.825 MHz (Bougeret et al. , 1995). The RPWS experiment is designed to measure the electric and magnetic fields of radio emissions and plasma waves across a broad

range of frequencies (Gurnett et al. , 2004). The high frequency receiver (HFR) enables the analysis of incoming radio waves by combining signals from three 10 m long antenna elements. The covered frequency bandwidth is from 3.5 to 16 MHz.

In this study, we combine the simultaneous observations of solar Type III radio bursts observed by the three spacecraft. We consider solar Type III bursts observed from September 2000 to March 2001. During this period, Wind and Cassini were nearly orbiting in the ecliptic plane, and Ulysses was in the Southern hemisphere at heliocentric latitude higher than −50°. The distances of Wind, Ulysses and Cassini spacecrafts, with regard to the Sun, were in the order of 1 AU, 2.4 AU and 4.5 AU, respectively (see Fig. 3).

3 Method of data investigation

We study more than 100 solar bursts simultaneously observed by the three spacecraft, from September 2000 to March 2001. The method of data investigations consists of the analysis of: (a) the variation of the observed frequency versus the time, (b) the intensity level versus the frequency, and (c) the Archimedean spiral at the origin of the solar burst. Hereafter we use the solar Type III burst observed on 6 October 2000 as an example for the data investigation.

In Fig. 1 we show the Type III burst observed by the three spacecraft on 6 October 2000. The blue, red and violet colors are associated, respectively, to Ulysses, Wind and Cassini observations. Despite the different spacecraft locations with regards to the Sun, the spectral shapes are found to be similar in the frequency range between 13 and 2 MHz, and look different at lower frequencies, less than 1 MHz. The frequency at about 15 kHz is the Langmuir frequency observed by Ulysses spacecraft. It is the lowest observed frequency. It is produced by thermal motions of the electrons in the vicinity of the spacecraft (Meyer-Vernet and Perche, 1989).

Figure 2 shows the variation of the intensity level (in dB) vs. the frequency (in kHz). The Wind observations (in red colors in Fig. 2) showed an increase of the intensity level emission from 10 to 20 dB between 60 and 100 kHz. Followed by a maximum of intensity of about 20 dB in the frequency range 100 kHz to 1 MHz. Then a clear decrease of the intensity level from 20 to 3 dB at high frequency (i.e. between 1 and 13 MHz). The Ulysses observations (blue color in Fig. 2) showed an increase from 5 to 10 dB at very low frequency (10 to 15 kHz). This is followed by a maximum about 10 dB in the frequency range from 15 kHz to about 900 kHz. The Cassini intensity level (in violet color in Fig. 2) has a maximum of about 6 dB between 30 and 200 kHz, followed by a decrease to about 2 dB in the frequency range 200 kHz to 2 MHz. We note that the intensity level decreases when the distance to the Sun increases. This explains the maximum of 20 dB observed by Wind, and the lowest level of 6 dB for Cassini. Instrumental effect seems to be at the origin of the increase of the Wind intensity level between 60 and 100 kHz. The two other spacecraft show a constant intensity level in the bandwidth 60–100 kHz. Also the decrease of the intensity above 200 kHz in the Cassini observations is due to the reception system. Wind and Ulysses show a constant level, and no decrease between 200 and 900 kHz. Furthermore the decrease of Wind intensity above 1 MHz can be explained by an instrumental effect because the RAD2 and RAD1 frequency bandwidth started and ended at about 1 MHz.

Further we analyze the geometry of the solar burst source with regards to the spacecraft locations. We first use the empirical formula which allows us to estimate the distance at which occur the observed frequency. We consider the empirical formula derived by Bougeret et al. (1984) using the Helios 1 and 2 in situ measurements. For this we use:

$$f_p = 22.5 r^{-1.05} \qquad (1)$$

where f_p is the local plasma frequency and r is the heliocentric distance expressed in AU. Then we consider that the electrons at the origin of the Solar Type III bursts follow the interplanetary magnetic field as shown in Fig. 3. The horizontal and vertical axes are expressed in Solar radii. The trajectory is an Archimedean spiral contained in the ecliptic plane. The equation of the curves in Fig. 3 is given by:

$$\phi = \frac{\Omega}{V_{SW}} \frac{r}{215} \qquad (2)$$

where, Ω, V_{SW}, and r are, respectively, the solar angular rotation (in rad s^{-1}), solar wind speed (in km s^{-1}) and the distance to the Sun in solar radii (in km). This equation includes the foot of the spiral on the Sun's surface, and gives the orientation of the spiral with respect to the Sun-Earth line. We have used a list of sunspot observations to find the location of the active regions on the Sun. This list is provided by the National Geophysical Data Center (NGDC) by FTP area (ftp.ngdc.noaa.gov). We have found two active regions on the Sun for the considered event of 6 October 2000, at about 00:50 UT, one in eastern part and other in the western one. Since this active region is simultaneously observed by

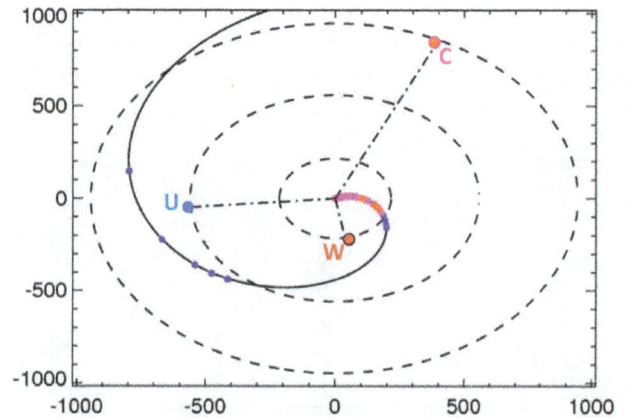

Figure 3. Archimedean spiral associated to the electrons at the origin of the solar type III bursts showed in Fig. 1. The higher frequencies (~ 10 MHz) are supposed to be generated in the solar corona and the lower one (~ 100 kHz) in the interplanetary medium. The dashed and dash-dotted lines indicated, respectively, the spacecraft orbits and the positions of Wind (W), Ulysses (U) and Cassini (C) on 6 October 2000.

Wind and Cassini spacecraft (both on the western part of the Sun), we have decided to selec the source region located at 45° W. In Eq. (2), we assume a constant solar wind speed of 400 km s^{-1}.

The spacecraft orbits are shown in Fig. 3 where their positions on 6 October 2000 are indicated with the letters "W" for Wind, "U" for Ulysses and "C" for Cassini. It is clear that the active region observed in the western part of the Sun is the real source location of the observed Type III burst. We can assume that the emission diagram is a cone with a symmetrical axis along the magnetic field line emerging from the Sun at about 45° W. The cone opening angle can be estimated as about 90°. At lower frequencies (less than 1 MHz) it is not evident to presume the shape of the emission diagram. The lower spectrum of the Type III burst was observed by the three spacecraft which means a large opening angle, more than 180°. However we note that the shapes are different, as recorded by each spacecraft.

4   Results and preliminary conclusion

We investigate more than 100 solar bursts observed from September 2000 to March 2001. We find that 44 % of the solar bursts are associated to active regions localized in the western part of the Sun which is visible to the Earth. A small part of about 15 % is linked to active regions in the eastern part of the Sun. And the rest of the bursts (41 %) can not be associated to active regions on the Sun. This means that the corresponding active regions, can't be observed from the Earth, and should be behind the Sun and on the western limb of our star. Only in such configurations Wind and Cassini can detect simultaneously the radio emission of the solar burst.

Also we find that the spectral shapes (as shown in Fig. 1) of the Solar bursts are similar in the frequency range between 13 and 2 MHz. This leads us to presume that the radio wave is not disturbed by the plasma environment in the solar corona and the interplanetary medium. This is not the case at frequencies lower than 1 MHz where we note different spectral shapes (as shown in Fig. 1). This is clear evidence of the presence of electron density inhomogeneity which alters the radio wave propagation in the interplanetary medium. Our result is in agreement with previous investigations (Dulk et al., 1985; Lecacheux et al., 1989). The origin of the different spectral shapes may be due to a high degree of scattering of radio waves originating at or near the local plasma frequency.

Further investigations should take into considerations other models of electron density (e.g. Mann et al., 1999). Also more bursts will be analyzed with the aim to combine our results to the works of Bonnin et al. (2008) and Krupar et al (2012). The future European mission Solar Orbiter (Boudjada et al., 2005) will give us the possibility to investigate the electron density at very close distances to the Sun during the nominal mission phase (duration of about 2.88 years), from about 0.2 to 0.9 AU (Marsch et al., 2001).

Acknowledgements. The authors are grateful to J. L. Bougeret (PI of the Wind experiment), D. A. Gurnett (PI of the RPWS experiment), and R. J. MacDowall (PI of the URAP experiment) for making the data available for this work.

Edited by: G. Mann
Reviewed by: two anonymous referees

References

Bonnin, X., Hoang, S., and Maksimovic, M: The directivity of solar type III bursts at hectometer and kilometer wavelengths: Wind-Ulysses observations, Astron. Astrophys., 489, 419–427, 2008.

Boudjada, M. Y., Macher, W., Rucker, H. O., and Fischer, G.: Solar Orbiter: Physical aspects towards a better knowledge of the solar corona, Adv. Space Res., 36, 1439, doi:10.1016/j.asr.2005.05.118, 2005.

Bougeret, J.-L., Kaiser, M. L., Kellogg, P. J., Manning, R., Goetz, K., Monson, S. J., Monge, N., Friel, L., Meetre, C. A., Perche, C., Sitruk, L., and Hoang, S.: WAVES: The radio and plasma wave investigation on the Wind spacecraft, Space Sci. Rev., 71, 231–263, 1995.

Bougeret, J. L., King, J. H., and Schwenn, R.: Solar radio burst and in situ determination of interplanetary electron density, Sol. Phys., 90, 401–412, 1984.

Dulk, G. A., Leblanc, Y., Robinson, P. A., Bougeret, J. L., and Lin, R. P.: Electron beams and radio waves of solar Type III bursts, J. Geophys. Res., 103, 17223–17233, 1998.

Dulk, G. A., Steinberg, J. L., Lecacheux, A., Hoang, S., and Mac-Dowall, R. J.: The visibility of type III radio bursts originating behind the sun, Astron. Astrophys., 150, L28–L30, 1985.

Fainberg, J. and Stone, R. G.: Type III solar radio burst storms observed at low frequencies, I. Storm morphology, Sol. Phys., 15, 222–233, 1970.

Ginzburg, V. L. and Zheleznyakov, V. V: On the Possible Mechanisms of Sporadic Solar Radio Emission – Radiation in an Isotropic Plasma, Astr. Zh., 35, 653–668, 1958.

Gurnett, D. A., Kurth, W. S., Kirchner, D. L., Hospodarsky, G. B., Averkamp, T. F., Zarka, P., Lecacheux, A., Manning, R., Roux, A., Canu, P., Cornilleau-Wehrlin, N., Galopeau, P., Meyer, A., Bostrom, R., Guastafsson, G., Wahlund, J.-E., Aahlen, L., Rucker, H. O., Ladreiter, H. P., Macher, W., Woolliscroft, L. J. C., Alleyne, H., Kaiser, M. L., Desch, M. D., Farrell, W. M., Harvey, C. C., Louarn, P., Kellogg, P. J., Goetz, K., and Pedersen, A.: The Cassini radio and plasma wave science investigation, Space Sci. Rev., 114, 395–463, 2004.

Hoang, S., Poquerusse, M., and Bougeret, J. L.: The directivity of solar kilometric Type III bursts: Ulysses-Artemis observations in and out of the ecliptic plane, Sol. Phys., 172, 307–316, 1997.

Krupar, V., Santolik, O., Cecconi, B., Maksimovic, M., Bonnin, X., Panchenko, M., and Zaslavsky, A.: Goniopolarimetric inversion using SVD: An application to type III radio bursts observed by STEREO, J. Geophys. Res., 117, A06101, doi:10.1029/2011JA017333, 2012.

Lecacheux, A., Steinberg, J. L., Hoang, S., and Dulk, G. A.: Characteristics of type III bursts in the solar wind from simultaneous observations on board ISEE-3 and Voyager, Astron. Astrophys., 217, 237–250, 1989.

Lin, R. P., Evans, L. G., and Fainberg, J.: Simultaneous observations of fast solar electrons and Type III radio burst emission near 1 AU, Astrophys. J., 14, 191–198, 1973.

MacDowall, R. J., Stone, R. G., and Kaiser, M. L.: Kilometric Type III burst directivity from simultaneous spacecraft observations, Bulletin of the American Astronomical Society, 14, 607–607, 1982.

Mann, G., Jansen, F., MacDowall, R. G., Kaiser, M. L., and Stone, R. G.: A heliospheric density model and type III radio bursts, Astron. Astrophys., 348, 614–620, 1999.

Marsch, E., Harrison, R., Pace, O. Antonucci, E., Bochsler, P., Bougeret, J.-L., Fleck, B., Langevin, Y., Marsden, R., Schwenn, R., and Vial, J.-C.: Solar Orbiter, a high resolution mission to the Sun and inner heliosphere, Physical aspects towards a better knowledge of the solar corona, In: Solar encounter. Proceedings of the First Solar Orbiter Workshop, edited by: Battrick, B. and Sawaya-Lacoste, H., ESA SP-493, Noordwijk: ESA Publications Division, ISBN 92-9092-803-4, 2001.

Meyer-Vernet, N. and Perche, C.: Tool kit for antennas in plasmas, J. Geophys. Res., 94, 2405–2415, 1989.

Poquérusse, M., Hoang, S., Bougeret, J. L., and Moncuquet, M.: Ulysses- ARTEMIS radio observation of energetic flare electrons, in Amer. Inst. Phys. Conf. Ser. 382, edited by: Winterhalter, D., Gosling, J. T., Habbal, S. R., Kurth, W. S., and Neugebauer, M., 62–65, 1996.

Stone, R. G., Pedersen, B. M., Harvey, C. C., Canu, P., Cornilleau-Wehrlin, N., Desch, M. D., de Villedary, C., Fainberg, J., Farrell, W. M., and Goetz, K.: ULYSSES radio and plasma wave observations in the Jupiter environment, Science, 257, 1524–1531, 1992.

16

Automated collection and dissemination of ionospheric data from the digisonde network

B. W. Reinisch, I. A. Galkin, G. Khmyrov, A. Kozlov, and D. F. Kitrosser

Environmental, Earth, and Atmospheric Sciences Department, Center for Atmospheric Research, University of Massachusetts, Lowell, USA

Abstract. The growing demand for fast access to accurate ionospheric electron density profiles and ionospheric characteristics calls for efficient dissemination of data from the many ionosondes operating around the globe. The global digisonde network with over 70 stations takes advantage of the Internet to make many of these sounders remotely accessible for data transfer and control. Key elements of the digisonde system data management are the visualization and editing tool SAO Explorer, the digital ionogram database DIDBase, holding raw and derived digisonde data under an industrial-strength database management system, and the automated data request execution system ADRES.

1 Introduction

Ionosondes used to be the workhorse of ionospheric radio research, and until the 1950s were the only supplier of reliable ionospheric data. Then in situ observations on rockets and satellites became available, remote sensing with UHF and VHF incoherent and coherent scatter radars provided great details at selected locations, and measurements on the ground and on satellites of radio signals from topside satellites determined the large scale electron density distribution in the ionosphere using tomography and limb-scanning techniques. Ionosondes continue to play an important role because they provide accurate vertical electron density profiles as function of time at many locations around the globe. This paper describes the networking of digisondes, the types of results obtained, their databasing, and the real time access. There are other digital ionosondes operating at various sites, some of them producing similar data as the Digisonde, like the KEL sounder, the Advanced Ionospheric Sounder or Dynasonde, and the CADI sounder. We describe the digisonde data collection and dissemination system as an example, because it is believed to be the largest and most advanced such system in routine operation.

Correspondence to: B. W. Reinisch
(bodo_reinisch@uml.edu)

2 Data from the ionosonde network

Global coverage of the ionosphere with ionosondes became reality with the International Geophysical Year in 1957–1958, and the necessity for standardized interpretation and scaling of ionograms became immediately apparent. URSI established a Worldwide Sounding Committee (WWSC), which issued the Handbook of Ionogram Interpretation and Reduction, prepared by Piggott and Rawer in 1961 and revised in 1972 (Piggott and Rawer, 1972). In 1969 the Ionosonde Network Advisory Group (INAG), with Roy Piggott as chairman, replaced the WWSC. Today INAG, currently under the chairmanship of Terence Bullett, is responsible for maintaining uniform ionogram reduction rules. For a long time the usefulness of the ionosonde network was handicapped by the limited data accessibility. Usually only the hourly ionograms were scaled, and the reduced values were hand recorded. The World Data Centers (WDC) prescribed a standard format for the monthly tabulation of ionospheric characteristics that allowed the systematic archiving of the data. While originally designed for the archiving of only hourly values, the URSI Ionospheric Informatics Working Group (IIWG), with Bodo Reinisch as the chairman, developed a more flexible format that permits the archiving of irregular time series data (Gamache and Reinisch, 1991; Reinisch, 1998a). This format is used today in the world data centers, where the data can be accessed with the Space Physics Interactive Data Resource (SPIDR) developed by the National Geophysical Data Center in Boulder, CO (Conkright, 1999). To facilitate the archiving and exchange of reduced data of individual ionograms the IIWG developed the Standard Archiving Output (SAO) format for the ionogram characteristics (Reinisch, 1998b). This format, recommended by URSI for all reduced ionogram data, is now widely used for data from different type ionosondes. The Digisonde network (Fig. 1) is using the SAO format for the automatically scaled data during the last 10 years.

Fig. 1. The global digisonde network. Data from unnamed stations currently are not available via Internet or DIDBase access.

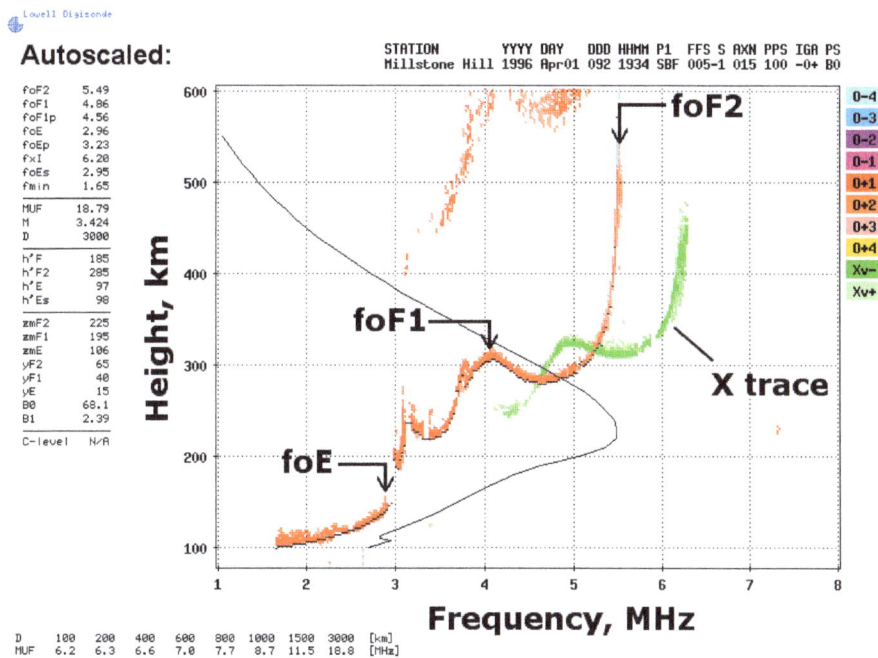

Fig. 2. Real time autoscaled daytime ionogram at Millstone Hill, listing the most important characteristics on the left.

3 Real time data

Many space weather applications require real time access to the ionospheric characteristics from a large set of ionosonde stations (Galkin et al., 1999). An early generation of digisondes, the "Digisonde 256" (Bibl and Reinisch, 1978) was designed with this goal in mind. The Digisonde 256 employed algorithms for automated onsite scaling of ionograms (Reinisch and Huang, 1983) and provided modem communications with remote data processing centers for delivery of the scaled data in real-time. In the late 1980s, the US Air Force Weather Service deployed ~20 Digisondes 256 in their DISS network to support the USAF Space Forecast Center operations (Reinisch, 1996a; Buchau et al., 1995)

feeding real time data to the PRISM and GAIM modeling projects (Daniell et al., 1995; Sojka et al., 2003) Another special network of 15 digisondes, using the second-generation Digisonde Portable Sounders (DPS) (Reinisch, 1996b) operates in Australia, providing ionospheric characteristics and electron density profiles to update the Real Time Ionospheric Models (RTIMs) of electron density distribution (Barnes et al., 2000; Reinisch et al., 1997). The RTIM densities are used by various Australian defense projects such as the Jindalee Over-the-horizon Radar Network (JORN) and Jindalee Facility at Alice Springs (JFAS). These two special digisonde networks are maintained by defense organizations that employ dedicated communication links and impose certain restrictions on public access to the data. Meanwhile, an increasing

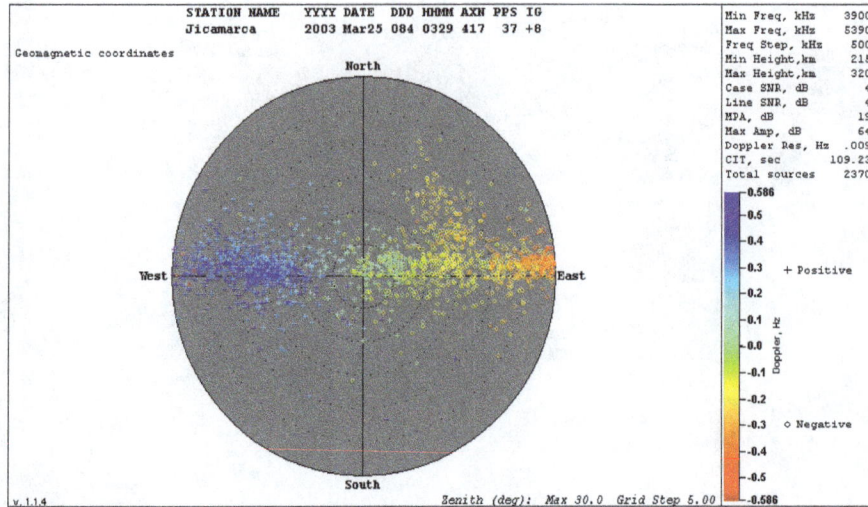

Fig. 3. Real time skymap at Jicamarca at 22:29 LT on 22 March 2003 showing the echo location and Doppler frequencies (color coded) for F-region echoes at frequencies between 3.9 and 5.4 MHz.

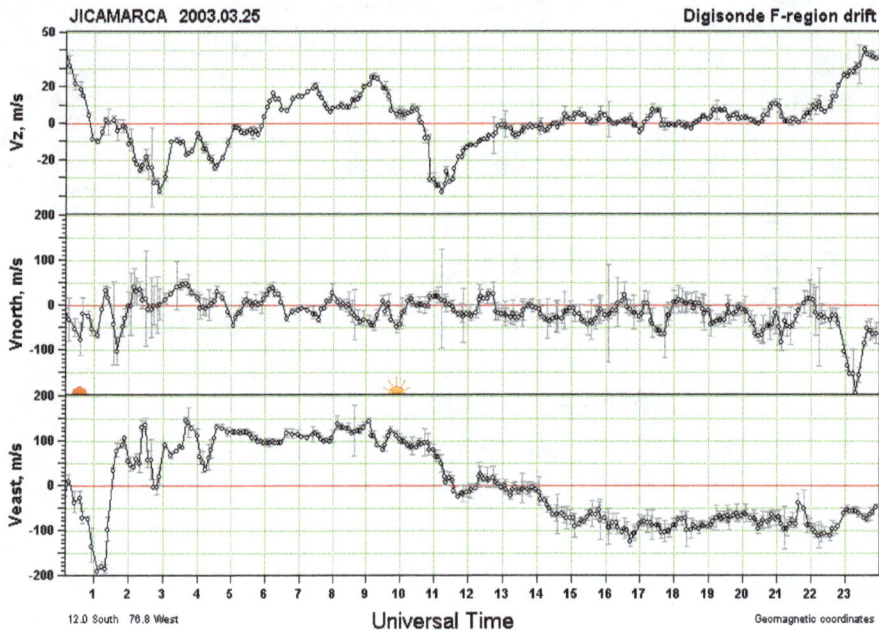

Fig. 4. Real time vertical and horizontal drift components as function of time at Jicamarca for 25 March 2003. The error bars are shown in gray. Sunset and sunrise are indicated in the central panel.

number of research digisondes discover advantages of the Internet as an inexpensive solution to the problem of rapid data dissemination. Using standard Internet protocols, it is possible not only to establish the real time data stream to remote processing centers, but also to publish both latest and retro data at the digisonde location using the WWW server that is included in the standard digisonde software since the late 1990s.

Modern online digisondes publish a variety of ionospheric data presentations in real-time on the WWW. Figures 2–5 show samples of the published "latest" images that are available within a minute after the sounding is complete; HTML

documents with the latest images are refreshed automatically on the local client computer so that they can be monitored at remote locations. Figure 2 shows a typical daytime midlatitude ionogram recorded at Millstone Hill, Massachusetts, together with the autoscaled ionospheric characteristics (left side) and a table of Maximum Usable Frequencies calculated for a number of preset path distances (bottom). Figure 3 presents an example of a latest skymap, a plot of calculated echo locations for an automatically selected subset of frequencies and heights in the F-region of the ionosphere. The skymap in Fig. 3 was taken at Jicamarca at 22:29 LT on 22 March 2003 during the night-time eastward drift of plasma

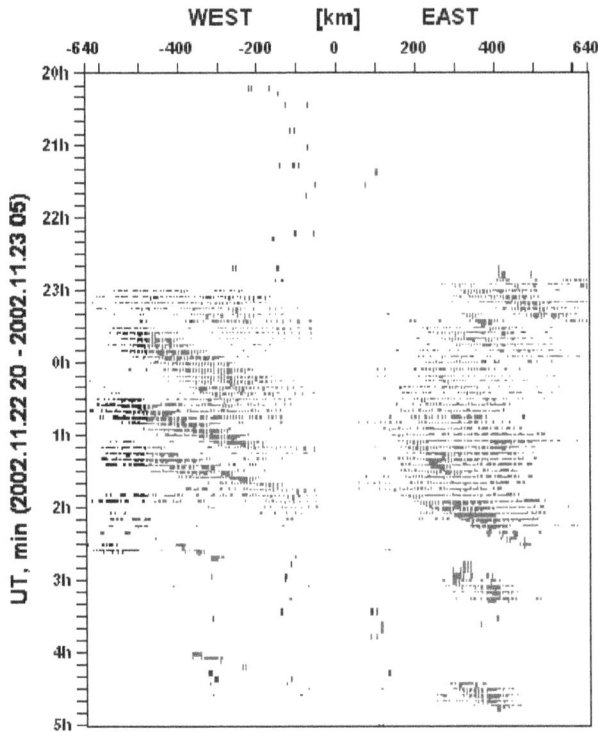

Fig. 5. Directogram at Cachimbo for 22/23 November 2004 shows as function of time the approximate location of irregularities seen by the multi-beam ionograms. Each horizontal line presents data from one ionogram.

irregularities along the magnetic equator; the color coding is used to indicate the Doppler shift of the sounding frequency. In addition to the skymaps indicating echo locations, the bulk motion of plasma across the station location can be calculated from the digisonde drift mode data. Figure 4 presents a sample of daily drift velocity plots (updated with the real-time data) showing vertical and horizontal components as a function of time. The error bars on the given values represent the degree to which the motions in the plasma can be considered as a bulk motion. Finally, Fig. 5 gives an example of a directogram, a daily plot of locations of the irregularities in the ionosphere derived from the multi-beam ionograms showing echo directions. Directograms are useful for quick identification of spread F conditions and characterization of the preferred motion of irregularities in the ionosphere.

4 Digisonde online databasing and quality control

With increasing number of applications relying on the "ground truth" extracted from ionograms, a staggering task of quality control and verification of the autoscaled data has emerged. While increasing attention is placed on development of techniques for automated quality control of autoscaling results (e.g. Conkright and McNamara, 1996), it is critical for many research and applied projects to verify and/or edit the derived ionospheric characteristics by human inspection. We placed the human verification process into

the framework of computer technologies that are convenient for both data verification experts (scalers) and the end users. Demanding the highest possible degree of accuracy for the ionogram-derived characteristics, this system now sets a new standard for management of ionosonde data.

Three key elements of the digisonde data management system are: (1) the "SAO Explorer", a digisonde data visualization and editing tool, (2) "DIDBase" (Digital Ionogram Database) holding raw and derived digisonde data under an industrial-strength database management system, and (3) "ADRES" (Automated Data Request Execution System) capable of accepting the "ground truth" requests for particular periods of time and locations and actively managing the process of acquiring, verifying, and reporting the requested data.

4.1 DIDBase

While the WDC SPIDR remains the major tool for accessing a great variety of geo, solar and space physics datasets, it was not designed to provide its users with the capability of interactively validating the autoscaled ionogram data. Arranging such validation process necessarily involves provision of raw ionogram data for reference, mating the database to an ionogram editing tool, and allowing online submissions of the edited/verified data back to the database. The need for a small, dedicated project was identified to establish an interactive, concurrent, read/write access to an archive of ionograms and scaled data. A pilot DIDBase project of this kind was started in 2001 at UMASS Lowell for the data collected from the network of 44 digisonde sounders (see Fig. 1 for the locations of contributing stations). Currently DIDBase holds ~80 GB of data and accepts real-time and near real-time submissions from 38 digisondes listed in Table 1.

The central piece of the DIDBase is a relational DBMS Firebird 1.0 chosen for its reliability and minimal administration expense. The database structure is designed to include the binary digisonde ionograms data together with multiple scaling records for each ionogram. By relaxing the conventional scenario "one ionogram – one set of derived characteristics", the DIDBase creates new possibilities for multiple trace identifications, storage of alternative ionogram interpretations, and a variety of comparative studies of the automated ionogram processing quality. As an example, the new technology makes it possible to store several simultaneously observed auroral E layers, which is of interest for high latitude ionospheric research. Although multiple scalings (interpretations) for a given ionograms are stored, the DIDBase maintains a single subset of preferably used ionospheric characteristics for each ionogram, which can be quickly accessed by the end users looking just for the "best" ionogram interpretation. A hierarchy of expert ranks and quality flags is used to automatically select one best value among existing versions.

Table 1. Online digisonde stations currently providing real time data to DIDBase.

URSI code	Station name	Latitude	Longitude	Model
AN438	ANYANG	37.4	127.0	D256
AS00Q	ASCENSION ISLAND	−8.0	345.6	DISS (D256)
AT138	ATHENS	38.0	23.5	DPS
BV53Q	BUNDOORA	−37.7	145.1	DPS
CO764	COLLEGE	64.9	212.0	DISS (D256)
CAJ2M	CACHOEIRA PAULISTA	−23.2	314.2	D256
DB049	DOURBES	50.1	4.6	D256
DS932	DYESS AFB	32.4	260.2	DISS (D256)
EG931	EGLIN AFB	30.4	273.2	DISS (D256)
EA036	EL ARENOSILLO	37.1	353.3	D256
FF051	FAIRFORD	51.7	358.5	DISS (D256)
FZA0M	FORTALEZA	−3.8	322.0	DPS
GA762	GAKONA	62.4	215.0	DPS
GSJ53	GOOSE BAY	53.3	299.7	DISS (D256)
HA419	HAINAN	19.4	109.0	DPS
GR13L	GRAHAMSTOWN	−33.3	26.5	DPS
JI91J	JICAMARCA	−12.0	283.2	DPS
JR055	JULIUSRUH	54.6	13.4	DPS
KS759	KING SALMON	58.4	203.6	DISS (D256)
LM42B	LEARMONTH	−21.8	114.1	DISS (D256)
LV12P	LOUISVALE	−28.5	21.2	DPS
MU12K	MADIMBO	−22.4	30.9	DPS
MHJ45	MILLSTONE HILL	42.6	288.5	DPS
NQJ61	NARSSARSSUAQ	61.2	314.6	DISS (D256)
SN437	OSAN AFB	37.1	127.0	DISS (D256)
PSJ5J	PORT STANLEY	−51.6	302.1	DPS
PQ052	PRUHONICE	50.0	14.6	DPS
PA836	PT ARGUELLO	34.8	239.5	DISS (D256)
THJ77	QAANAAQ	77.5	290.6	D256
PRJ18	RAMEY AFB	18.5	292.9	DISS (D256)
RO041	ROME	41.9	12.5	DPS
EB040	ROQUETES	40.8	0.3	D256
SAA0K	SAO LUIS	−2.6	315.8	D256
T139	SAN VITO	40.6	17.8	DISS (D256)
SMJ67	SONDRESTROM	67.0	309.1	DPS
TR169	TROMSO	69.6	19.2	DPS
TUJ2O	TUCUMAN	−26.9	294.6	D256
WP937	WALLOPS ISLAND	37.9	284.5	DISS (D256)

4.2   SAO explorer, interactive ionogram analysis tool

SAO-Explorer is a digisonde data analysis tool used for manual verification and editing of autoscaled digisonde ionograms, as well as a variety of visual presentations of ionograms and derived ionospheric characteristics. It serves best for the in-depth study of particular periods of time or locations where background ionograms are required to aid with data interpretation. The SAO-X workstations are granted both read and write access to the DIDBase allowing full scale, platform-independent, concurrent, remote operations with the archived data over the Internet. Any ionogram interpretation expert can register for write permission and verify/edit the autoscaled data from remote locations. In addition, the SAO-X workstations connect to the SPIDR database to read retrospective ionospheric data that are not available in the DIDBase or locally. Some examples of typical SAO-X outputs are shown in Figs. 6 and 7. The "profilogram" in Fig. 6 shows the vertical electron density distribution as function of time (Reinisch et al., 1994) for one day in October 2003 at the magnetic equator, Cachimbo, Brazil (Abdu et al., 2003) derived from the N(h) profiles calculated for each ionogram (see Fig. 2). Figure 7 displays three selected (of the 49 available) characteristics at Cachimbo for three consecutive days: foF2, hmF2, and frequency spread QF.

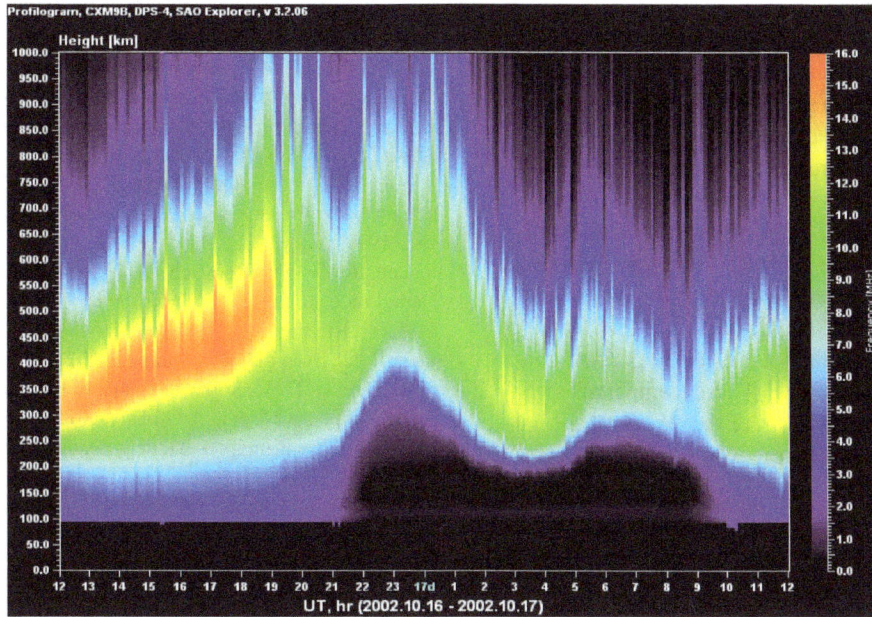

Fig. 6. Profilogram for 16 October 2002 at Cachimbo, Brazil. The plasma frequencies (proportional to $N^{1/2}$) are color-coded.

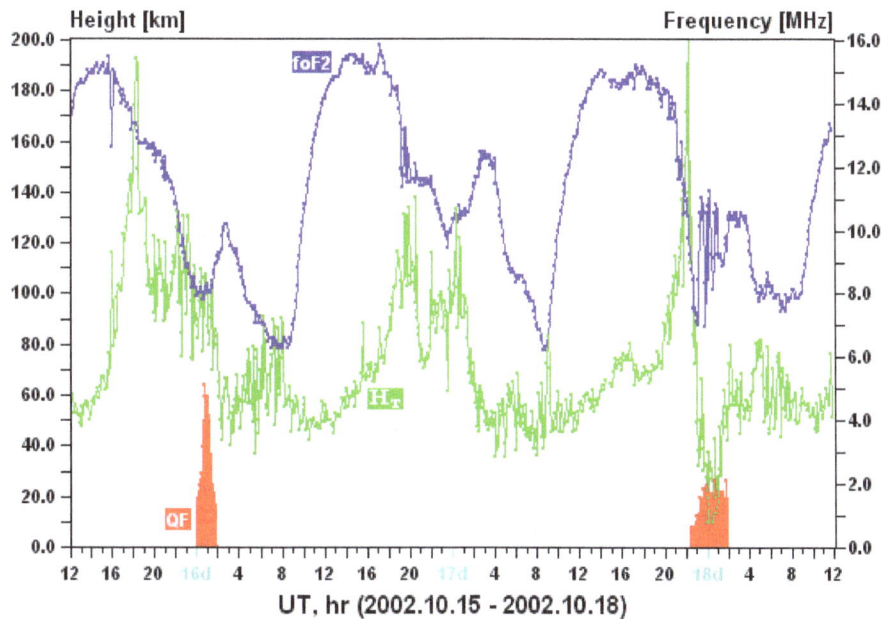

Fig. 7. Ionospheric characteristics as function of time at Cachimbo for three consecutive days in October 2002: foF2, topside scale height HT, and range spread QF in km. The vertical axis on the left shows the heights in km, and on the right the frequencies in MHz.

4.3 ADRES

Automated Data Request Execution Subsystem (ADRES) is an integrated working environment capable of accepting a data request and taking the necessary steps to return the appropriate data, and also to remotely control the sounder schedule, data acquisition, alerting scalers, and generating the final report.

If the requested data are already in DIDBase and are validated, the ADRES generates the report immediately. A pro-

vision is made to manage the requests for special modes of ionosonde operation (e.g. high ionogram rate during a satellite pass over the station, a coordinated campaign, or an event of interest) and for data that are not available in DIDBase or are not manually validated. The ADRES has a mechanism to automatically read incoming requests to adjust programs and schedules of the Internet-enabled digisondes. To acquire digisonde data, the ADRES maintains a list of FTP servers where the ionogram data can be found. A number of digisonde stations deliver their real time data directly to a

WDC and to DIDBase. As soon as the data are ingested, a message is generated to the SAO-X operators to validate/edit the autoscaling results. When the quality control procedure is completed, the final report is generated and delivered to the requesting party. Each step of the request execution is monitored, and the status of each request is available for remote access, just like any other data stored in the database.

Applications of the ADRES subsystem include the calibration and validation of space-borne UV sensors measuring ionospheric electron density profiles, and of the ionosonde-derived total electron content (ITEC) (Reinisch and Huang, 2001), by comparing it with the total electron content data from the TOPEX measurements. These campaigns provide a first opportunity to demonstrate the power of global ionosonde networking for quick access to ionospheric electron density distributions.

5 Summary

The establishment of a global Internet-connected digital ionosonde network with standardized data formats was the first gigantic step to make ionosonde data user-friendly. The new information system introduced in this paper was the missing second step. The new system, which includes the Digital Ionogram Data Base (DIDBase), the expert ionogram scaling tool "SAO Explorer", and the Automated Data Request Execution Subsystem (ADRES), provides organized freedom to ionogram data management and remote digisonde station control.

Acknowledgement. This work was in part supported by Air Force Contract No. AF19628-02-C-0092 and NASA Grant No. NAG5-13387.

References

Abdu, M. A., Batista, I. S., Reinisch, B. W., de Souza, J. R., de Paula, E. R., Sobral, J. H. A., and Bullett, T. W.: Equatorial Spread F and Ionization Anomaly Development as diagnosed from Conjugate Point Observation (COPEX) in Brasil, XXIII General Assembly of IUGG, Sapporo, Japan, 30 June–11 July, 2003.

Bibl, K. and Reinisch, B. W.: The Universal Digital Ionosonde, Radio Sci., 13, 519–530, 1978.

Conkright, R. O.: SPIDR on the Web: Space Physics Interactive Data Resource Online Analysis Tool, Proc. URSI XXVI GA, Toronto, Canada, 491, August 1999.

Conkright, R. O. and McNamara, L. F.: Quality control of automatically scaled vertical incidence ionogram data, Proc. Ionosph. Effects Symp. IES-96, 99–103, May 1996.

Galkin, I. A., Kitrosser, D. F., Kecic, Z., and Reinisch, B. W.: Internet access to ionosondes, J. Atmos. Solar-Terr. Phys., 61, 181–186, 1999.

Gamache, R. R. and Reinisch, B. W.: Databasing of Scientific Data, Proceedings of the Workshop on Geophysical Informatics, Moscow, Russia, WDC-A Report UAG-99, 102–108, 1991.

Daniell, R. E., Brown, L. D., Anderson, D. N., Fox, M. W., Doherty, P. H., Decker, D. T., Sojka, J. J., and Schunk, R. W.: Parameterized Ionosphere Model: A Global Ionospheric Parameterization based on First Principles Models, Radio Sci., 30, 1499–1510, 1995.

Piggott, W. R. and Rawer, K. (eds.): URSI Handbook of Ionogram Interpretation and Reduction, 2nd edition, Report UAG-23, WDC-A for STP, NOAA, Boulder, Colorado, 1972.

Reinisch, B. W. and Huang, X.: Automatic Calculation of Electron Density Profiles from Digital Ionograms, 3, Processing of Bottomside Ionograms, Radio Sci., 18, 477, 1983.

Reinisch, B. W., Anderson, D., Gamache, R. R., Huang, X., Chen, C. F., and Decker, D. T.: Validating Ionospheric Models with Measured Electron Density Profiles, Adv. Space Res., 14, 12, 67–70, 1994.

Reinisch, B. W.: Ionosonde, in Upper Atmosphere, edited by Dieminger, W., Hartmann, G. K., and Leitinger, R., Springer, 370–381, 1996.

Reinisch, B. W.: Modern Ionosondes, in Modern Ionospheric Science, edited by Kohl, H., Ruster, R., and Schlegel, K., European Geophysical Society, 37191 Katlenburg-Lindau, Germany, 440-458, 1996.

Reinisch, B. W.: CHARS: URSI IIWG format for archiving monthly ionospheric characteristics, INAG Bulletin No. 62, WDC-A for STP, Boulder, CO, 38–46, 1998a.

Reinisch, B. W.: SAO (Standard ADEP Output) format for ionogram scaled data archiving, INAG Bulletin No. 62, WDC-A for STP, Boulder, CO, 47–58, 1998b.

Reinisch, B. W. and Huang, X.: Deducing Topside Profiles and Total Electron Content from Bottomside Ionograms, Adv. Space Res., 27, 1, 23–30, 2001.

Sojka, J. J., Thompson, D. C., Schunk, R. W., Eccles, V., Makela, J. J., Kelley, M. C., Gonzoles, S. A., Aponte, N., and Bullett, T. W.: Ionospheric data assimilation: recovery of strong mid-latitudinal density gradients, JASTP, 65, 2087–1097, 2003.

17

GPS TEC and ITEC from digisonde data compared with NEQUICK model

J.-C. Jodogne[1]**, H. Nebdi**[1]**, and R. Warnant**[1][*]

[1]Institut Royal Météorologique, 3 Avenue Circulaire, B-1180 Bruxelles, Belgium
[*]Royal Observatory of Belgium

Abstract. At the Dourbes station, a digisonde 256 is co-located with a Turbo Rogue GPS receiver. Real time processing of the digisonde data gives the electron density profile and the ITEC value (SAO file) for each sounding. The GPS receiver produces data that are treated at the Royal Observatory in order to extract a vertical TEC. Running the well-known NeQuick ionospheric model allows to compute vertical TEC values. Comparisons of the results obtained in 1996 and 2001 by these different approaches are shown.

1 Introduction

The use of empirical models for the Total Electron Content (TEC) appears to increase in different applications. We benefit of the co-location of two systems able to provide TEC estimations to compare with such a model.

NeQuick model is a one adopted by the COST 251 Action and updated during the succeeding COST 271 Action till now (Leitinger et al., 2002). To be short, the CCIR coefficients are used for foF2 and we run the software program with the monthly averaged solar flux for each hour at the Dourbes location. The profilers for E-, F1- and F2-regions are Epstein functions with different parameters for bottom and top parts. The topside ionosphere is simply a semi-Epstein layer. One of the features of the NeQuick model is to compute the electron content between any starting point (lat, long and height) and ending point in straight line. We limit ourselves to vertical direction. Two graphs for respectively 2001 (high solar activity) and 1996 (low solar activity) displayed at Fig. 1 give an idea of the model behaviour.

A Turbo Rogue receiver was installed at the top of the main building of the Dourbes station for derivation of GPS TEC values (Warnant and Jodogne, 1998). We remove all data from GPS satellite whose elevation is less than 88.5°.

At the Dourbes station a digisonde 256 with the Artist software produces hourly ionograms. The SAO output file gives the usual characteristics and an ITEC (Ionospheric TEC) up to infinity (Huang and Reinisch, 2001). It is such a value automatically produced that was used for this work.

2 NeQuick compared with GPS TEC

In order to see the data compared with the model we choose several months. GPS TEC was produced each quarters of the hour. We compute the median, the lower and upper quartiles, the maximum and the minimum for each quarter of hour of the month. The graphs present these values. We show the hourly NeQuick values as diamonds on the graphs. Typical season's months are displayed (January, March, June and September) for the years 1996 and 2001 (Figs. 2 and 3). To easily see the contrast between the two years we put four graphs on one panel.

When the solar activity was low the model gives higher values except for September 1996. For high solar activity values are quite good for January and especially for March but too small during June and September 2001.

3 NeQuick compared with Ionospheric Total Electron Contents

As for the GPS data, we compute the same statistical parameters for the year 2001. However GPS TEC were produced each quarter of hour but we record hourly ionograms only. Again we show NeQuick values as diamonds on the graphs (Fig. 4).

The median's data from January and March are always below the model's values while they are higher in June during day's hours. In September the agreement between model and experimental values is very good, especially during the first half of the day.

Correspondence to: J.-C. Jodogne
(jodogne@oma.be)

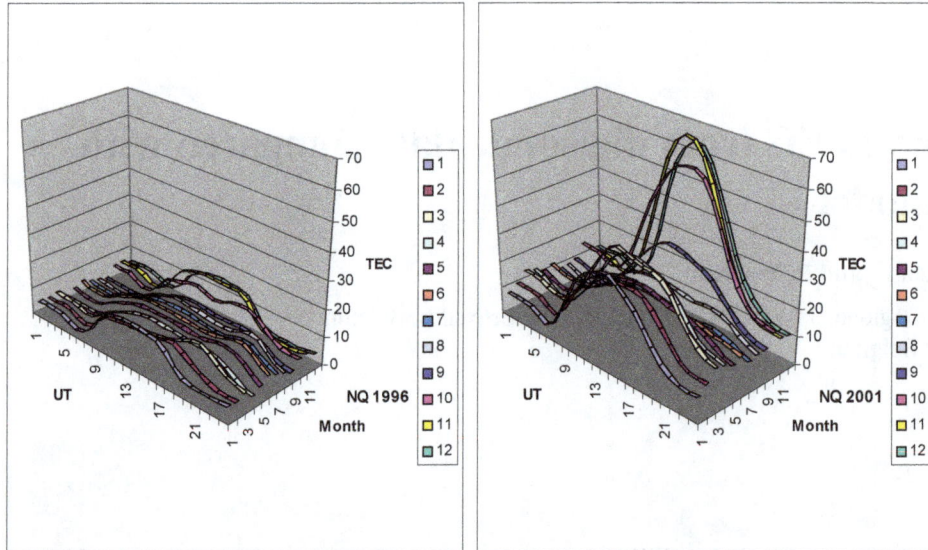

Fig. 1. TEC values up to 20 000 km from the NeQuick model for 1996 and 2001.

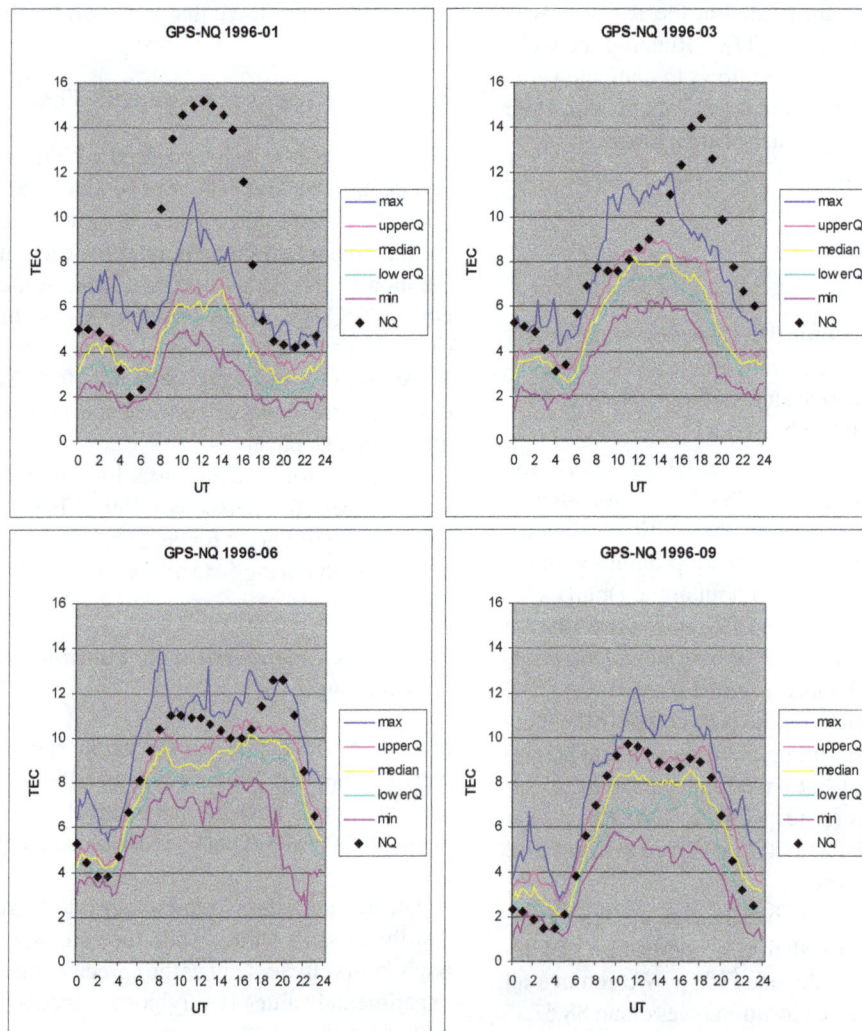

Fig. 2. GPS TEC versus TEC from model for typical season's months of 1996.

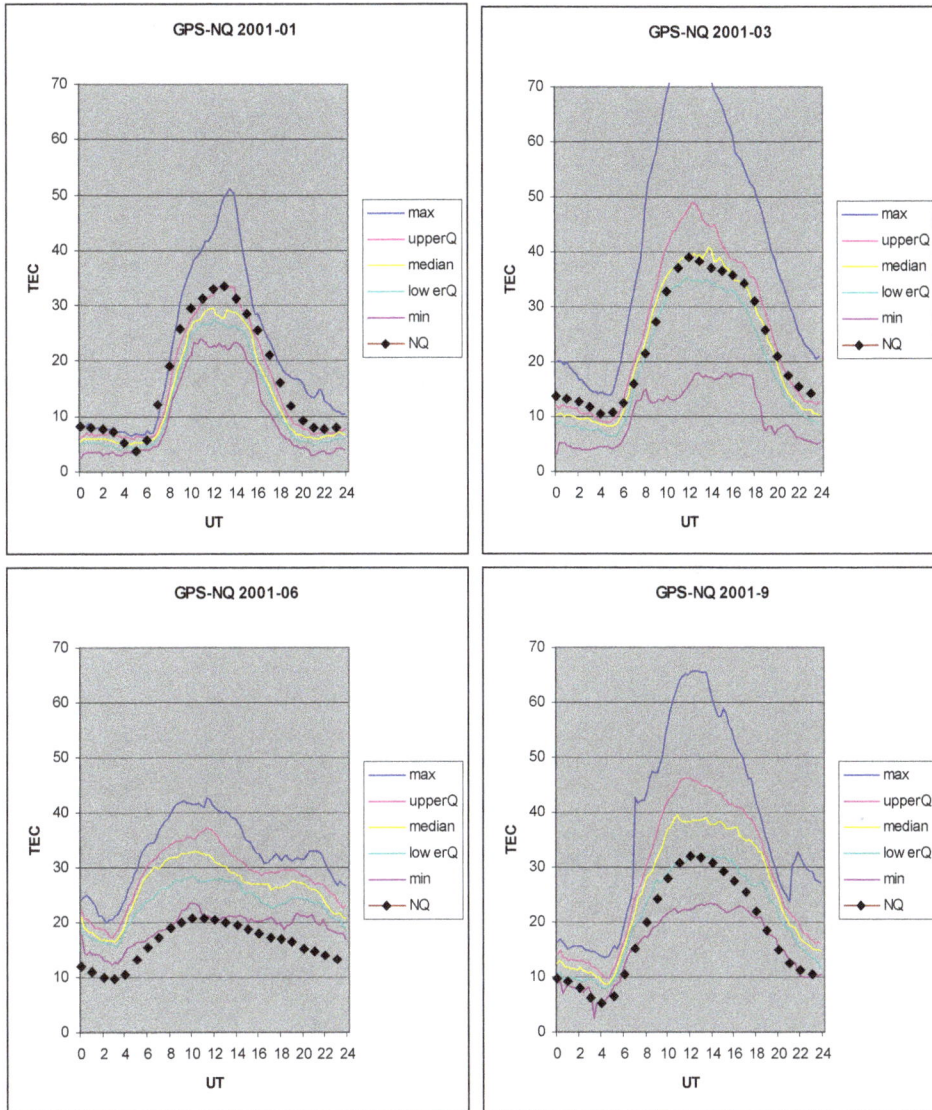

Fig. 3. Same as Fig. 2 but for year 2001.

In March 2001 we run the digisonde each 30 min. When these data are taken into account we get the graph of Fig. 5. The influence of waves in the ionosphere appears clearly. For some hours, the upper quartiles, medians and lower quartiles have nearly the same values despite the wavy structure (2:00, 3:00, 8:30, 10:30, 12:30, 13:30, 19:30, 20:30, 22:30). For GPS TEC during this month, the difference between the upper quartile and the lower quartile is small and the model fits quite well the data.

4 Ionospheric Total Electron Contents compared with GPS TEC

We display the medians for GPS as curves and for digisonde as points (Fig. 6). The shapes of both estimations are quite similar. As well known the GPS values are always larger than those from the digisonde.

5 Conclusions

As TEC estimations from GPS used in this work are means from about 30 samples during 15 min, it is understandable that the scatter of the GPS data is lesser. The night's data of the two experimental systems seems to be closer (except for June) to the model's values than those of the day. During daytime the discrepancies can reach more than 50% (January 1996) for GPS.

As this study is made with limited data series we don't want to give final conclusions. However the order of magnitude of the model's values are good, in particular for January and March 2001 were medians of GPS TEC and model's values almost coincide.

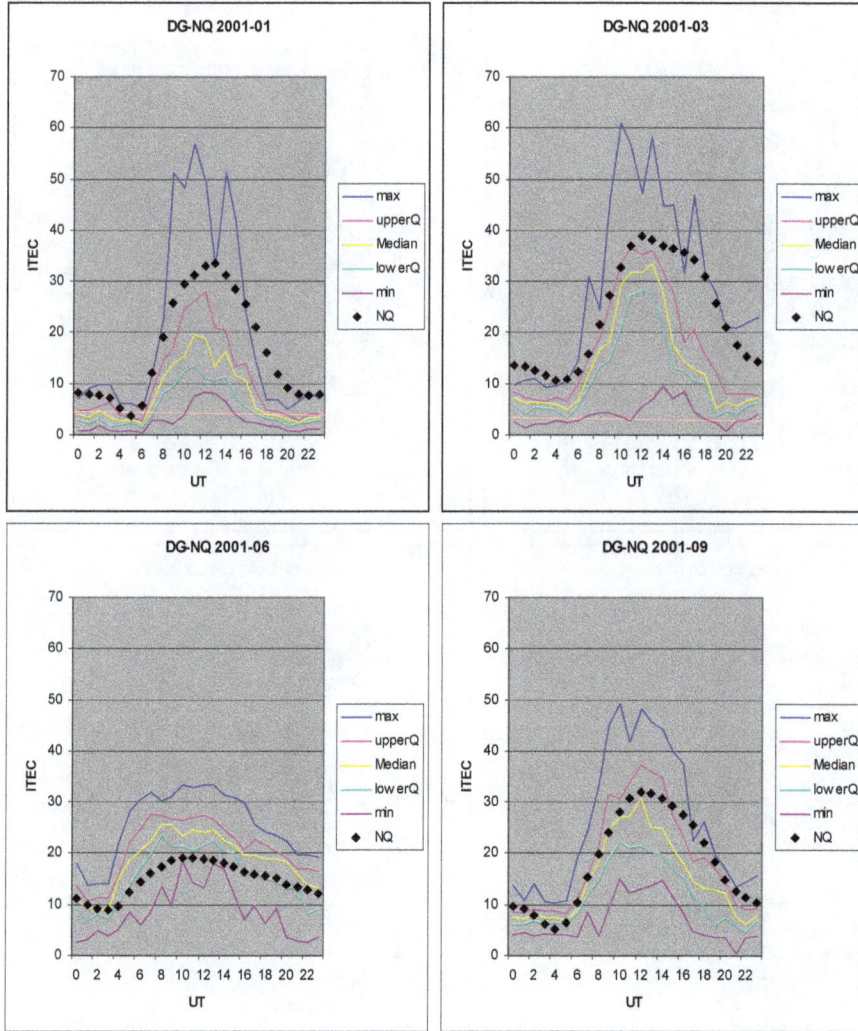

Fig. 4. Same as Fig. 3 but ITEC values from SAO files of the digisonde.

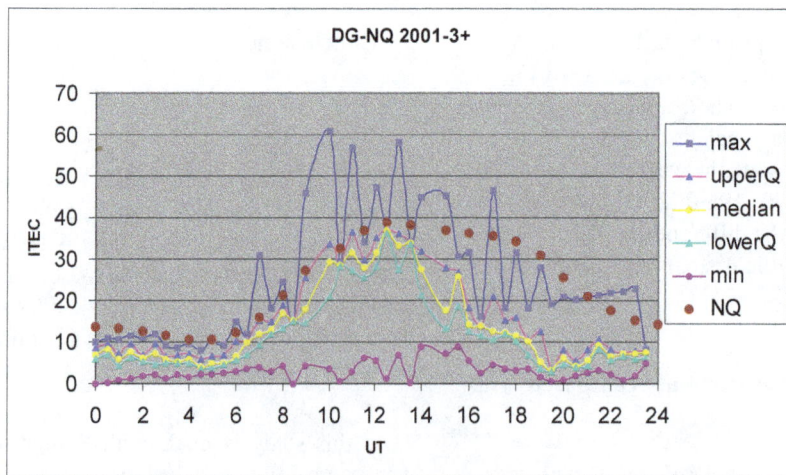

Fig. 5. ITEC data for each 30 min sounding and March 2001.

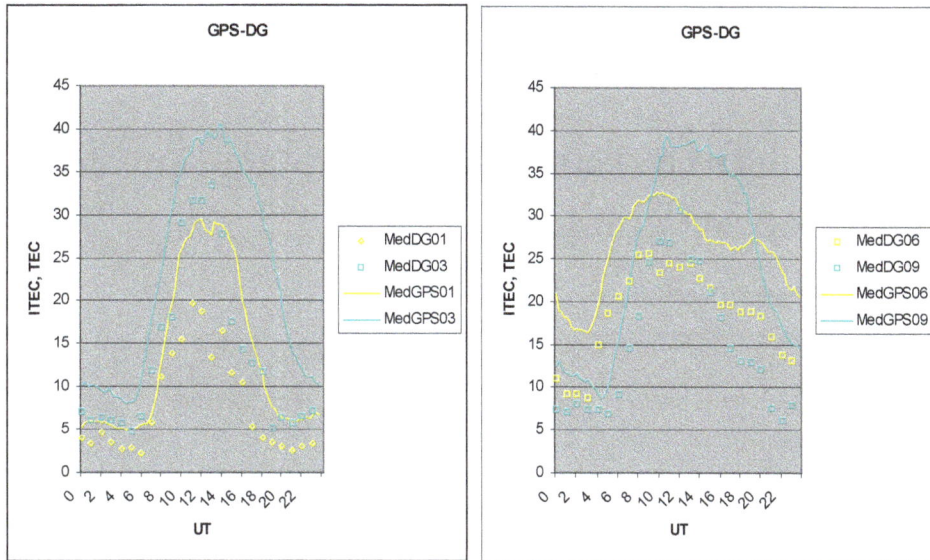

Fig. 6. Medians for digisonde ITEC (points) and GPS TEC (curves) during January and March (left panel) or June and September (right panel).

References

Huang, X. and Reinisch, B. W.: Vertical electron content from iono-grams in real time, Radio Science, 36, 2, 335–342, 2001.

Leitinger, R., Radicella, S., and Nava, B.: Electron density models for assessment studies – new developments, Acta Geodet. Geophys. Hung. 37, 183–193, 2002.

Warnant, R. and Jodogne, J.-C.: A comparison between TEC com-puted using GPS and Ionosonde measurements, Acta Geodet. Geophys. Hung. 33, 147–153, 1998.

18

Using scale heights derived from bottomside ionograms for modelling the IRI topside profile

B. W. Reinisch[1], **X. Huang**[1], **A. Belehaki**[2], **and R. Ilma**[3]

[1]Environmental, Earth and Atmospheric Sciences Department, Center for Atmospheric Research, University of Massachusetts Lowell, USA.
[2]Institute for Space Applications and Remote Sensing, National Observatory of Athens, Greece
[3]Jicamarca Radio Observatory, Lima, Peru

Abstract. Groundbased ionograms measure the Chapman scale height H_T at the F2-layer peak that is used to construct the topside profile. After a brief review of the topside model extrapolation technique, comparisons are presented between the modeled profiles with incoherent scatter radar and satellite measurements for the mid latitude and equatorial ionosphere. The total electron content TEC, derived from measurements on satellite beacon signals, is compared with the height-integrated profiles ITEC from the ionograms. Good agreement is found with the ISR profiles and with results using the low altitude TOPEX satellite. The TEC values derived from GPS signal analysis are systematically larger than ITEC. It is suggested to use H_T, routinely measured by a large number of Digisondes around the globe, for the construction of the IRI topside electron density profile.

1 Introduction

In a number of recent publications we have shown that the topside electron density profiles can be derived with good accuracy from the ionograms of groundbased ionosondes (Reinisch and Huang, 2001; Belehaki et al., 2003). An α-Chapman function with constant scale height H_T is assumed for the topside electron density distribution,

$$C(h)=NmF2 \cdot \exp\left[\frac{1}{2}\left(1-z-e^{-z}\right)\right];$$
$$z=\frac{h-hmF2}{H_T}. \qquad (1)$$

The scale height H_T is derived from the measured bottomside profile, which can be represented in terms of α-Chapman functions with a scale heights $H(h)$ that vary with height (Rishbeth and Garriott, 1969):

Correspondence to: B. W. Reinisch
(bodo_reinisch@uml.edu)

$$N(h)=N_m \left(\frac{H_m}{H(h)}\right)^{\frac{1}{2}} \exp\left\{\frac{1}{2}\left[1-y-\exp(-y)\right]\right\};$$
$$y=\int_{h_m}^{h} \frac{dh}{H(h)}. \qquad (2)$$

The subscript m refers to the values at the layer peak. The value H_m at the F2-layer peak can be calculated from the known function $N(h)$ (Huang and Reinisch, 2001). Figure 1 illustrates the process of constructing the topsisde profile. The bottomside profile on the left is derived from a daytime ionogram recorded in Kokobunji, Japan, on 8 March 1999. The center panel shows the height variation of the calculated scale height $H(h)$ with a maximum in the F1-region and minimal variation near the F2 peak. It seems therefore reasonable to assume that $H(h>hmF2)\approx H(hmF2)$ for a few hundred km above $hmF2$. The topside profile is then calculated with $H_T=H(hmF2)$ (right panel).

2 Validation of topside extrapolation technique

The best validation of the topside extrapolation technique is obtained at locations where an incoherent scatter radar (ISR) and an ionosonde measure vertical profiles simultaneously. Figures 2 and 3 show comparisons of such measurements at a mid latitude site (Millstone Hill, Massachusetts, 42° N) and the magnetic equator (Jicamarca, Peru). The Millstone Hill data in Fig. 2, showing the integrated electron content up to 800 km for four seasons in 1990 for all days for which ISR profile measurements were made, demonstrate the very good agreement between the two techniques. In Fig. 3, time averaged ISR profiles (red) at Jicamarca are compared with the hourly Digisonde profiles (green) for the available ISR observations from 19:00 LT on 11 June to 04:00 LT on 12 June 2002. While there is mostly very good agreement up to 800 km altitude in the evening and early night hours, noticeable differences occur above 600 km at 03:00 and 04:00 LT

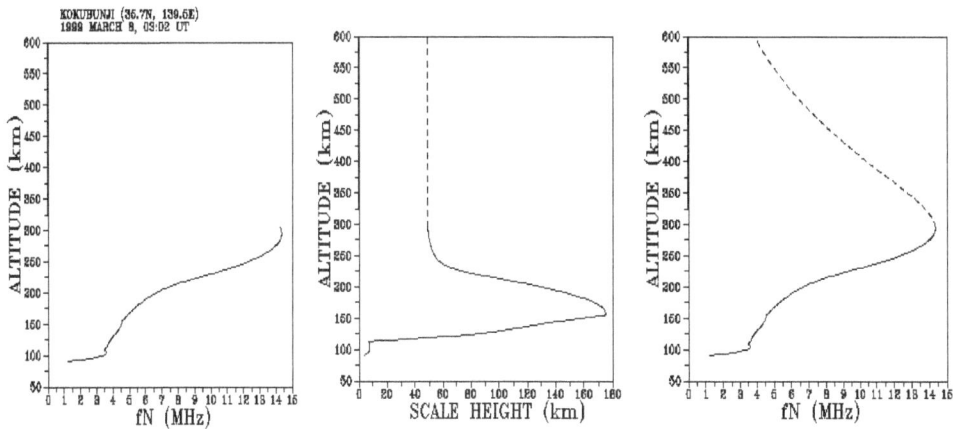

Fig. 1. Construction of the topside profile (right, dotted curve) from the measured bottomside profile (left) using the derived Chapman scale height (center).

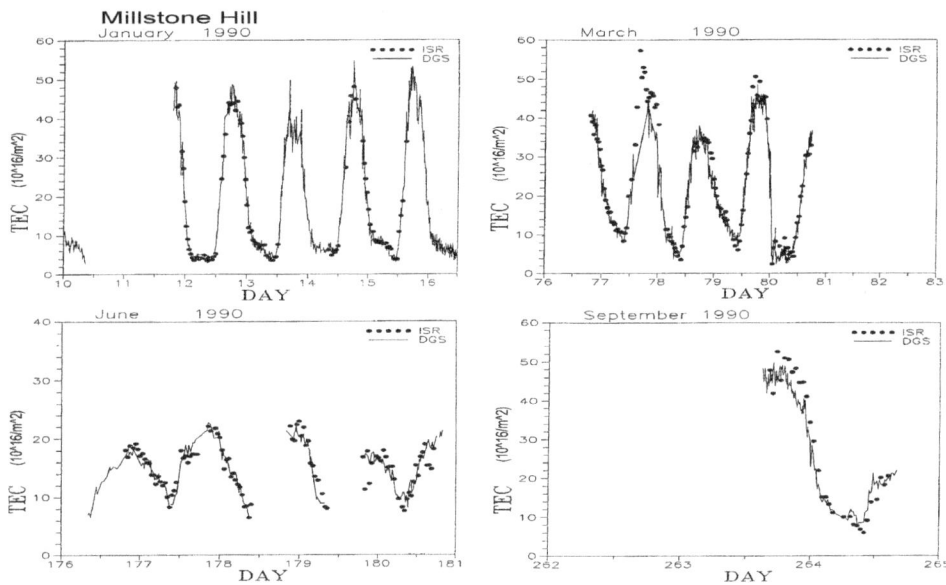

Fig. 2. Diurnal variations of the height-integrated electron density profiles at Millstone Hill derived from ISR (dotted) and Digisonde (thin line) measurements for January, March, June, and September 1990.

(08:00 and 09:00 UT). When more ISR profiles become available, we will find out whether the ionosonde technique systematically underestimates H_T at late night hours, and how it performs during the daytime in the equatorial ionosphere.

The height-integrated ionosonde profiles represent the ionospheric total electron content "ITEC", and comparisons with TEC measurements on satellite beacon signals can be made. Best agreement should be expected from vertical TOPEX observations when the satellite orbit passes over the ionosonde site. Jicamarca is close to the Pacific Ocean, which is important since TOPEX TEC measurements are only possible over ocean surfaces, and suitable TOPEX passes (A. Komjathy, personal communication) were selected for comparison.

For March and April 1998 TOPEX had 9 passes close to Jicamarca. Figure 4 shows the TOPEX TEC values (red) during the 10 s of closest approach to Jicamarca, superimposed on the diurnal Digisonde ITEC variations. The lengths of the TOPEX bars indicate the TEC variation observed during 10 s. Until 21 March the TOPEX passes occurred at nighttime (LT = UT–5 h) confirming the agreement between the techniques that was shown in Fig. 3. The measurements on daytime passes on and after 28 March show also good agreement during daytime.

Vertical TEC data derived from oblique GPS signals (Sardon et al., 1994; Jakowski, 1996) contain the plasmaspheric electron content (Lund et al., 1999), since GPS satellites orbit at ~20 000 km altitude. Comparing GPSTEC with ITEC should therefore show a systematic difference. Figures 5 and 6 show GPSTEC and ITEC data for Athens, Greece (38° N).

Hourly average of electron density profile over Jicamarca — ISR vs DPS
Day: 12-Jun-2002

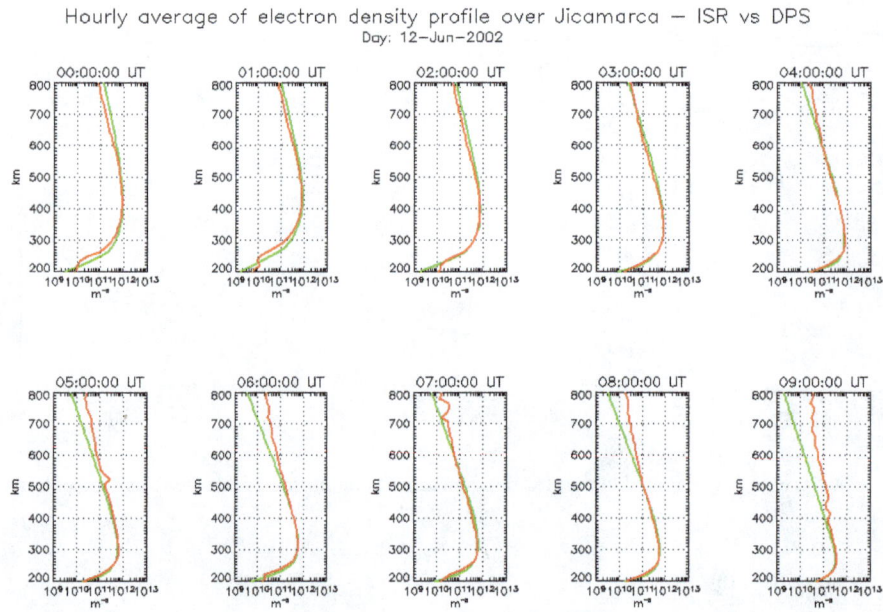

Fig. 3. Hourly nighttime F-layer profiles from ISR (red) and Digisonde (green) measurements at Jicamarca.

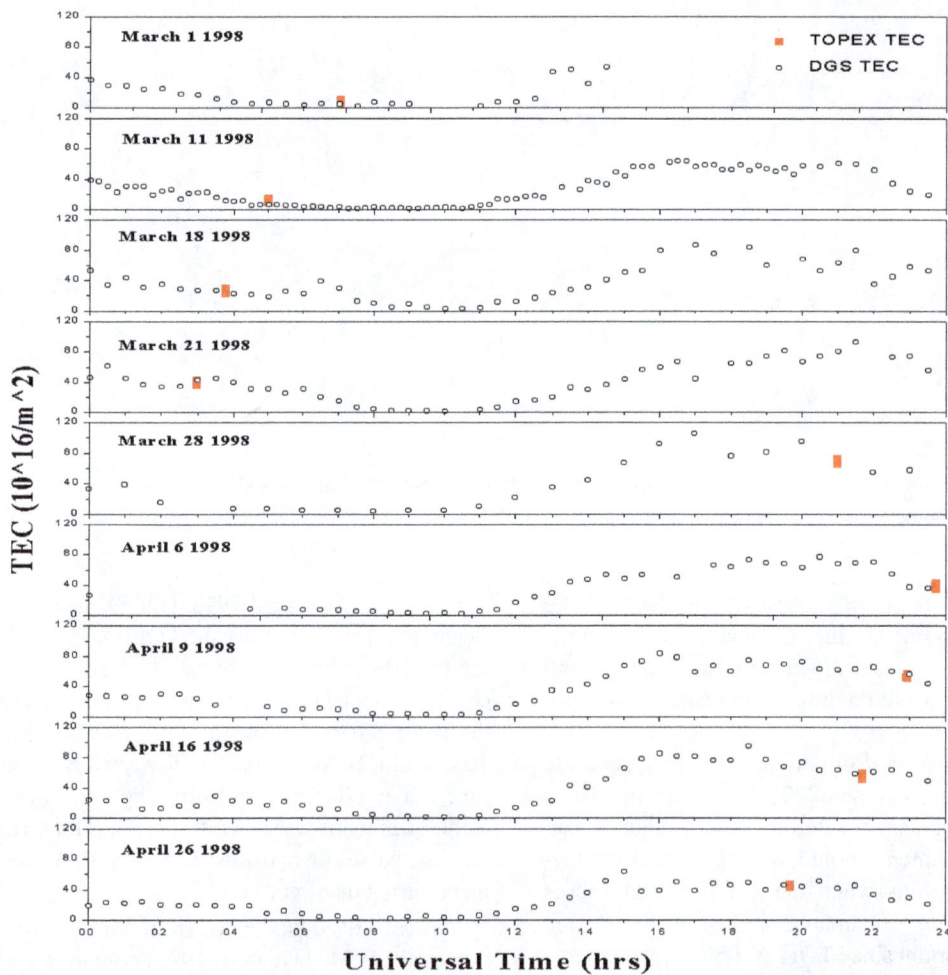

Fig. 4. Diurnal variation of Digisonde ITEC at Jicamarca compared with TOPEX TEC (red) results.

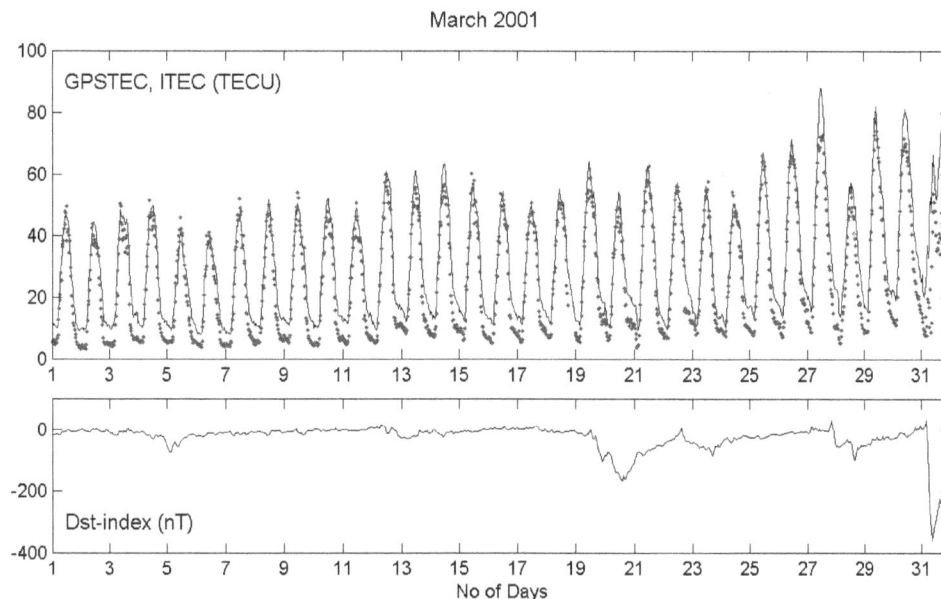

Fig. 5. Diurnal variation of GPSTEC (solid line) and ITEC (dotted line) at Athens, Greece, and the Dst-index (lower panel) during March 2001.

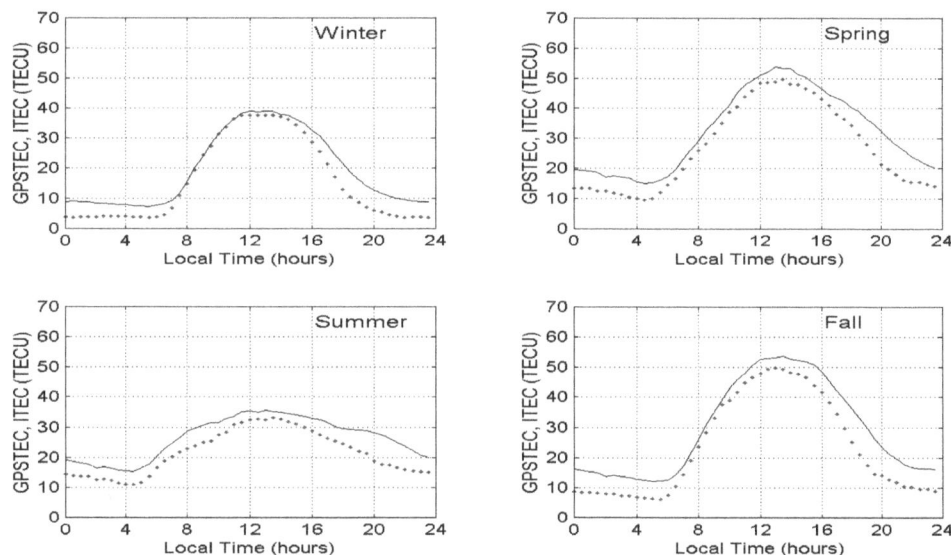

Fig. 6. Diurnal variation of GPSTEC (solid lines) and ITEC (dotted lines) for the four seasons, using the mean of the monthly median values for Athens, Greece, 2000/2001.

The diurnal variations in March 2001 again confirm the good correlation between the techniques, even during geomagnetic storms (Belehaki et al., 2004). The mean of the monthly median values for the four seasons (Fig. 6) clearly reveals the systematic difference between GPSTEC and ITEC, interpreted by Belehaki et al. (2004) as the diurnal and seasonal variations of the plasmaspheric contribution to GPSTEC.

3 Discussion

Evidence has been presented that the topside profiles at mid and low latitudes can be derived from the bottomside profiles measured by groundbased ionosondes. Only three characteristics are required, $NmF2$, $hmF2$, and H_T, all of them automatically determined with modern ionosondes. We therefore suggest that the scale heights H_T, routinely determined in the global Digisonde network and from other suitable ionosondes, be statistically analyzed and used as an input for the construction of the IRI topside electron density model.

Acknowledgement. The UML authors were supported by the AF Research Laboratory under Contract Number F19628-02-C-0092.

References

Belehaki, A., Jakowski, N., and Reinisch, B. W.: Comparison of ionospheric ionization measurements over Athens using ground ionosonde and GPS derived TEC values, Radio Sci., 38, 6, 1105, 2003.

Huang, X. and Reinisch, B. W.: Vertical total electron content from ionograms in real time, Radio Science, 36, 36, 2, 335–342, 2001.

Jakowski, N.: TEC Monitoring by Using Satellite Positioning Systems, Modern Ionospheric Science, (Eds. H. Kohl, R. Rüster, K. Schlegel), EGS, Katlenburg-Lindau, ProduServ GmbH Verlagsservice, Berlin, 371–390, 1996.

Lunt, N., Kersley, L., and Baily, G. J.: The influence of the protonosphere on GPS observations: Model simulations, Radio Sci., 34, 2, 725–732, 1999.

Reinisch, B. W. and Huang, X.: Deducing topside profiles and total electron content from bottomside ionograms, Adv. Space Res., 27, 1, 23–30, 2001.

Rishbeth, H. and O.K Garriott, *Introduction to ionospheric physics*, Academic Press, New York, 1969.

Sardon, E., Rius, A., and Zarraoa, N.: Estimation of the transmitter and receiver differential biases on the ionospheric total electron content from Global Positioning System observations, Radio Sci., 29, 577–586, 1994.

19

Enhanced sporadic E occurrence rates during the Geminid meteor showers 2006–2010

C. Jacobi[1]**, C. Arras**[2]**, and J. Wickert**[2]

[1]Institute for Meteorology, University of Leipzig, Stephanstr. 3, 04103 Leipzig, Germany
[2]German Research Centre for Geosciences GFZ, Potsdam, Department Geodesy & Remote Sensing, Telegrafenberg, 14473 Potsdam, Germany

Correspondence to: C. Jacobi (jacobi@uni-leipzig.de)

Abstract. Northern Hemisphere midlatitude sporadic E (E_s) layer occurrence rates derived from FORMOSAT-3/COSMIC GPS radio occultation (RO) measurements during the Geminid meteor showers 2006–2010 are compared with meteor rates obtained with the Collm (51.3° N, 13.0° E) VHF meteor radar. In most years, E_s rates increase after the shower, with a short delay of few days. This indicates a possible link between meteor influx and the production of metallic ions that may form E_s. There is an indication that the increase propagates downward, probably partly caused by tidal wind shear. However, the correlation between E_s rates and meteor flux varies from year to year. A strong correlation is found especially in 2009, while in 2010 E_s rates even decrease during the shower. This indicates that additional processes significantly influence E_s occurrence also during meteor showers. A possible effect of the semidiurnal tide is found. During years with weaker tidal wind shear, the correlation between E_s and meteor rates is even weaker.

1 Introduction

Sporadic E (E_s) layers are thin vertical regions of enhanced electron density in the lower ionosphere. Their origin is generally accepted to be vertical ion drift convergence driven by vertical shears in the horizontal tidal winds, with the long-living ions needed for the layers to be provided by meteors (Whitehead, 1960). There have been long discussions about how sporadic E layers are linked to meteor rates. The similarity of the seasonal cycles of both meteor rates and E_s occurrence rates or strength suggested a cause-and-effect explanation for the sporadic E layer seasonal dependence (Haldoupis et al., 2008).

However, not all features of the seasonal E_s cycle can be found in meteor rates, too. One reason may be that meteor radars, which are usually utilised to provide meteor rate seasonal cycles, only detect part of the incoming meteor flux. Another possible reason is that metallic ions are relatively long-lived and some details of short-term variability are thus not visible in E_s. Nevertheless, it is of interest whether short-period meteor events, especially meteor showers, may influence E_s rates.

The Geminids are a major meteor shower, which forms every year between December 4–17 with its peak activity on December 13 (at solar longitude $\lambda = 262°$). Its parent body is the asteroid 3200 Phaeton. Geminid shower meteors are relatively slow with a geocentric velocity of about 35 km/s (e.g., Stober et al., 2011a). Consequently, they burn at comparatively low altitudes and are thus well visible in the altitude range accessible to standard meteor radars (about 80–100 km). The Geminid meteor shower is the major shower visible in radio detections, while other showers are less well visible at least if the analysis is not focused on altitudes above about 100 km.

In this paper we present E_s occurrence rates detected by the GPS (Global Positioning System) radio occultation method using F3C (FORMOSAT-3/COSMIC, FORMOsa SATellite mission-3/Constellation Observing System for Meteorology, Ionosphere and Climate) data during the Geminid meteor showers 2006–2010, and compare these with meteor rates observed with VHF meteor radar. In Sects. 2 and 3 the methods are briefly introduced. In Sect. 4, we present time series of E_s and meteor count rates, which are discussed in Sect. 5. Section 6 concludes the paper.

2 Sporadic *E* analysis using FORMOSAT-3/COSMIC radio occultations

The F3C constellation was launched on 14 April 2006 and consists of six satellites (Anthes et al., 2008). The main scientific instrument aboard each satellite is a GPS receiver, which applies the GPS radio occultation technique (e.g., Kursinski et al., 1997) for vertical atmosphere sounding on a global scale.

To obtain information on the E_s occurrence we use signal-to-noise ratio (SNR) profiles of the 50 Hz GPS L1 signal below 120–140 km according to Arras et al. (2008). Strong changes in the vertical electron density gradients, as it is usual in presence of a sporadic *E* layer, appear as strong fluctuations in the SNR above about 85 km altitude. These disturbances are caused by signal divergence/convergence which leads to a decrease/increase of the signal intensity at the receiving antenna. The fluctuations are extracted from the background by applying a band pass filter which only accepts disturbances covering altitude intervals between 1.0 and 12.5 km. If the standard deviation of the SNR in a 2.5 km interval exceeds the threshold of 0.2, the disturbance in the SNR profile is regarded as significant. Since E_s are very thin layers, the standard deviation should rise abruptly. Consequently, a second criterion is introduced defining that the standard deviation has to rise suddenly by more than 0.14 between two adjacent intervals. In order to avoid using disturbances resulting from other effects than sporadic *E*, all profiles are excluded from further investigation, if the standard deviation exceeds the threshold of 0.2 in more than five intervals. The maximum deviation from the mean profile represents the approximate altitude of the sporadic *E* layer (e.g., Arras et al., 2008, 2009). Note that this method does not provide information about the strength of the E_s layer (amplitude of related electron density) but only on the occurrence rates in a given time and space interval.

GPS RO data are not uniformly distributed around the globe (see also Arras et al., 2009). In the left panel of Fig. 1, the total number of occultations per 5 degree latitude interval between 4 and 17 December, taken as the average of the years 2006–2010, are presented as black line. In the lower part of the panel, the occurrence rates, defined as the number of detected E_s divided by the number of occultations in a 5 km height and 5 degrees latitude gate, are presented. The majority of E_s is found in the summer hemisphere, but there is also considerable E_s activity at lower winter latitudes. The maximum number of E_s is found at altitudes slightly above 100 km. There is a tendency for lower altitudes in the winter hemisphere. In the right panel of Fig. 1, the 20–60° N mean occurrence rates per 1 km height interval are presented. Most E_s are found between 90 and 110 km. Note, however, that this result is partly due to the upper limit of the GPS RO profiles, which is set to around 120 km for F3C.

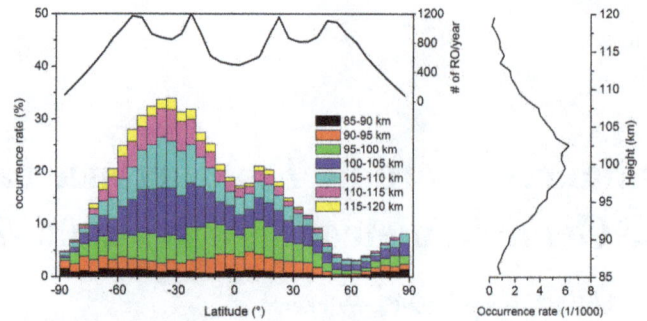

Fig. 1. Left panel: 2006–2010 mean sporadic *E* occurrence rates per 5 km height interval for different 5° latitude bands for the time interval 4–17 December of each year. Right panel: Mean occurrence rates per 1 km height interval for all latitudes between 20° N and 60° N and for the same time interval.

3 Collm meteor radar measurements

At Collm, Germany (51.3° N, 13.0° E), a SKiYMET all sky meteor radar is operated at 36.2 MHz since summer 2004. The antenna system consists of one 3-element Yagi transmitting antenna and five 2-element Yagi receiving antennas, forming an interferometer. Peak power is 6 kW. Pulse repetition frequency is 2144 Hz, but effectively only 536 Hz, due to 4-point coherent integration. The sampling resolution is 1.87 ms. The angular and range resolutions are $\sim 2°$ and 2 km, respectively. The pulse width is 13 μs, the receiver bandwidth is 50 kHz (see also Stober et al., 2011b).

We consider zenith angles between 0° and 70°, and distances of up to 400 km from the transmitter. Meteor count rates are taken every 2 h, and running 24 h means are calculated. In the following, we consider the Geminid meteor shower as one that is visible in each altitude interval, and which considerably influences meteor count rates. We analyse count rates at altitudes between 75 and 105 km. The radar is also used for wind measurements, using Doppler frequency shift of the reflected radio wave from meteor trails and minimising the squared difference between radial winds and the half-hourly horizontal winds projected on these. Mean winds and tidal parameters are calculated using least-squares fitting (e.g., Jacobi, 2011).

4 Results

As an example, in Fig. 2 the 24 h mean meteor count rates and E_s occurrence rates are shown for the year 2010. After the minimum in late winter/early spring there is an increase and maximum E_s and meteor rates in summer. This behaviour has led to the conclusion that the annual cycle of meteor rates is responsible for the seasonal cycle of E_s (Haldoupis et al., 2008). After midsummer however, the meteor rates remain fairly high until about November while E_s

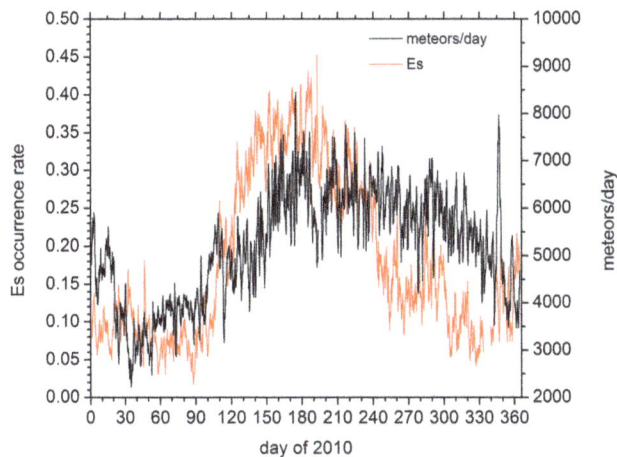

Fig. 2. 24 h mean Collm meteor count rates and 20–60° N mean sporadic E occurrence rates in 2010.

occurrences decrease. In December there is usually an increase in E_s occurrence, which is not present in all years.

Figure 2 shows that the seasonal cycle of meteor rates only partly explains the seasonal cycle of E_s. On shorter time scales, however, some peculiarities are found when E_s and meteor rates behave similarly. One of them is the maximum of E_s rates during the Geminid meteor showers in December. This may indicate that strong meteor showers could lead to enhanced E_s rates owing the increasing mass flux and thus ion production rate.

Two examples of E_s and meteor rates during the Geminid meteor showers in different years are presented in Fig. 3. We added the 2-hourly mean meteor rates multiplied by 12, to give an impression how the 24 h means are obtained. Meteor rates have a distinct diurnal cycle with maximum rates in the early morning. This may influence the trends of the 24 h means presented and definitely makes it difficult to detect the exact time of a meteor shower peak. The E_s rates are taken over all longitudes and thus do not show the diurnal cycle. In 2006, the E_s rates increase with some delay after the time of increasing meteor rates. Meteor rates after solar longitude $\lambda = 256°$ show a double-peak structure, which is also represented in E_s rates. In 2010, however, the picture is not that clear. E_s rates undergo an oscillation not very clearly linked to the Geminid shower. However, in most years an E_s increase is preceded by an increase in meteor rates with a time delay of 2 to 3 days, although there is no quantitative connection between the respective maxima.

On the left panel of Fig. 4, 5 yr averages of E_s occurrence and meteor rates are presented together with the standard error. Owing to the small number of years included, the error is partly large due to interannual variability. One can see that on average, E_s rates maximise about 2.5 days after the meteor rates. Note that there is an E_s maximum also shortly before the meteor rate maximum, however, this is preceded by a weak enhancement of meteor rates, too. In the right panel

of Fig. 4, the cross-correlation functions, taken from data of the days #335–355 of each year, are presented. One can see that the E_s-meteor rate correlation in respective years behave in different manners, but there is a tendency for the cross-correlation to maximise at a lag of few days, except for 2007, when the correlation is low at a lag of few days.

5 Discussion

Considering the standard error bars of the 5 year mean E_s rates, it can be concluded that the enhancement of E_s after the Geminid meteor shower is hardly significant. In part this can be due to the small number of years considered, but definitely there is considerable interannual variability of the E_s behaviour. As is the case in 2007, during some years E_s does not seem to be strongly influenced by the meteor shower, while in other years a rather strong correlation is found. Clearly, other influencing factors must play a role.

The wind shear theory predicts that at midlatitudes E_s are formed at the convergence nodes of vertical ion drift owing mainly to vertical shear of the zonal wind. Comprehensive overviews on this effect has been presented, e.g., by Haldoupis (2011, 2012). Generally, E_s layers form at altitudes of 120–130 km, which is at the upper limit or slightly above the region covered by the used RO. The main contribution to wind shear is by the semidiurnal tide (SDT), such that the SDT signature is clearly visible in E_s phases (e.g. Arras, 2009). Figure 5 presents SDT zonal wind amplitudes as measured at Collm during December 2006–2010. One can see that the amplitudes are smaller in 2007, 2008, and 2010 compared to 2006 and 2009. Comparing this with Fig. 4 reveals that these are the years when the cross-correlation function at lag up to about 5 days does not exceed values of 0.5, while in 2006 and 2009 larger values are found. Although the number of years considered is too small to draw more substantial conclusions, and the SDT amplitudes are only a proxy for the wind shear, this nevertheless indicates that SDT wind shear variability may modulate the E_s reaction on meteor showers.

It has to be noted that possible enhancement of E_s after meteor showers should be a rather indirect process. To date there is no proof that after the Geminid shower the concentration of metallic ions is really enhanced, and measurements showed inconclusive and partly contradicting results. For example, Dunker et al. (2013) found a decrease in sodium column abundance during the 2010 Geminid meteor shower, which is consistent with the decrease in E_s rates shown in the right panel of Fig. 3. One reason may be that the mass influx of the Geminids is small compared with the sporadic background (e.g. Ceplecha et al., 1998). The Collm radar is not sensitive to meteors above about 105 km, and the meteor rates provided here are therefore qualitative when they are used to describe the total meteor flux. Haldoupis et al. (2008) also pointed out that they only found a poor

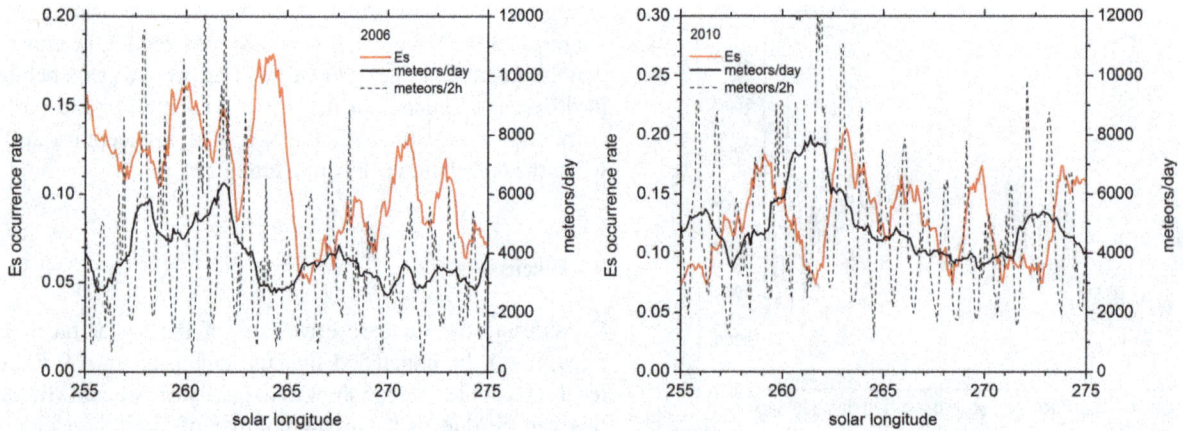

Fig. 3. Daily sporadic E layer occurrence rates in 2006 (left) and 2010 (right). Daily mean meteor rates are added, as well as 2-hourly meteor rates multiplied by 12.

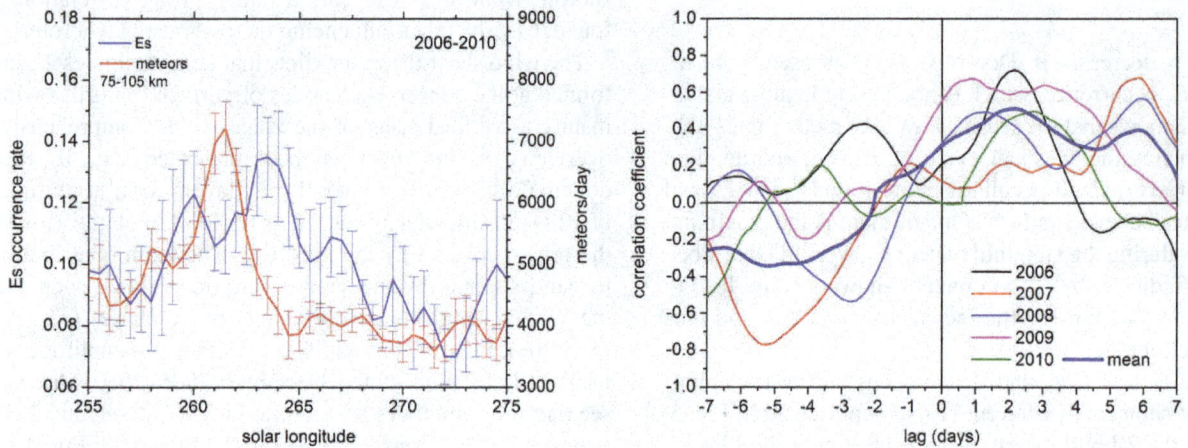

Fig. 4. Left panel: 2006–2010 mean sporadic E occurrence rates (blue) and 24 h mean meteor count rates (red). Standard errors are added. Right panel: cross-correlation functions between sporadic E occurrence rates and meteor count rates. Positive lag denotes meteor rate changes heading E_s ones.

correlation between the Geminid meteor rates and E_s critical frequencies. Moreover, other slow meteor showers like the Quadrantids in January do not seem to have considerable influence on the E_s rates (see Fig. 2).

Another open question refers to the time delay between meteor shower and E_s increase. To shed more light on this, we present in Fig. 6 cross-correlation functions between E_s occurrence rates and meteor count rates in 2009 at different altitudes. The curves are shifted by 0.2 against each other to visualize the height dependence. One can clearly see that there is a vertical shift of the delay of E_s rates with respect to the meteor rates, so that at lower altitudes the E_s rates increase later. This means that, although the Geminid meteors are found in every height accessible to the Collm radar, ions probably first form at larger altitudes and are then transported downwards. In total, this leads to a time delay of the overall E_s occurrence rates. However, the descent speed of E_s layers in winter is usually of the order of $2\,\mathrm{km\,h^{-1}}$ at altitudes above

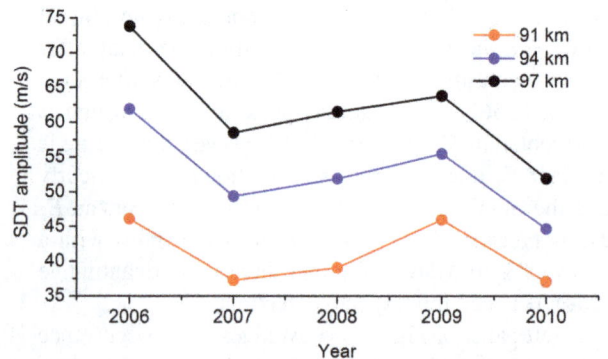

Fig. 5. December mean semidiurnal tidal zonal wind amplitudes over Collm.

about 100 km (Arras et al., 2009), so that there is another reason for delay of the layer formation. Generally, the E_s layer

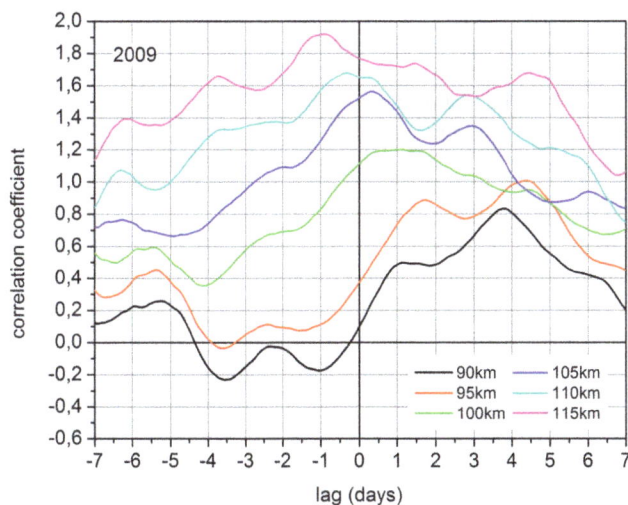

Fig. 6. Cross-correlation functions between sporadic E occurrence rates and meteor count rates in 2009 for different 10 km E_s height gates centered at the heights given in the legend. Positive lag denotes meteor rate changes heading E_s ones. The curves are shifted by 0.2 against each other to visualize the height dependence.

descent follows the phase speed of the SDT convergent node at altitudes well above about 100 km, but slows down owing to enhanced ion-neutral collisions and also since the diurnal tide with a shorter wavelength provides the downward transport (Haldoupis, 2012; Haldoupis et al., 2006; Christakis et al., 2009). Indeed, Arras et al. (2009, their Fig. 6) have found a delay of E_s phase downward progression with respect to SDT wind shear ones, but these gave rise to only few hours delay and would not explain a delay of more than one day. Other mechanisms, like a possible mean downward wind, could be the reason for the longer delay in some cases.

6 Conclusions

Our results indicate that there is a tendency for sporadic E layer occurrence rates to increase after the Geminid meteor shower. This increase is observed with an average time delay of 2.5 days. This effect is variably strong pronounced in different years, in particular during 2007 it appeared to be quite weak. Comparison with SDT amplitudes indicates a possible relationship with changing wind shear magnitudes.

Taking into account the small number of years considered so far and the comparatively large error bars of the mean effect, conclusions should be drawn with care. This is also the case, since supporting evidence of increasing metal concentration after meteor showers is not available. Further experimental and modelling studies are required to substantiate the results. For example, ionosonde measurements will be helpful to increase the upper boundary of the observational data base and will shed more light on the layer formation and downward propagation effect.

Acknowledgements. This study was partly supported by Deutsche Forschungsgemeinschaft under JA 836/22-1. We acknowledge UCAR (Boulder, US) and NSPO (Taiwan) for the provision of FORMOSAT-3/COSMIC data and related support. Solar longitudes have been calculated based on those provided by the International Meteor Organisation (IMO) on http://www.imo.net/data/solar.

References

Anthes, R. A., Bernhardt, P. A., Chen, Y., Cucurull, L., Dymond, K., F., Ector, D., Healy, S. B., Ho, S.-P., Hunt, D. C., Kuo, Y.-H., Liu, H., Manning, K., McCormick, C., Meehan, T. K., Randel, W. J., Rocken, C., Schreiner, W. S., Sokolovskiy, S. V., Syndergaard, S., Thompson, D. C., Trenberth, K. E., Wee, T.-K., Yen, N. L., and Zeng, Z.: The COSMIC/FORMOSAT-3 Mission: Early Results, B. Am. Meteorol. Soc., 83, 313–333, doi:10.1175/BAMS-89-3-313, 2008.

Arras, C., Wickert, J., Beyerle, G., Heise, S., Schmidt, T., and Jacobi, Ch.: A global climatology of ionospheric irregularities derived from GPS radio occultation, Geophys. Res. Lett., 35, L14809, doi:10.1029/2008GL034158, 2008.

Arras, C., Jacobi, Ch., and Wickert, J.: Semidiurnal tidal signature in sporadic E occurrence rates derived from GPS radio occultation measurements at midlatitudes. Ann. Geophys., 27, 2555–2563, doi:10.5194/angeo-27-2555-2009, 2009.

Ceplecha, Z., Borovicka, J., Elford, W. G., Revelle, D. O., Hawkes, R. L., Porubcan, V., and Simek, M.: Meteor phenomena and bodies, Space Sci. Rev., 84, 327–471, 1998.

Christakis, N., Haldoupis, C., Zhou, Q., and Meek, C.: Seasonal variability and descent of mid-latitude sporadic E layers at Arecibo, Ann. Geophys., 27, 923–931, doi:10.5194/angeo-27-923-2009, 2009.

Dunker, T., Hoppe, U.-P., Stober, G., and Rapp, M.: Development of the mesospheric Na layer at 69° N during the Geminids meteor shower 2010, Ann. Geophys., 31, 61–73, 2013, http://www.ann-geophys.net/31/61/2013/.

Haldoupis, C.: A tutorial review on sporadic E layers, in Aeronomy of the Earth's Atmosphere and Ionosphere. IAGA Special Sopron Book Series 2, 381–394, doi:10.1007/978-94-007-0326-1-29, 2011.

Haldoupis, C.: Midlatitude sporadic E. A typical paradigm of atmosphere-ionosphere coupling, Space Sci. Rev., 168, 441–461, doi:10.1007/s11214-011-9786-8, 2012.

Haldoupis, C., Meek, C., Christakis, N., Pancheva, D., and Bourdillon, A.: Ionogram height-time-intensity observations of decending sporadic E layers at mid-latitudes, J. Atmos. Solar Terr. Phys., 68, 539–557, doi:10.1016/j.jastp.2005.03.020, 2006.

Haldoupis, C., Pancheva, D., Singer, W., Meek, C., and Mac-Dougall, J.: An explanation for the seasonal dependence of midlatitude sporadic E layers, J. Geophys. Res., 112, A06315, doi:10.1029/2007JA012322, 2008.

Jacobi, Ch.: Meteor radar measurements of mean winds and tides over Collm (51.3° N, 13° E) and comparison with LF drift measurements 2005–2007, Adv. Radio Sci., 9, 335–341, doi:10.5194/ars-9-335-2011, 2011.

Kursinski, E. R., Hajj, G. A., Schofield, J. T., Linfield, R. P., and Hardy, K. R.: Observing earth's atmosphere with radio occultation measurements using the global positioning system, J. Geophys. Res., 102, 23429–23465, 1997.

Stober, G., Jacobi, Ch., and Singer, W.: Meteoroid mass determination from underdense trails. J. Atmos. Solar-Terr. Phys., 73, 895–900, doi:10.1016/j.jastp.2010.06.009, 2011a.

Stober, G., Singer, W., and Jacobi, Ch.: Cosmic radio noise observations using a mid-latitude meteor radar, J. Atmos. Solar-Terr. Phys., 73, 1069–1076, doi:10.1016/j.jastp.2010.07.018, 2011b.

Whitehead, J. D.: Formation of the sporadic E layer in the temperate zones, Nature, 188, 567–567, 1960.

20

35 years of International Reference Ionosphere – Karl Rawer's legacy

D. Bilitza

Raytheon ITSS/SSDOO, GSFC, Code 632, Greenbelt, MD 20771, USA

Abstract. This presentation is given in honor of Prof. Karl Rawer's 90th birthday. It looks back at 35 years of research and development in the framework of the International Reference Ionosphere (IRI) project. K. Rawer initiated this international modeling effort and was the first Chairman of the IRI Working Group. IRI is a joint project of the Committee on Space Research (COSPAR) and the International Union of Radio Science (URSI) whose goal it is to establish an international standard model for the ionospheric densities, temperatures and drifts. This year we are celebrating Karl Rawer's 90th birthday and also the 35-year anniversary of the IRI effort. My talk will review the close involvement of Karl Rawer in all stages of the development and improvement of this international standard from early on and his still very active participation in this effort.

1 A brief history of Rawer's IRI

In the mid-sixties it became clear that an international standard model for the ionosphere was needed similar to the successfully established COSPAR International Reference Atmosphere (CIRA) for the thermospheric parameters (CIRA, 1961). Such models are required for the specifications of environmental parameters in the thermosphere and ionosphere for the design of space-based instrument, for satellite orbit determination and control, for analysis of radioastronomy data and satellite altimetry data, and many more applications. Foreseeing the need for such a model COSPAR in 1968 initiated the International Reference Ionosphere (IRI) project and asked Karl Rawer to become its first Chairman. Since than Karl Rawer has been closely involved with the IRI effort and has been the main reason for the great success and broad application of this international standard representation of the ionosphere. It is therefore quire appropriate that we are celebrating at this occasion Karl Rawer's 90th birthday as well

as IRI's 35th anniversary. Table 1 lists all the major IRI milestones throughout these past 35 years and highlights Karl Rawer's involvement. It also shows that the IRI group under the guidance of Karl Rawer has always tried to keep up with the rapidly changing computer and network environment, to be able to provide the user community with fast and easy access to the IRI model and its parameters.

It is interesting to note that the original IRI charter had only asked "...to provide vertical profiles of the main ionospheric parameters for suitable chosen locations, hours, seasons, and levels of solar activity; representing monthly median conditions based on experimental evidence." The IRI group, of course, quickly moved past the limitations in time and space and its first major release (Rawer et al., 1978) already included global coverage for the electron density. COSPAR's Terms of Reference for the IRI project now states: "The Task Group was established to develop and improve a standard model of the ionospheric plasma parameters. The model should be primarily based on experimental evidence using all available ground and space data sources; theoretical considerations can be helpful in bridging data gaps and for internal consistency checks. Where discrepancies exist between different data sources the IRI team should promote critical discussion to establish the reliability of the different databases. IRI should be updated as new data become available and as old data sources are fully evaluated and exploited. IRI is a joint working group of COSPAR and URSI. COSPAR's prime interest is in a general description of the ionosphere as part of the terrestrial environment for the evaluation of environmental effects on spacecraft and experiments in space. URSI's prime interest is in the electron density part of IRI for defining the background ionosphere for radiowave propagation studies and applications.

Correspondence to: D. Bilitza
(bilitza@gsfc.nasa.gov)

2   IRI members and meetings

K. Rawer used four very important ingredients in establishing the IRI success story:

1. Working group members who provided a good and balanced cross-section in terms of the representation of different countries and continents (see Fig. 1) as well as in terms of the representation of different measurement techniques. This turned out to be great asset in getting the IRI effort access to all essential ground and space data sets for ionospheric parameters.

2. Annual IRI Workshops (see Table 2) that became the many venue for discussing improvements and enhancements of the model and that became the catalyst for a multitude of international collaborations whose goal it was to improve specific aspects of the model. A trademark of this quite informal meetings was (and is) the "Final Discussion" session during which the IRI team decides on the improvements and additions to be included in the next version of the model and during which "volunteers" are enlisted to investigate new data sources and specific modeling questions for future model updates.

3. Publication of the workshop papers first in Space Research, and then in Advances in Space Research (Table 2) has resulted in an excellent record of the IRI activity and has produced a unique series documenting the international efforts in ionospheric modeling. As the main editor of these publications K. Rawer has helped countless non-English speaking colleagues to correct their language and grammar and to publish their important results sometimes for the first time in an English-language journal.

4. Newest technology to make the IRI model quickly and easily accessible to its wide user community and its broad spectrum of applications. IRI was one of the first international projects to distribute its model in the form of computer codes in a number of different computer languages. Following closely in step with the computer revolution of the eighties and nineties, the programs were first provided on 9-track tapes, than on punched tapes and cards, than on floppy disks, than online, and the latest incarnation is the IRIWeb, an interface that lets user compute and plot IRI parameters online (http://nssdc.gsfc.nasa.gov/space/models/iri.html).

3   Applications

The great legacy of Karl Rawer's push for an International Reference Ionosphere is best documented by the many applications that depend on this model:

a) Standard for Engineering Applications

- IRI is used as the standard in NASA Guidelines (Anderson, 1994).

- IRI is the standard ionospheric model in the System Engineering Handbook of the European Cooperation for Space Standardization (ECSS, 1997).

- IRI is under consideration to become the ISO standard for the ionospheric parameters.

- IRI is the ionospheric model used in ESA/ESTEC's Space Environment Information System (SPENVIS) http://www.spenvis.oma.be/spenvis/ and in MSFC's Space Environment Effects (SEE) web interface.

b) Visualization Tool for Educational Applications

- IRI Total Electron Content (TEC) world maps (U. Leicester, U. K.) http://ion.le.ac.uk/remote_sensing/models/tec.html

- 3-D electron density visualization using AVS (Watari et al., 2003)

c) Ionospheric Correction for Single-Frequency Satellite Altimeters

- Longtime data record of sea surface heights (Pathfinder Project); updating IRI with ionosonde data (Bilitza et al., 1997)

- ERS Quick-look data (ESA)

- Work with Geosat Follow On (GFO) data (Zhao et al., 2002)

d) Background ionosphere for Evaluation of Data Mapping Techniques

- Testing algorithms that convert GPS measurements into global TEC maps (Hernandez-Pajares et al., 2002)

- TEC from NNSS Doppler measurements (Ciraolo and Spalla, 2002)

- Reliability of tomographic methods.

4   Conclusion

On behalf of the IRI Working Group and the many IRI users, I would like to say, "Thank you Karl" and congratulations on your 90th Birthday and on the 35th Birthday of your IRI model.

Table 1. IRI Milestones.

Year	Event	Description	Media	Image
1968	COSPAR establishes IRI WG	K. Rawer, Chair		
1969	URSI joins in			
1972	Prelimenary set of Tables (Rawer and Ramakrishnan)	IRI parameters at selected locations	Report	
1973	COSPAR Symposium, Konstanz, Germany (K. Rawer, Organizer) [2]	Guidelines established for data that should be used for D-region modeling		
1978	URSI Special Report: IRI-79 [3]	Global coverage for densities, CCIR Maps for foE, foF1, foF2, and M(3000)F2	Report ALGOL and FORTRAN code on punched tape and cards	
1981	World Data Center A for Solar-Terrestrial Physics Report: IRI-79 [4]		Graphs and tables of IRI parameters	
1986	IRI-86 on floppy disk for use on Personal Computers (PCs)	Global coverage for temperatures based on AE-C,-D,-E and AEROS-A,-B data	Floppy disk with DOS interactive program	
1990	National Space Science Data Center (NSSDC) Report [5]	URSI maps for foF2	Retrievable from NSSDC's archive via Anonymous ftp and available for online computation as part of NSSDC's Online Data and Infromation Service (NODIS)	
1995	IRI-95 online (IRIWeb)	Improvements at low magnetic latitudes	IRIWeb: compute and plot IRI parameters on the internet.	
1999	URSI Resolution	IRI recognized as the international standard for the ionosphere		
2001	IRI-2001 with many improvements and new parameters [6]	Improvements: D- F1-region, STORM and Intercosmos Te model New paramts.: F1prob., equat. vert. ion drift		**IRI-2001**

Table 2. IRI Workshops and Publications.

Year	Location	Topic	Publication
1971[†]	Seattle, USA		Space Res. XII, 1229-1335, 1972
1973	Konstanz, FRG	Measurements and Results of Lower Ionosphere	Akademie-Verlag, Berlin, 1974
1974[†]	Sao Paulo, Brazil		Space Res. XV, 295- 334, 1975
1980[†]	Budapest, Hungary	International Reference Ionosphere – IRI-79	WDC-A-STP, UAG-90, 1984
1982[†]	Ottawa, Canada	The Upper Atmosphere of the Earth and Planets	Adv. Space Res. (ASR) 2(10) 1982
1983	Stara Zagora, Bulgaria	Towards an Improved IRI	ASR 4(1) 1984
1984[†]	Graz, Austria	Models of the Atmosphere and Ionosphere	ASR 5(7) 1985
1985	Louvain, Belgium	IRI - Status 1985/86	ASR 5(10) 1985
1986[†]	Toulouse, France	IRI - Status 1986/87	ASR 7(6) 1987
1987	Novgorod, Russia	Ionospheric Informatics	ASR 8(4) 1988
1988[†]	Espoo, Finland	Ionospheric Informatics and Empirical Modelling	ASR 10(8) 1990
1989	Abingdon, UK	Development of IRI-90	ASR 10(11) 1990
1990[†]	The Hague, Netherlands	Enlarged Space and Ground Data Base for	ASR 11(10) 1991
1991	Athens, Greece	Adv. in Global/Reg. Descript. of Ionospheric Parameter	ASR 12(7) 1992
1992[†]	Washington, DC, USA	Ionospheric Models	ASR 13(3) 1993
1993	Trieste, Italy	Off Median Phenomena and IRI	ASR 14(12) 1994
1994[†]	Hamburg, FRG	The High Latitudes in the IRI	ASR 16(1) 1995
1995	New Dehli, India	Low and Equat. Latitudes in IRI	ASR 18(6) 1996
1996[†]	Birmingham, UK	Descript. of Ionospheric Storm Effects and Irregularities	ASR 20(9) 1997
1997	Kühlungsborn, Germany	New Develops. in Ionospheric Modeling and Prediction	ASR 22(6) 1998
1998[†]	Nagoya, Japan	Lower Ionosphere: Measurements and Models	ASR 25(1) 2000
1999	Lowell, MA, USA	IRI- Workshop 1999	ASR 27(1) 2001
2000[†]	Warsaw, Poland	Modelling the Topside Ionosphere and Plasmasphere	ASR 29(6) 2002
2001	Sao Jose dos Campos, Brazil	Description of the Low Latitude Ionosphere in the IRI	ASR 31(3) 2003
2002[†]	Houston, Texas, USA	Improved Ionosphere Specification and Forecast	ASR in press
2003	Grahamstown, South Africa	Quantifying ionospheric variability	ASR in preparation

[†] IRI session during the General Assembly of the Committee on Space Research.

P. Bradley, M. Rycroft (U.K.),
K. Rawer, W. Singer (Germany),
A. Alcayde, R. Hanbaba (France),
B. Zolesi, S. Radicella (Italy),
M. Friedrich (Austria),
E. Kopp (Switzerland),

L. Triskova, V. Truhlik (Czech Rep)
I. Kutiev (Bulgaria)
I. Stanislawska (Poland),
S. Kouris (Greece)

A. Danilov
V. K. Depuev
T. Gulyaeva
G. Ivanov-Kholodny
E. Kazimirovsky
A. Mikhailov
S. Pulinets

B. Reinisch
D. Bilitza
T. Fuller-Rowell
K. Bibl
X. Huang
J. Sojka
D. Anderson
V. Wickwar

K. Oyama
K. Igarashi
S. Watanabe

K. Mahajan
A.P. Mitra
S.P. Gupta

M. Abdu

R. Ezquer
M. Mosert de Gonzalez

J. Adeniyi   A. Poole, L.-A. McKinnell

P. Wilkinson
P. Tyson
B. Ward

IRI Working Group Members

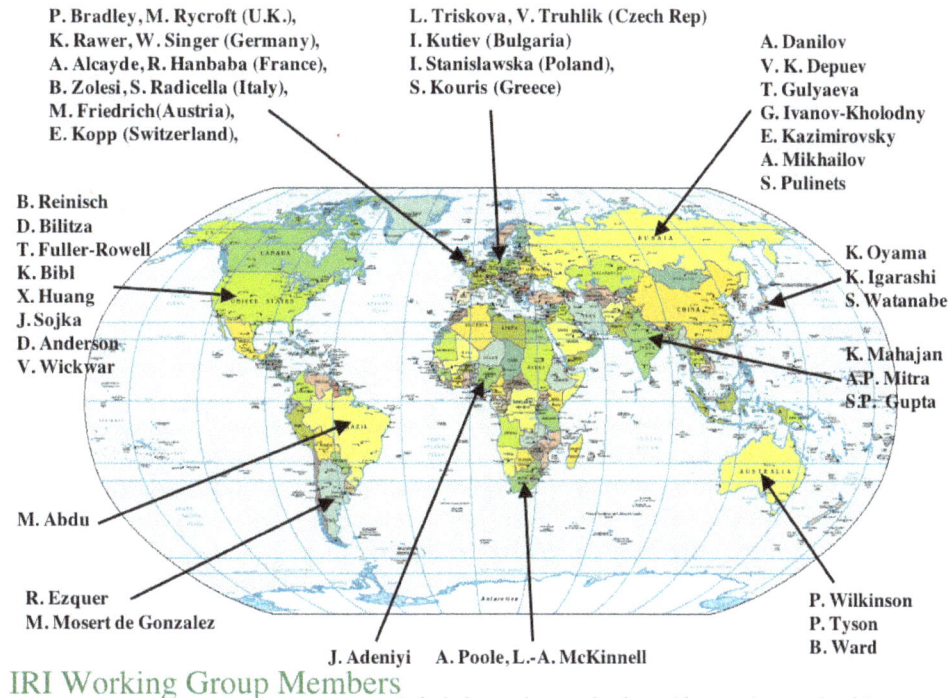

Fig. 1. Global distribution of members of the URSI/COSPAR Working Group on the International Reference Ionosphere.

References

Anderson, B. J. (Ed.): Natural orbital environment definition for use in aerospace vehicle development, NASA Tech Memo, NASA-TM-4527, 1994.

Bilitza, D.: International Reference Ionosphere 1990, National Space Science Data Center, Report 90-22, Greenbelt, Maryland, USA, 1990.

Bilitza, D., Bhardwaj, S., and Koblinsky, C.: Improved IRI predictions for the GEOSAT time period, Adv. Space. Res. 20, 9, 1755–1760, 1997.

Bilitza, D.: International Reference Ionosphere 2000, Radio Sci., 36, 261–275, 2001.

CIRA, COSPAR International Reference Ionosphere, North-Holland Publications, Amsterdam, The Netherlands, 1961.

Ciraolo, L. and Spalla, P.: TEC analysis of IRI simulated data, Adv. Space Res., 29, 6, 959–966, 2002.

ECSS, European Cooperation for Space Standardization, System Engineering: Space Environment, Handbook, Noordwijk, The Netherlands, 1997.

Hernandez-Pajares, M., Juan, J., Sanz, J., and Bilitza, D.: Combining GPS measurements and IRI model values for Space Weather specification, Adv. Space Res. 29, 6, 949–958, 2002.

Rawer, K. (Ed): Methods and Measurements and Results of Lower Ionosphere Structure, Akademie-Verlag, Berlin, GDR, 1974.

Rawer, K., Ramakrishnan, S., and Bilitza, D.: International Reference Ionosphere 1978, URSI, Brussels, 1978.

Rawer, K., Lincoln, J., and Conkright, R. (Eds): International Reference Ionosphere, IRI-79, Report UAG-82, World Data Center A for Solar-Terrestrial Physics, boulder, Colorado, USA, 1981.

Watari, S., Iwamoto, I., Igarashi, K., Isogai, M., and Arakawa, Y.: 3-D visualization of the IRI model, Adv. Space Res., 31, 3, 781–784, 2003.

Zhao, C., Bilitza, D., Shum, C., Schaer, S., Beutler, G., and Ge, S.: Evaluation of IRI95 ionosphere model for radar altimeter applications, Adv. Space Res. 29, 6, 967–976, 2002.

Meteor radar measurements of mean winds and tides over Collm (51.3° N, 13° E) and comparison with LF drift measurements 2005–2007

C. Jacobi

Institute for Meteorology, University of Leipzig, Stephanstr. 3, 04103 Leipzig, Germany

Abstract. An all-sky VHF meteor radar (MR) has been continuously operated at Collm (51.3° N, 13° E) since summer 2004. The radar measures meteor parameters, diffusion coefficients, and horizontal winds in the mesopause region. There exists a temporal overlap of the MR wind measurements with co-located low-frequency (LF) ionospheric drift measurements until 2007. Comparison of MR and LF semidiurnal tidal phases allows to empirically determine the virtual height overestimation of LF reflection heights due to the group retardation of LF waves. LF reference heights have to be reduced by up to 20 km to match real heights. Correction of LF heights for group retardation allows to determine the wind underestimation by the LF method compared with meteor radar measurements and opens the possibility to continue long-term trend analysis using mesosphere/lower thermosphere winds.

1 Introduction

As background information for linear models, for validation purposes, and for estimation of further derived parameters there is a need for empirical models of wind parameters such as prevailing winds and tidal parameters in the mesosphere/lower thermosphere (MLT) region. The MLT is of special interest because its structure is mainly dynamically driven and represents to a certain degree the boundary between middle and upper atmosphere. To accomplish global empirical models, there is a need to construct and updates models, e.g., from radar measurements at single sites.

Operational radars used for MLT wind measurements in general either use the Doppler shift of the reflected radio wave, or apply the spaced receiver (D1) method. While the former is the usual method for meteor radars (MR), the latter has been used for the conventional analysis of medium frequency (MF) radars as well as for the low-frequency (LF) method applied, e.g., with earlier measurements over Collm, Germany. MR and D1 wind comparisons have provided hints to systematic differences between the results of the two methods (Hocking and Thayaparan, 1997; Manson et al., 2004; Hall et al., 2005). Jacobi et al. (2009) compared LF, MR and MF winds of one year, and found that MF and LF winds are smaller than MR ones and the differences increase with height. Literature results in all cases indicate that generally winds measured using MF and LF are smaller than those from MR, while the reasons and details of these differences are still under discussion.

Over Collm, Germany, mesopause region wind have been measured since the late 1950s using the LF D1 method (referring to 52° N, 15° E), and climatologies of the prevailing wind and the semidiurnal tide (SDT) have been constructed from these data (e.g., Kürschner and Jacobi, 2005). The data have also been used for the analysis of long-tem trends (Jacobi and Kürschner, 2006). In 2004, these measurements have been replaced by a VHF meteor radar, with some years of overlapping data. This opens the possibility to analyse MR winds together with LF winds in nearly the same volume during the years 2005–2007. This provides a climatology of MR winds together with a long-term comparison of MR and LF with correction terms for the LF wind underestimation, in order to more correctly interpreting earlier climatologies from the LF wind measurements and to possibly extending long-term MLT wind time series for trend analysis.

Correspondence to: C. Jacobi
(jacobi@uni-leipzig.de)

2   Measurements

2.1   Collm meteor radar

A VHF all-sky MR is operated at Collm (51.3° N, 13° E) on 36.2 MHz since summer 2004. Pulse repetition frequency is 2144 Hz, but is effectively reduced to 536 Hz, due to 4-point coherent integration. Power is 6 kW and a 3-element Yagi is used as transmitting antenna. The sampling resolution is 1.87 ms. The pulse width is 13 μs and the receiver bandwidth is 50 kHz. The angular and range resolutions are $\sim 2°$ and 2 km, respectively. The wind measurement principle is the detection of the Doppler shift of the reflected radio waves from ionised meteor trails, which delivers radial wind velocity along the line of sight of the radio wave. An interferometer, consisting of five 2-element Yagi antennas arranged as an asymmetric cross is used to detect azimuth and elevation angle from phase comparisons of individual receiver antenna pairs. Together with range measurements this enables the meteor trail position detection. The raw data collected consist of azimuth and elevation angle, wind velocity along the line of sight, and meteor height. The data collection procedure is also described in detail by Hocking et al. (2001).

The individual meteor trail reflection heights roughly vary between 75 and 110 km, with a maximum around 90 km (Stober et al., 2008). In standard configuration, the data are binned in 6 different not overlapping height gates centred at 82, 85, 88, 91, 94, and 98 km. Individual radial winds calculated from the meteors are collected to form half-hourly mean values using a least squares fit of the horizontal wind components to the raw data under the assumption that vertical winds are small (Hocking et al., 2001). An outlier rejection is added.

2.2   Collm LF lower ionospheric drifts

At Collm, MLT winds have also been obtained by D1 LF radio wind measurements from 1959-2008, using the ionospherically reflected sky wave of three commercial radio transmitters. The data were combined to half-hourly zonal and meridional mean wind values. The virtual reflection heights, varying between about 80 and 120 km, have been estimated between late 1982 and 2007 using measured travel time differences between the separately received ionospherically reflected sky wave and the ground wave (Kürschner et al., 1987). These heights may drastically overestimate the real heights owing to group retardation of the LF radio wave. The LF reference heights vary systematically in the course of one day (e.g. Kürschner et al., 1987). In addition, due to increased absorption during daylight hours, regular daily data gaps are present, which are particularly long in summer. Therefore, obtaining the means through simple averaging of data of one height gate is not possible, and a special regression analysis based on Groves (1959) is necessary (Kürschner and Schminder, 1986).

2.3   Prevailing wind and tidal analysis

For the estimation of LF monthly mean wind parameters, a multiple regression analysis with quadratically height-dependent coefficients is used to determine the monthly mean prevailing wind and the semidiurnal tidal wind from the half-hourly wind components (Kürschner and Schminder, 1986; Jacobi et al., 1999; Kürschner and Jacobi, 2005), i.e. modeling the winds at time t as:

$$v_z(t) = \sum_{k=0}^{p}\left\{ h^k a_{k,z} + b_{k,z}h^k \sin\omega t + c_{k,z}h^k \cos\omega t \right\} + \varepsilon, \quad (1)$$

$$v_m(t) = \sum_{k=0}^{p}\left\{ h^k a_{k,m} + b_{k,m}h^k \sin\omega t + c_{k,m}h^k \cos\omega t \right\} + \varepsilon, \quad (2)$$

with $p = 2$ for a quadratic height dependence, $v_z(t)$ and $v_m(t)$ as the zonal and meridional measured half-hourly winds, h as the (virtual) height and $\omega = 2\pi/12$ h as the frequency of the SDT. Then, the zonal (v_{0z}) and meridional (v_{0m}) mean winds are estimated as

$$v_{0z}(h) = \sum_{k=0}^{p}a_{k,z}h^k, \quad v_{0m}(h) = \sum_{k=0}^{p}a_{k,m}h^k, \quad (3)$$

and the estimates of the amplitudes v_{2z} and zonal phases T_{2z} of the SDT are:

$$v_{2z}(h) = \sqrt{\left(\sum_{k=0}^{p}b_{k,z}h^k\right)^2 + \left(\sum_{k=0}^{p}c_{k,z}h^k\right)^2}, \quad (4)$$

$$T_{2z}(h) = \frac{1}{\omega}\text{atan}\left\{ \frac{\sum_{k=0}^{p}b_{k,z}h^k}{\sum_{k=0}^{p}c_{k,z}h^k} \right\}. \quad (5)$$

Meridional amplitudes v_{2m} and phases T_{2m} are calculated accordingly. In order to improve the separation and the spectral selectivity of the evaluation of the tidal components in the presence of data gaps, circular polarization of the SDT horizontal components is assumed, thus

$$b_{k,m} = -c_{k,z}, \quad \text{and} \quad c_{k,m} = b_{k,z}, \quad (6)$$

(Kürschner, 1991), which of course results in $v_{2m} = v_{2z}$ and $T_{2m} = T_{2z}$-3 h. This assumption is justified at higher midlatitudes (Jacobi et at., 1999).

Prevailing winds from the MR are calculated accordingly, however, since there are no regular daytime data gaps, regression with height-dependent coefficients is not necessary and Eqs. (1–6) are applied with $p = 0$ for each height gate separately. Note also that assumption of circular polarization of the SDT components is not necessary for the MR data analysis and in addition to the SDT also diurnal and terdiurnal components can be estimated. In order to ensure comparability of the MR and LF winds, circular polarisation was assumed and thus Eq. (6) was applied also to the MR data.

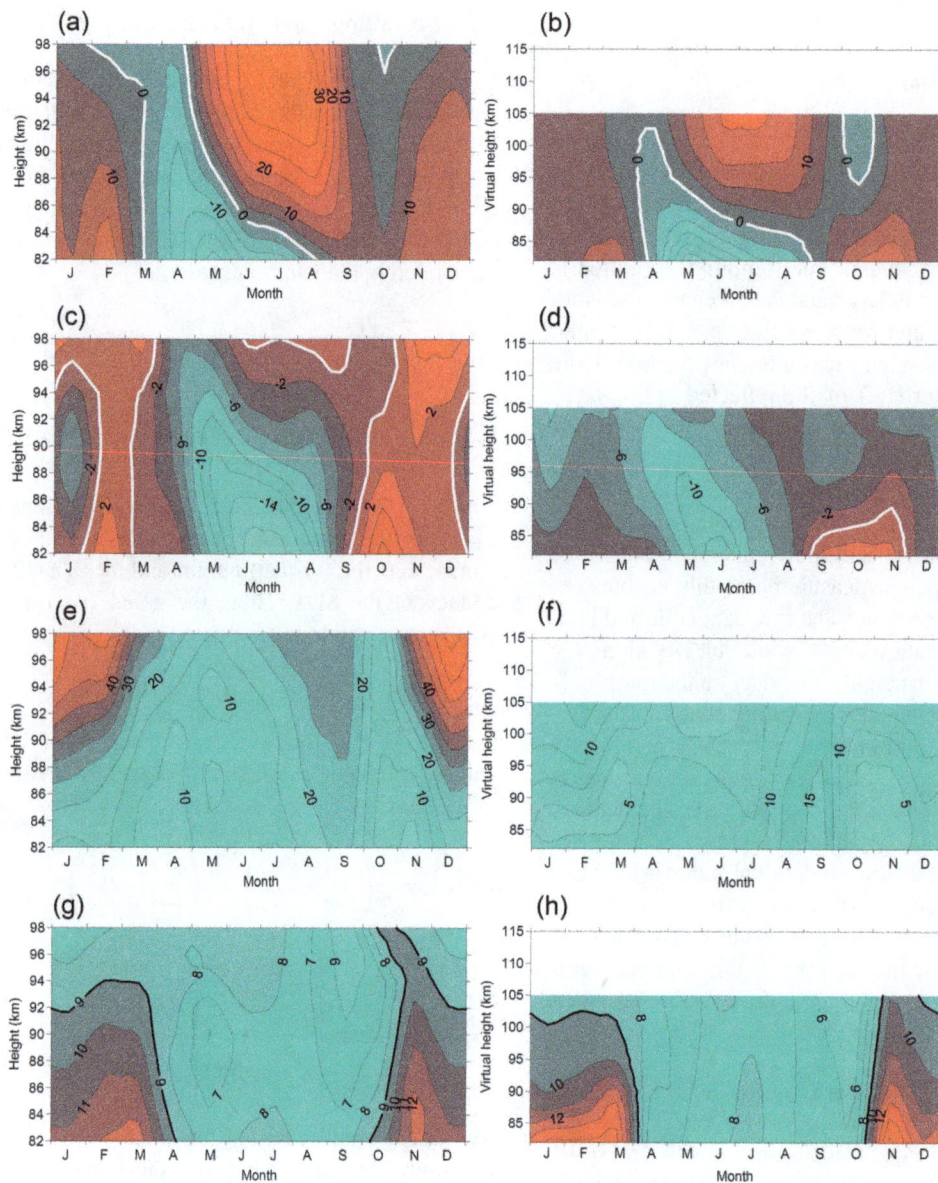

Fig. 1. MLT wind parameters as measured with MR (left) and LF (right). Parameters are zonal prevailing wind (**a**), (**b**), meridional prevailing wind (**c**), (**d**), SDT amplitude (**e**), (**f**), and zonal SDT phase (**g**), (**h**). Values are given in m/s, except for the phases given in LT. Height-time cross-sections are based on monthly mean winds and tidal parameters.

3   2004–2007 mean winds measured by meteor radar and LF

3-year mean monthly mean prevailing winds, SDT amplitudes and zonal phases are shown in Fig. 1. MR winds are presented on the left, LF winds on the right hand side. Note that the height scaling of the LF winds differs from the MR one in order to achieve approximate correspondence of the wind systems. The gross features of both prevailing wind and SDT seasonal cycle are similar with both measurements, although some details differ as, e.g., southerly meridional winds in winter (except for January) measured

by MR throughout the height range considered, while the LF measurements show southward winds. Differences are also seen in the summer SDT; while LF amplitudes decrease with height, the MR ones partly increase with altitude. The summer mean winds at the lowermost height gates correspond well between LF and MR, however, due to the short length of the night during summer, only few data are available by LF during summer and then the summer wind parameters there are essentially extrapolated from upper heights and should be interpreted with care.

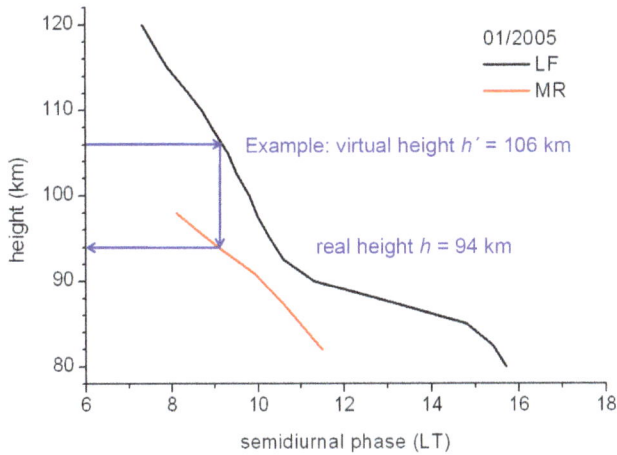

Fig. 2. Example of SDT vertical phase profiles observed by MR and LF. The blue arrows demonstrate the method of real height estimation.

The most striking feature is the strong underestimation of LF tidal amplitudes and, less strongly expressed, zonal prevailing winds in summer. This is clearly an effect of the spaced receiver bias. To quantify this effect, however, a height correction has to be applied first. This correction is not possible on a physical basis, since the actual D region electron density profile is not known. Therefore an empirical statistical approach is used, which is described in the following Sect. 4.

4   LF virtual height correction

Owing to the combined effect of wind underestimation and height overestimation of LF measurements, a direct comparison of wind amplitudes is not possible, because one does not know the real reference height of the LF measurements. The tidal phase, however, being defined as the time of eastward wind maximum, is not affected by wind underestimation and thus may be used for virtual height correction. The method is demonstrated in Fig. 2 for the monthly mean phase profiles of January 2005. It consists of selecting some virtual height h', looking for the LF SDT phase at this virtual height, and then searching for the real height h where the MR SDT phase has the same value as the LF one. This procedure is repeated for each virtual height, which provides a height correction profile for the month under consideration. The correction estimation is then repeated for each of the monthly phase profiles.

These data are shown in Fig. 3, simply presenting MR (real) heights h vs. LF virtual heights h', and a quadratic fit is added. Note that this fit was not produced by forcing the line to pass through some selected point, but the regression shows that – on a long-term average – the group retardation is negligible below 81 km. At greater heights, it may result in an overestimation of the height by up to 20 km. Owing to the

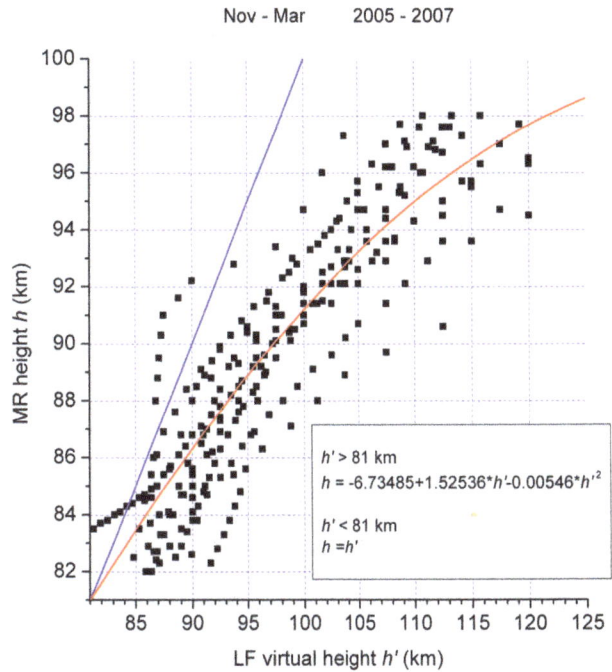

Fig. 3. Real heights observed by MR vs. LF virtual heights for November–March. The blue line represents $h' = h$, while the black line is calculated by a 2nd order least-squares fit.

small phase shift with height during summer (Fig. 1g, h), the described method can only be applied for the data collected during the winter months. Therefore the correction is, strictly speaking, only valid for November–March, however, as one may see from the following Figs. 4–7, leads to reasonable results for each month including the summer ones.

Figure 4 presents 3-year winter (November–March) and summer (May–September) mean vertical profiles of the zonal and meridional prevailing wind and the SDT amplitude. LF data are shown without and with virtual height correction, and the MR profiles are added for comparison. After correction, the LF prevailing wind vertical structure well corresponds with the MR one especially with respect to the vertical gradients, although there is a clear underestimation of the amplitudes, and the winter LF meridional winds show a bias with respect to the MR ones. The improvement is visible especially for summer profiles, because the winter vertical gradients are small and therefore the effect necessarily is small, too. This shows that, although the summer data have not been included into the phase analysis, the correction may be useful for the entire year. The winter LF SDT amplitudes after height correction have a height structure that is in qualitative agreement with the MR one. For summer, LF analyses do not see the amplitude increase at greater heights, mainly owing to the short nights and thus reduced temporal coverage there. Thus, mean profiles differ significantly and height correction will not lead to a substantial improvement of the mean profile.

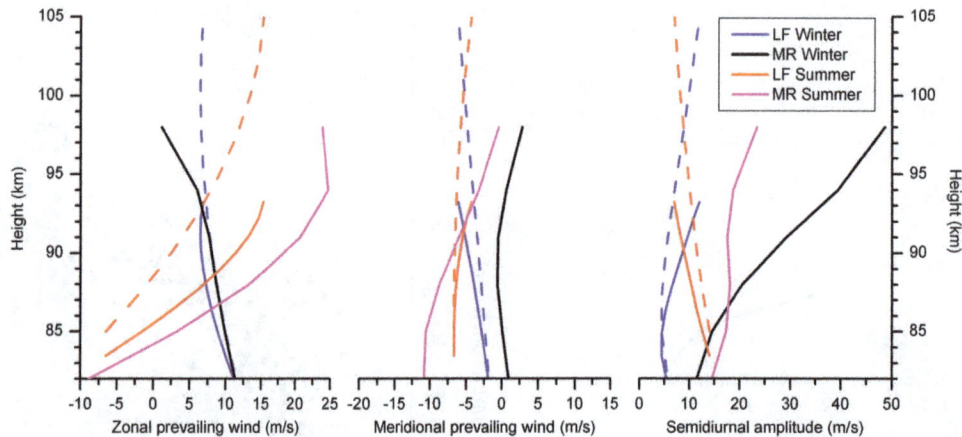

Fig. 4. Winter (December–February) and summer (June–August) mean zonal prevailing wind (left panel), meridional prevailing wind (middle panel) and semidiurnal amplitude (right panel) as measured by LF without (dashed blue/red) and after virtual height correction (solid blue/red)and by MR (solid black/magenta).

Table 1. Results of a linear best-fit analysis of wind parameters measured by LF vs. MR according to $v_{LF} = a + b \times v_{MR}$ before (respective upper row) and after (respective lower row) virtual height correction. For the SDT the intercept a is set to zero. Correlation coefficients r are also added. In the table "winter" refers to November–March and "summer" refers to May–September.

Height	Season	Zonal prevailing wind			Meridional prevailing wind			SDT tidal amplitude		
		a	b	r	a	b	r	a	b	r
85 km	Year	−4.55	1.39	0.57	−0.35	0.88	0.56	−	0.64	0.35
		−0.23	0.75	0.68	−2.14	0.58	0.60		0.58	0.46
85 km	Winter	7.22	0.28	0.14	−2.46	1.21	0.26	−	0.28	0.49
		6.83	0.21	0.28	−2.96	0.77	0.45		0.30	0.74
85 km	Summer	−10.70	1.28	0.65	0.84	1.39	0.67	−	0.85	0.29
		−3.92	0.76	0.69	4.31	0.99	0.73		0.77	0.46
88 km	Year	−0.74	0.50	0.31	−1.42	0.78	0.57	−	0.45	0.22
		2.01	0.44	0.83	−3.49	0.44	0.66		0.44	0.57
88 km	Winter	6.99	0.21	0.23	−2.64	0.50	0.31	−	0.21	0.49
		4.63	0.30	0.61	−3.60	0.60	0.31	−	0.21	0.49
		1.45	0.42	0.85	−0.71	0.65	0.74		0.58	0.73
91 km	Year	2.44	0.21	0.39	−2.86	0.71	0.58	−	0.32	0.15
		1.45	0.56	0.85	−4.18	0.45	0.48		0.37	0.71
91 km	Winter	6.34	0.12	0.31	−3.31	0.45	0.39	−	0.19	0.78
		2.54	0.50	0.68	−4.85	0.59	0.61		0.31	0.67
91 km	Summer	−3.65	0.37	0.56	−0.71	0.97	0.72	−	0.61	0.57
		2.89	0.49	0.79	−2.77	0.46	0.36		0.48	0.90

Results of linear fits of LF winds vs. MR winds after $v_{LF} = a + b \times v_{MR}$ are presented in Table 1 for different seasons and 3 different heights. Correlation coefficients r are also given. With few exceptions, correlation coefficients increase after height correction, which is valid both for winter and for summer. Note that this is also the case for the summer SDT, notwithstanding the long-term mean gradient differences between LF and MR. Correlation coefficients for the zonal prevailing wind are generally larger for summer than for winter, which is due to the small mean gradients in winter, and the resulting relatively important role of interannual variability. Note the large negative values of the intercept a in summer for the uncorrected LF heights, which are due to the partial extrapolation of the zonal wind at low altitudes and thus are not reliable. Owing to this fact the summer slopes b decrease after height correction. Meridional prevailing winds are generally weaker, and the correlation coefficients do not show a clear tendency after height correction. The SDT slope is always smaller than unity, indicating underestimaton of amplitudes by the LF method. However,

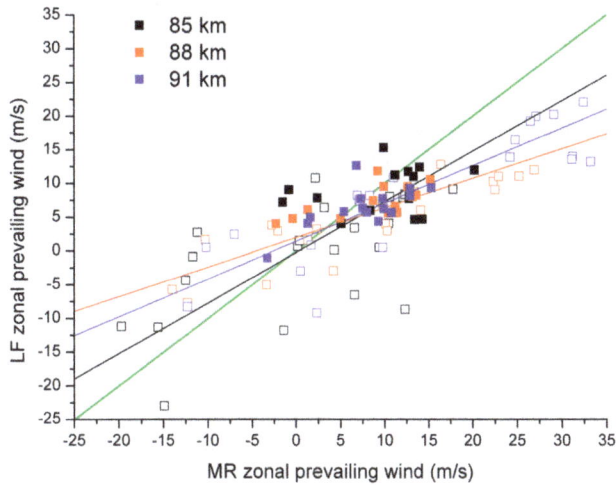

Fig. 5. LF vs. MR zonal prevailing wind at 85, 88, and 81 km after virtual height correction of the LF data. Linear best fit curves for the whole data set are added. The heavy solid green line denotes $y = x$.

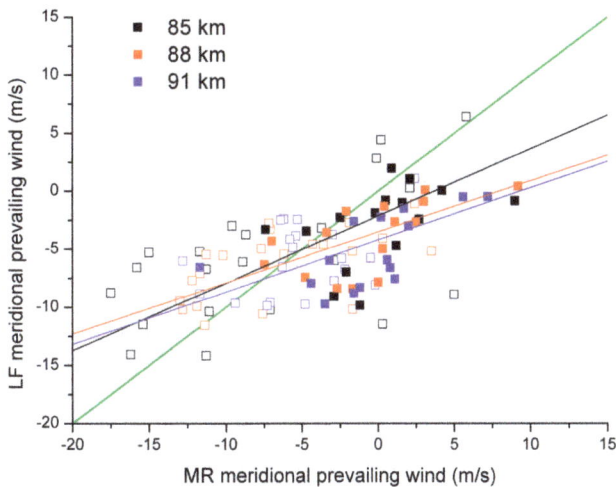

Fig. 7. As in Fig. 4, but for the SDT amplitude. The linear fits were performed with prescribing the intercept as $a = 0$.

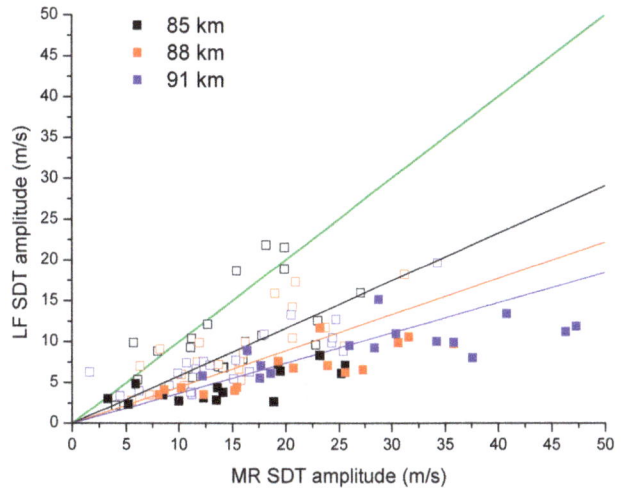

Fig. 6. As in Fig. 4, but for the meridional prevailing wind.

while after virtual height correction the slope increases for winter, it decreases for summer, the later being the effect of the different sign of the vertical gradients.

5   Mean wind and SDT amplitude comparison after height correction

Figs 1 and 4 show that, in addition to the height difference, LF measurements underestimate the winds with respect to the MR. The height-corrected LF zonal prevailing winds for 3 radar height gates are shown in Fig. 5, together with linear regression results after $v_{0z,\text{LF}} = a + b \times v_{0z,\text{MR}}$ (see also Table 1). Solid symbols show winter (November–March here) data. LF underestimation is somewhat weaker in the lower-

most height gate considered, however, one has to take into account that some of the strong westward winds at 85 km in summer may be a result of extrapolation if there are few LF data only, so that the slope (coefficient b) may be slightly too small. Otherwise the summer data do not stand out against the other ones so that one may use the wind LF height correction during the whole year as a first approximation.

The meridional component (Fig. 6) is underestimated by LF in a similar manner as the zonal one, but the correspondence is generally worse, and the underestimation even stronger, an effect that has already been noted by Jacobi et al. (2009). Also, the intercept of the regression lined differs somewhat from zero. As is the case with the zonal wind, there is no clear height effect visible.

The strongest effect is seen with the winter SDT amplitude (Fig. 7). The LF measurements underestimate the MR amplitudes by about 50%, and the effect increases with height, as in Jacobi et al. (2009) with the hourly means. Note, however, that there are some large amplitudes mainly in late summer at lower altitudes (black open squares), when winds are not underestimated by LF. This is an effect that has already been shown by Jacobi et al. (2009) and leads to differences of the seasonal cycle of SDT amplitudes measured by LF, MR, or MF (e.g., Jacobi et al., 1999, their Fig. 7). Apart from this summer anomaly, one may conclude that the increase with height of the underestimation of hourly winds by LF is mainly valid for the SDT amplitudes, while a general tendency of wind underestimation is seen in all wind parameters.

6 Discussion and conclusions

An overlap of about three years of MR and LF wind measurements in the MLT allows the analysis of necessary virtual height correction of LF reference heights as well as the estimation of LF mean wind and SDT amplitude underestimation. The virtual height effect is small near 80 km, but increases to more than 20 km difference between virtual and real height at 120 km virtual height. The virtual height correction is performed using an empirical statistical approach. In future study, empirical D region electron densities from the IRI model (Bilitza, 2001; Bilitza and Reinisch, 2008) should be considered for a more physical approach, but it has to be taken into account that the data base for the D-region in the IRI model is small and thus IRI results are uncertain as well.

The analysis of LF winds has to be carried out assuming circularly polarized SDT components. Analysis of monthly mean tidal parameters using the MR measurements without assuming circular polarization lead to a mean phase difference of 2.94 ± 0.57 h and a zonal-meridional amplitude difference of -0.52 ± 2.87 m/s, confirming earlier results using other midlatitude radars (Jacobi et al., 1999). Thus, the applied LF phase determination is justified. This, however, is only proved for the latitude range considered. At other latitudes, phase polarization may be different at least during part of the year (e.g., Zhao et al., 2005).

The LF winds generally underestimate the MR ones by about 50%. The effect is broadly independent of height for the prevailing wind, but increase with height for the SDT amplitudes. It has to be taken into account, however, that the mean winds are generally weaker than the SDT amplitudes, and clearer effects may be masked by the general variability. The reason for wind underestimation by LF is not fully understood. It may be an indirect effect of gravity waves, which may provide some signature in a spaced receiver analysis as is employed by the LF measurements. A similar effect is seen in MF winds (Manson et al., 2004; Jacobi et al., 2009). Since gravity wave activity minimizes around the equinoxes (e.g., Placke et al., 2011), this may also explain the weaker effect on late summer SDT amplitudes.

Acknowledgements. This study was supported by Deutsche Forschungsgemeinschaft under grant JA 836/22-1.

Topical Editor Matthias Förster thanks Jens Taubenheim and Dora Pancheva for their help in evaluating this paper.

References

Bilitza, D.: International Reference Ionosphere 2000, Radio Sci., 36, 261–275, 2001.

Bilitza, D. and Reinisch, B.: International Reference Ionosphere 2007: Improvements and new parameters, Adv. Space Res., 42, 599–609, 2008.

Groves, G. V.: A theory for determining upper-atmosphere winds from radio observations on meteor trails, J. Atmos. Terr. Phys., 16, 344–356, 1959.

Hall, C. M., Aso, T., Tsutsumi, M., Nozawa, S., Manson, A. H., and Meek, C. E.: A comparison of mesosphere and lower thermosphere neutral winds as determined by meteor and medium-frequency radar at 70° N. Radio Sci., 40, RS4001, doi:10.1029/2004RS003102, 2005.

Hocking, W. K. and Thayaparan, T.: Simultaneous and collocated observations of winds and tides by MF and meteor radars over London, Canada (43° N, 81° W), during 1994–1996, Radio Sci., 2, 833–865, 1997.

Hocking, W. K., Fuller, B., and Vandepeer, B.: Real-time determination of meteor-related parameters utilizing modern digital technology, J. Atmos. Solar-Terr. Phys., 63, 155–169, 2001.

Jacobi, C., Portnyagin, Y. I., Solovjova, T. V., Hoffmann, P., Singer, W., Fahrutdinova, A. N., Ishmuratov, R. A., Beard, A. G., Mitchell, N. J., Muller, H. G., Schminder, R., Kürschner, D., Manson A. H., and Meek, C. E.: Climatology of the semidiurnal tide at 52° N–56° N from ground-based radar wind measurements 1985–1995, J. Atmos. Solar-Terr. Phys., 61, 975–991, 1999.

Jacobi, C., Arras, C., Kürschner, D., Singer, W., Hoffmann, P., and Keuer, D.: Comparison of mesopause region meteor radar winds, medium frequency radar winds and low frequency drifts over Germany, Adv. Space Res., 43, 247–252, 2009.

Jacobi, C. and Kürschner, D.: Long-term trends of MLT region winds over Central Europe, Phys. Chem. Earth, 31, 16–21, 2006.

Kürschner D.: Ein Beitrag zur statistischen Analyse hochatmosphärischer Winddaten aus bodengebundenen Messungen, Z. Meteorol., 41, 262–266, 1991.

Kürschner, D. and Schminder, R.: High-atmosphere wind profiles for altitudes between 90 and 110 km obtained from D1 LF measurements over central Europe in 1983/1984, J. Atmos. Terr. Phys., 48, 447–453, 1986.

Kürschner, D., Schminder, R., Singer, W., and Bremer, J.: Ein neues Verfahren zur Realisierung absoluter Reflexionshöhenmessungen an Raumwellen amplitudenmodulierter Rundfunksender bei Schrägeinfall im Langwellenbereich als Hilfsmittel zur Ableitung von Windprofilen in der oberen Mesopausenregion, Z. Meteorol., 37, 322–332, 1987.

Kürschner, D. and Jacobi, C.: The mesopause region wind field over Central Europe in 2003 and comparison with a long-term climatology, Adv. Space Res., 35, 1981–1986, 2005.

Manson, A. H., Meek, C. E., Hall, C. M., Nozawa, S., Mitchell, N. J., Pancheva, D., Singer, W., and Hoffmann, P.: Mesopause dynamics from the scandinavian triangle of radars within the PSMOS-DATAR Project, Ann. Geophys., 22, 367–386, doi:10.5194/angeo-22-367-2004, 2004.

Placke, M., Hoffmann, P., Becker, E., Jacobi, C., Singer, W., and Rapp, M.: Gravity wave momentum fluxes in the MLT – Part II: Meteor radar investigations at high and midlatitudes in comparison with modeling studies, J. Atmos. Solar-Terr. Phys., doi:10.1016/j.jastp.2010.05.007, 2011.

Stober, G., Jacobi, C., Fröhlich, K., and Oberheide, J.: Meteor radar temperatures over Collm (51.3° N, 13° E). Adv. Space Res., 42, 1253–1258, 2008.

Zhao, G., Libo, L., Wan, W., Ning, B., Xiong, J.: Seasonal behavior of meteor radar winds over Wuhan, Earth Planets Space, 57, 61–70, 2005.

22

Structural changes in lower ionosphere wind trends at midlatitudes

C. Jacobi[1]**, E. G. Merzlyakov**[2]**, R. Q. Liu**[1]**, T. V. Solovjova**[2]**, and Y. I. Portnyagin**[2]

[1]Institute for Meteorology, University of Leipzig, Stephanstr. 3, 04103 Leipzig, Germany
[2]Research and Production Association "Typhoon", Institute for Experimental Meteorology, 4, Pobeda Str., 249038 Obninsk, Russia

Abstract. Long-term variability of the mesosphere/lower thermosphere (lower E region ionosphere) since 1970 has been analyzed using wind data series obtained at Collm (52° N, 15° E) using the LF drift method and at Obninsk (55° N, 37° E) applying VHF meteor radar. Applying piecewise linear trend analysis with a priori unknown number and positions of breakpoints shows that trend models with breakpoints are generally to be preferred against straight lines. There is a strong indication for a change of trends in wind parameters around 1975–1980. Similar changes are also found in the lower atmosphere, e.g., in tropospheric temperatures. This indicates a coupling between atmospheric layers at time scales of decades.

1 Introduction

A general question discussed in the atmospheric community is, to which degree the global climate is changing, and whether there are long-term trends present in different climate parameters. While it is widely accepted that anthropogenic greenhouse warming of the atmosphere is increasing, additional variability, e.g. due to the interplay of atmospheric circulation patterns, may give rise to periods of stronger or smaller temperature increase superposed to the general trend (Swanson and Tsonis, 2009). Even in a generally warming climate, there may be periods of several years with no temperature increase (Easterling and Wehner, 2009) due to natural climate variability, which can be confused with a change in secular trends. It is therefore of general interest to analyse climate time series not only with respect to linear trends, but also to possible changes of trend parameters,

Correspondence to: C. Jacobi
(jacobi@uni-leipzig.de)

and to define breakpoints, when parameters of linear trend models are changing.

Time series of middle atmosphere parameters have been proposed to be useful for climate trend detection (e.g. Thomas, 1996) due to comparatively large signals, and owing to the fact that the middle and upper atmosphere is coupled with the troposphere (e.g., Kazimirovsky et al., 2003). Of particular interest in this connection have been radar measurements of the mesosphere/lower thermosphere (MLT) horizontal wind, essentially because these data are, for a few selected stations, available for several decades now. The earliest continuous measurements have started in 1953 over UK (Greenhow and Neufeld, 1961). Very long records are available from measurements at Kühlungsborn and Collm, Germany (Bremer et al., 1997) dating back to the 1950s and at Obninsk, Russia (Portnyagin et al., 2006; Merzlyakov et al., 2009) beginning in 1964. Long-term measurements over Arctic and Antarctic sites dating back to the 1960s have been presented by Portnyagin et al. (1993a, b).

Analyses of possible linear long-term trends in MLT time series have been performed, e.g., by Bremer et al. (1997). The general results were that the meridional and zonal prevailing winds were decreasing until the early 1990s, and the semidiurnal tidal (SDT) amplitudes were decreasing as well. Later, Jacobi et al. (2005) reported a linear trend in SDT phases also. The question, whether these trends are really linear, has been discussed as well. Jacobi et al. (1997) found that linear trends in Collm wind time series are changing with time. Later, applying piecewise linear trend fitting, Portnyagin et al. (2006) found trend breaks in the Obninsk and Collm wind time series around the late 1970s and the early 1990s.

Portnyagin et al. (2006) and later Merzlyakov et al. (2009) have used seasonal mean winds and tidal parameters to detect possible structural changes in long-term trends. Since the wind field, especially during the equinoxes, may vary rapidly, in this paper we will make an attempt to analyze

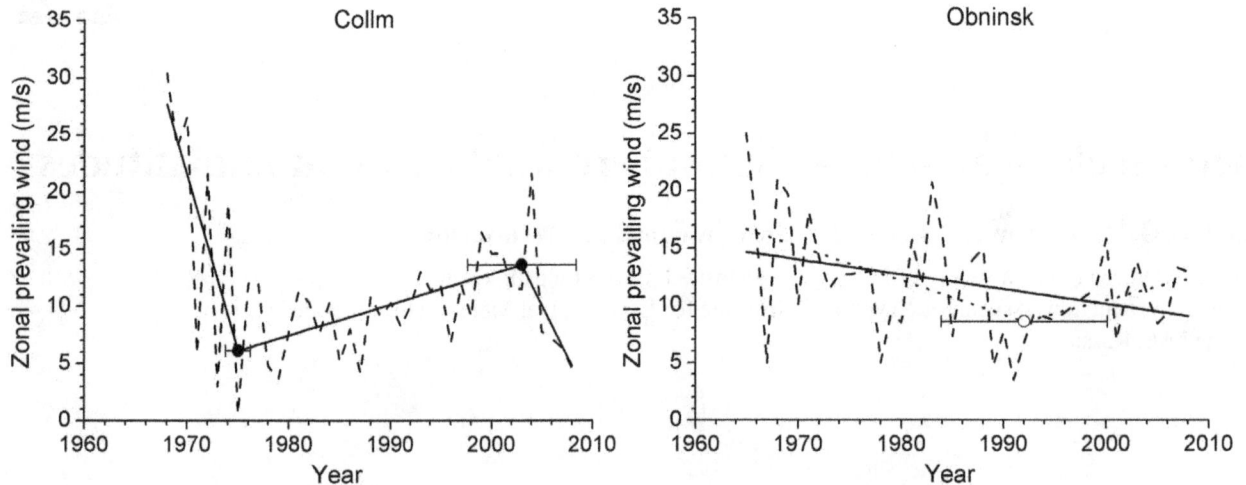

Fig. 1. January mean zonal prevailing winds (positive eastward) over Collm (left panel) and Obninsk (right panel). Results for the primary solution (solid line and symbols) and secondary solution (dashed line and open symbol) according to BIC of piecewise trend analyses are added.

long-term trends based on monthly mean winds over Collm and Obninsk. The measurements are shortly described in Sect. 2. In Sect. 3 the method for piecewise linear trend fitting is outlined. Results are presented in Sect. 4, and Sect. 5 concludes the paper.

2 Measurements

Two of the longest lower ionospheric wind time series presently available are used, namely winds measured at Collm (52° N, 15° E) during 1968–2008 and at Obninsk (55° N, 37° E) during 1964-2007. Both stations are located at similar middle Northern Hemisphere latitudes, at a distance of only approximately 1620 km. The wind measurements were carried out by meteor radar without height finding at Obninsk and by the LF closely spaced receiver method at Collm.

2.1 Collm LF lower ionospheric drifts

At Collm Observatory, MLT winds have been obtained by D1 LF radio wind measurements from 1959–2008, using the ionospherically reflected sky wave of three commercial radio transmitters. The data are combined to half-hourly zonal and meridional mean wind values. A multiple regression analysis is used to determine the daily prevailing wind and the semidiurnal tidal wind from the half-hourly wind components assuming clockwise circularly polarized tidal wind components (Kürschner, 1991). The same method is applied to calculated monthly wind parameters using monthly median half-hourly winds (Jacobi and Kürschner, 2006). The data are attributed to the mean reflection height at about 90 km.

The virtual reflection heights have been estimated since late 1982 using measured travel time differences between the separately received ionospherically reflected sky wave and the ground wave (Kürschner et al., 1987). This information is not used here, in order to avoid possible artifacts due to different analysis procedures before and after 1982. Data measured before 1968 have not been used, since before that date measurements have only been taken during evening hours, which do not allow an accurate discrimination between the semidiurnal tide and the mean wind. Other major changes in the measurements include a switch from manual data evaluation to automatic measurements in 1972 and the averaging of the results over three measuring paths since late 1978. Both of these changes are connected with a decrease of the statistical uncertainty, such that short-period variations are likely to be affected. However, the long-term mean wind analysis should not be affected. The Collm wind data set represents an extended dataset which has been used in Jacobi et al. (2009), however, there focus was laid on January through May means in order to compare with ozone variability. Here, we present results based on monthly mean values.

2.2 Obninsk meteor radar winds

Wind measurements over Obninsk using the meteor radar technique have been carried out since 1964. Frequency used is 33.3 MHz. The radar transmitting system consists of four antennas pointing in the directions N, E, S and W. For the present investigation focusing on monthly means and large-scale circulation patterns the mean values of the two respective beams are considered to improve the statistical significance. The wind values are combined to hourly means. Height determination is not available and the wind values are ascribed to a height of 90 km. The monthly mean values of

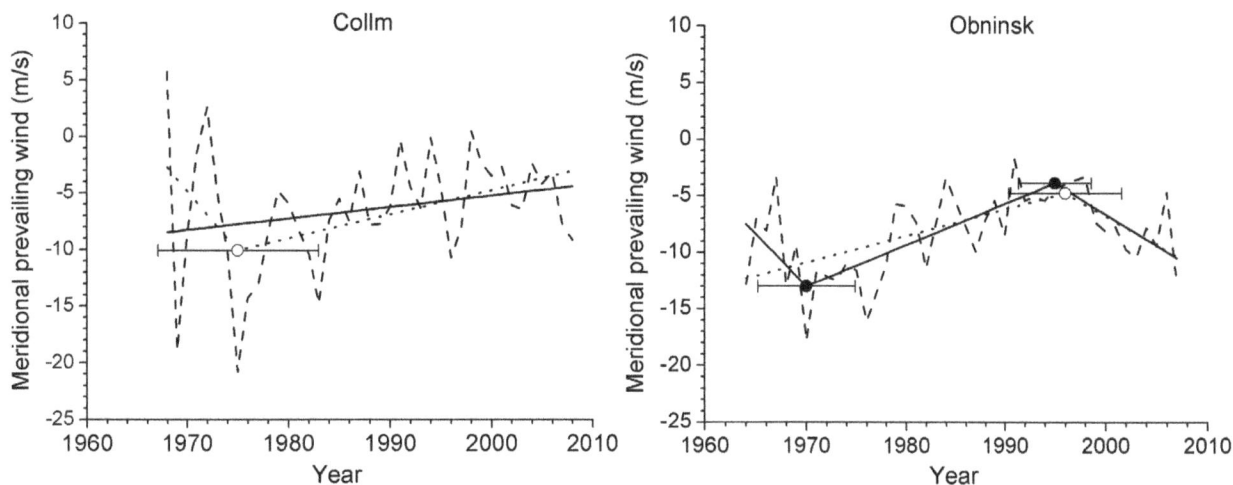

Fig. 2. As in Fig. 1, but for the June meridional prevailing winds (positive northward).

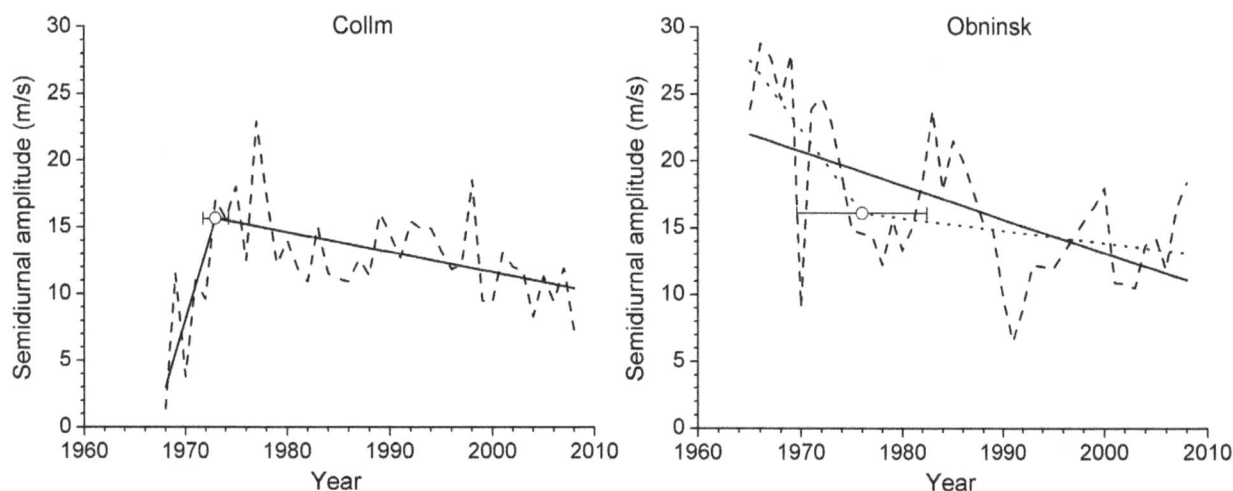

Fig. 3. As in Fig. 1, but for the January SDT amplitude.

the prevailing winds and tidal parameters are obtained from the hourly mean winds by applying a least square fitting with harmonics of 48 h, 24 h, 12 h, and 8 h. The root mean square errors of these means are about 1–2 m/s except for the first four years, when they are about 3–4 m/s.

A modification of the radar in 1968 led to a significant increase of the number of meteor echoes. In 1973 the data processing procedure was modified. In addition, instrumentation and software modifications were made in the first half of the 1980s and in 1999. Each time the measurements were modified, parallel measurements using the old and the modified techniques were carried out for several months to ensure that there was no artificial shift in the data. The Obninsk wind dataset is the same as has been presented in Merzlyakov et al. (2009) and Jacobi et al. (2009), but differs in part from that one analyzed by Portnyagin et al. (2006). Some months

of old wind data have been added and several monthly means have been re-evaluated. Some gaps in seasonal mean data have thus been filled.

3 Piecewise trend analysis of time series

In this study piecewise linear trends with a priori unknown years of breakpoints were estimated. The practical algorithm used is partly based on Tomé and Miranda (2004). The analysis was performed for different numbers of breakpoints (up to 2, and including a linear trend model as a possibility as well) by minimizing the sum of squared residuals. To avoid edge effects, breakpoints are only considered if they are located at least 4 points from the beginning/end of the time series. Then each of the three basic trend assessments were augmented with first- and second-order autoregressive

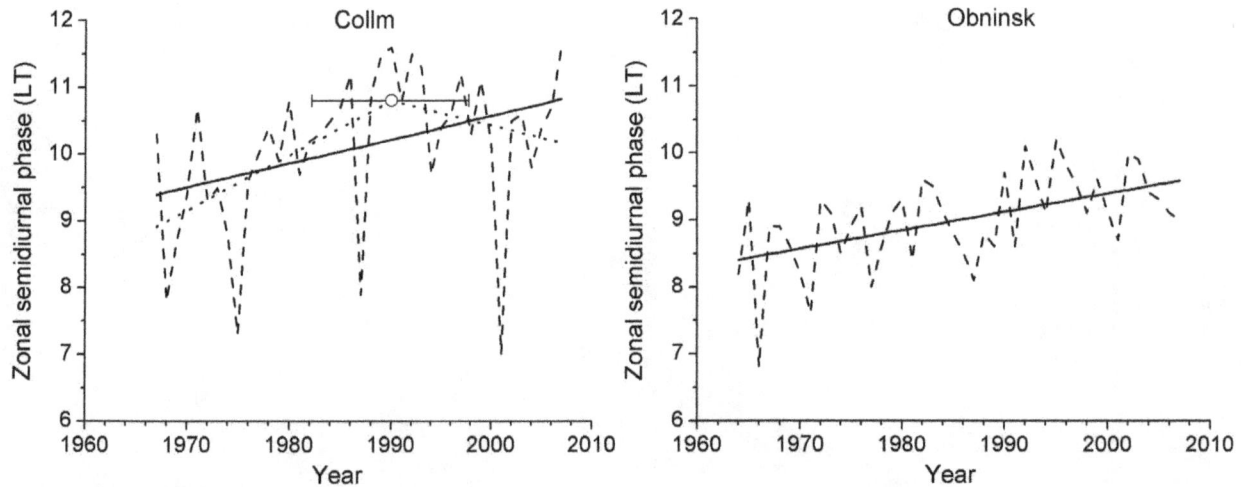

Fig. 4. As in Fig. 1, but for the November zonal SDT phase (time of maximum eastward wind in LT).

components (AR(1) and AR(2), e.g. Seidel and Lanzante, 2004). Correspondingly, the goodness of fit of the residuals to a Gaussian distribution was assessed by using the Anderson-Darling (e.g. Romeu, 2003) statistic to test the null hypothesis of normally distributed residuals. This was done with removal of the AR(1) and AR(2) behavior and with the model that does not include autoregressive behavior in the residuals (AR(0)). Any model for which the null hypothesis is rejected at the 10% significance level (Stephens, 1974) was eliminated from further consideration. The mean value and the standard lag-one and -two autocorrelations of each accepted normally distributed residual series were calculated to check whether the residuals can be regarded as a realization of white noise. Finally, to obtain the best model, the standard form of the Schwarz-Bayesian Information Criterion (BIC, e.g. Ng and Perron, 2005; Portnyagin et al., 2006) was employed by minimizing the BIC, provided that the residuals are accepted as one-dimensional normally distributed white noise. The uncertainty of trend parameters was estimated using the Monte-Carlo simulation approach.

Note that our restriction to a maximum of 2 breakpoints is arbitrary. In the case that more breakpoints are allowed, the model tends to analyze the effect of possible decadal variations, including the 11-year solar cycle effect (e.g. Jacobi and Kürschner, 2006), which is not the topic of this analysis. The use of piecewise linear trends does not allow to consider structural changes in general (as, for example, in Merzlyakov et al., 2009), but this approach is a natural extension of the linear regression model and can be readily performed. We present results with minimum BIC and secondary solutions with second smallest BIC. Note that ranking the possible solutions according to BIC is only a statistical procedure. Other solutions may be physical as well, and the primary solutions by BIC may be non-physical, e.g. in the presence of artifacts. Therefore, the interpretation of the results has to be made with special care.

4   Results

In this section we present examples of time series including breakpoint analyses to provide an impression of how breakpoints are distributed. The selection of the months presented is arbitrary. Note that there are many combinations of month of year and wind parameter, where no breakpoint is found according to minimum BIC. In total, only about 10% of all time series under consideration contain breakpoints. This fact is above all owing to the relatively large variability from year to year, which is present in monthly mean time series. Analyses of seasonal means (Portnyagin et al., 2006) reveal more stable results.

In Fig. 1 January zonal mean winds over Collm and Obninsk are shown. Some similarities are visible, e.g. the decreasing trend in the early years of the measurements, which has also been reported by Bremer et al. (1997), and an increasing trend after 1990. Differences between the two time series in Fig. 1, however, are also visible, e.g. the stronger zonal winds seen in the Obninsk January time series approximately between 1980 and 1990 are visible in the Collm time series, too, but they are much weaker expressed there. Results for the primary and secondary solutions according to BIC of piecewise trend analyses are added in Fig. 1. The secondary solution for Collm is one with the same breakpoints as the primary one, but different autoregression of the residuals, so that only one trend curve is shown. Breakpoints for Collm are found in 1975 and 2003, while for Obninsk only a secondary breakpoint is found in 1992. However, if one would allow solutions with three breakpoints (not shown), the primary solution for Collm would include a third breakpoint in 1987, with error bars that overlap with the ones for the Obninsk secondary solution.

Examples for the meridional prevailing wind are shown in Fig. 2. Generally the year-to-year variability over Collm is larger than over Obninsk. This may be owing to larger

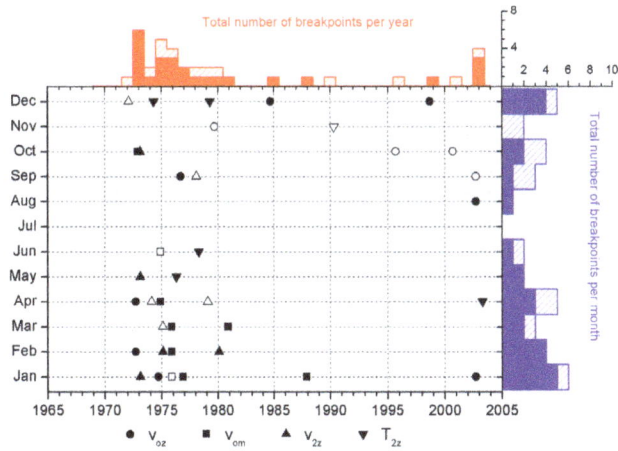

Fig. 5. Summary of breakpoints analyzed in Collm prevailing zonal (V_{oz}) and meridional (V_{om}) prevailing winds and tidal amplitudes (V_{2z}) and phases (T_{2z}). Symbols show year and month of the respective breakpoints. In the upper panel, the sum of breakpoints, for all the 4 parameters analyzed together is shown for each year. On the right hand panel the total number of breakpoints for each month is given.

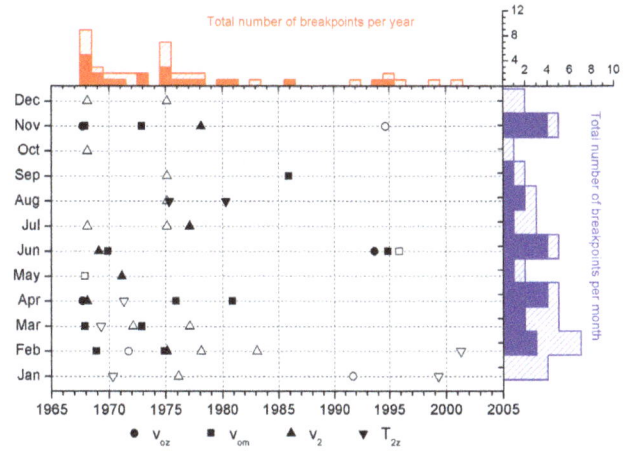

Fig. 6. As in Fig. 5, but for Obninsk winds.

uncertainty of Collm monthly means due to the long daily data gaps in summer. However, the long-term tendencies for both time series are similar. These include increase of the southward wind in the early years (note that the winds are generally negative, i.e. southward directed), a long-term decrease after ~1975, and again an increase after 1995. The latter, however, is not so clearly expressed over Collm in relation to the overall variance, so that it is not among the first two solutions selected according to minimum BIC.

The SDT phases, given as the time of eastward zonal wind maximum, generally show an increasing trend (towards later phases), which has already been noted, e.g., by Jacobi et al. (2005) and Jacobi and Kürschner (2006). As an example, the November mean phases are given in Fig. 4. Note that there is a mean difference of mean phases, which is present during the whole winter (Jacobi et al., 2005) and which may indicate that the November SDT is not only due to the migrating (westward traveling with wavenumber 2) STD, but other SDT components may be present. In the case of the November phases there is one secondary breakpoint in the Collm time series.

A summary of breakpoints seen in the Collm data series is presented in Fig. 5, while the corresponding results for Obninsk are presented in Fig. 6. A symbol is plotted in the figure each time, when a breakpoint appears in any of the time series of the zonal prevailing wind, meridional prevailing wind, SDT amplitude or SDT phase. The choice of parameters used is arbitrary, and, in order not to cram the figure with too many lines, we did not present error bars as was done in Figs. 1–4. Therefore the information that can be taken from these summaries is qualitative only. The ma-

jority of breakpoints is found in or after 1975. There is also a larger number of breakpoints for Collm in 1973, however, this year is very close to the change in data analysis at Collm, and may be owing to that. This idea is corroborated by the fact that in 1973 there is only one breakpoint for Obninsk. On the other hand, the 1970 breakpoints for Obninsk may be owing to artifacts in connection with the radar modification in 1968. These artifacts can be evaluated if one would take into account changes in the data variance. This might lead to a possible removal of these breakpoints. This would be a subsequent extension of the trend model. There is also some less clear clustering of breakpoints around 1995 (Obninsk only) and after 2000 (Collm only). The former possible breakpoint coincides with a change in middle atmosphere ozone, ozone laminae, and possibly temperature trend (Reinsel et al., 2005; Križan and Laštovička, 2005; Bremer and Peters, 2008). The latter breakpoint would be in correspondence with indications that the global climate has recently shifted (Swanson and Tsonis, 2009).

Only about 10% of the curves contain breakpoints according to the piecewise analysis. Taking into account that the analyses of seasonal means result in breakpoints in the majority of cases (Portnyagin et al., 2006; Merzlyakov et al., 2009), this gives rise to the conclusion, that the interannual variability of monthly means is large, and only the strongest and most robust breakpoints are found. In this case, we may conclude that there is one primary breakpoint in the time series around 1975–1980. This appears at the same time as a major change in global temperatures (e.g. Swanson and Tsonis, 2009).

For most months, large interannual variability is present, which in most cases lead to a linear (or no) trend as the best solution in the framework of piecewise linear trends. Therefore, breakpoints which have been identified in earlier analyses (Portnyagin et al., 2006; Merzlyakov et al, 2009) are not necessarily visible here. This is particularly the case with breakpoints around 1990, which are not visible in analyses using monthly data. We may conclude that owing to the strong variability of the MLT wind field, the use of seasonal means or more sophisticated trend models may be superior to using monthly means and piecewise linear trends, if weaker breakpoints like the one around 1990 are sought. On the other hand, especially for the equinoxes, seasonal means are not always meaningful, as they consist of averages over very different circulation patterns. This means, breakpoints that are visible in monthly means notwithstanding the large variability of these datasets, can be considered as robust.

The analysis shows that the majority of breakpoints are found roughly between 1975 and 1980. This is time interval, when a major atmospheric structural change has been observed, which is, for example, clearly visible in global temperature records. This suggests that this global change influences the dynamics of the middle atmosphere.

Acknowledgements. This study was supported by Deutsche Forschungsgemeinschaft under JA 836/22-1 and by RFBR under grant 08-05-91950. Topical Editor M. Förster thanks E. Kazimirovsky and J. Taubenheim for their help in evaluating this paper.

References

Bremer, J., Schminder, R., Greisiger, K. M., Hoffmann, P., Kürschner, D., and Singer, W.: Solar cycle dependence and long-term trends in the wind field of the mesosphere/lower thermosphere, J. Atmos. Solar-Terr. Phys., 59, 497–509, 1997.

Bremer, J. and Peters, D.: Influence of stratospheric ozone changes on long-term trends in the meso- and lower thermosphere, J. Atmos. Solar-Terr. Phys., 70, 1473–1481, 2008.

Easterling, D. R, and Wehner, M. F.: Is the climate warming or cooling? Geophys. Res. Lett., 36, L08706, doi:10.1029/2009GL037810, 2009.

Greenhow, J. and Neufeld, E. L.: Winds in the upper atmosphere, Q. J. Roy. Meteorol. Soc., 87, 472–489, 1961.

Jacobi, C., Schminder, R., Kürschner, D., Bremer, J., Greisiger, K. M., Hoffmann, P., and Singer, W.: Long-term trends in the mesopause wind field obtained from D1 LF wind measurements at Collm, Germany, Adv. Space Res., 20, 2085–2088, 1997.

Jacobi, C., Portnyagin, Y. I., Merzlyakov, E. G., Solovjova, T. V., Makarov, N. A., and Kürschner, D.: A long-term comparison of mesopause region wind measurements over Eastern and Central Europe, J. Atmos. Solar-Terr. Phys., 67, 227–240, 2005.

Jacobi, C. and Kürschner, D.: Long-term trends of MLT region winds over Central Europe, Phys. Chem. Earth, 31, 16–21, 2006.

Jacobi, C., Hoffmann, P., Liu, R. Q., Križan, P., Laštovièka, J., Merzlyakov, E. G., Solovjova, T. V., and Portnyagin, Y. I.: Midlatitude mesopause region winds and waves and comparison with stratospheric variability, J. Atmos. Solar-Terr. Phys., 71, 1540–1546, 2009.

Kazimirovsky, E., Herraiz, M., and De la Morena, B. A.: Effects on the ionosphere due to phenomena occurring below it, Surveys in Geophys., 24, 139–184, 2003.

Križan, P. and Laštovička, J.: Trends in positive and negative ozone laminae in the Northern Hemisphere, J. Geophys. Res., 110, D10107, doi:10.1029/2004JD005477, 2005.

Kürschner D.: Ein Beitrag zur statistischen Analyse hochatmosphärischer Winddaten aus bodengebundenen Messungen, Z. Meteorol., 41, 262–266, 1991.

Kürschner, D., Schminder, R., Singer, W., and Bremer, J.: Ein neues Verfahren zur Realisierung absoluter Reflexionshöhenmessungen an Raumwellen amplitudenmodulierter Rundfunksender bei Schrägeinfall im Langwellenbereich als Hilfsmittel zur Ableitung von Windprofilen in der oberen Mesopausenregion, Z. Meteorol., 37, 322–332, 1987.

Merzlyakov, E. G., Jacobi, C., Portnyagin, Y. I., and Solovjova, T. V.: Structural changes in trend parameters of the MLT winds based on wind measurements at Obninsk (55° N, 37° E) and Collm (52° N, 15° E), J. Atmos. Solar-Terr. Phys., 71, 1547–1557, 2009.

Ng, S. and Perron, P.: Practitioners' corner: a note on the selection of time series models, Oxford Bulletin of Economics and Statistics, 67, 115–134, 2005.

Portnyagin, Y. I., Forbes, J. M., Fraser, G. J., Vincent, R. A., Avery, S. K., Lysenko, I. A., and Makarov, N. A.: Dynamics of the Antarctic and Arctic mesosphere and lower thermosphere regions – I. The prevailing wind, J. Atmos. Terr. Phys., 55, 827–841, 1993a.

Portnyagin, Y. I., Forbes, J. M., Fraser, G. J., Vincent, R. A., Avery, S. K., Lysenko, I. A., and Makarov, N. A.: Dynamics of the Antarctic and Arctic mesosphere and lower thermosphere regions – II. The semidiurnal tide, J. Atmos. Terr. Phys., 55, 843–855, 1993b.

Portnyagin, Y. I., Merzlyakov, E. G., Solovjova, T. V., Jacobi, C., Kürschner, D., Manson, A., and Meek, C.: Long-term trends and year-to-year variability of mid-latitude mesosphere/lower thermosphere winds, J. Atmos. Solar-Terr. Phys., 68, 1890–1901, 2006.

Reinsel, G. C., Miller, A. J., Weatherhead, E. C., Flynn, L. E., Nagatani, R. M., Tiao, G. C., and Wuebbles, D. J.: Trend analysis of total ozone data for turnaround and dynamical contributions, J. Geophys. Res., 110, D16306, doi:10.1029/2004JD004662, 2005.

Romeu, J. L.: Anderson-Darling: a goodness of fit test for small samples assumptions. Reliability Analysis Center, Selected Topics in Assurance Related Technologies, 10, 1–6, 2003.

Seidel, D. J. and Lanzante, J. R.: An assessment of three alternatives to linear trends for characterizing global atmospheric temperature changes, J. Geophys. Res., 109, D14108, doi:10.1029/2003JD004414, 2004.

Stephens, M. A.: EDF Statistics for goodness of fit and some comparisons, J. Am. Stat. Assoc., 69, 730–737, 1974.

Swanson, K. L. and Tsonis, A. A.: Has the climate recently shifted?, Geophys. Res. Lett., 36, L06711, doi:10.1029/2008GL037022, 2009.

Thomas, G. E.: Is the polar mesosphere the miner's canary of global change?, Adv. Space Res., 18, 49–58, 1996.

Tomé, A. R. and Miranda, P. M. A.: Piecewise linear fitting and trend changing points of climate parameters, Geophys. Res. Lett., 31, L02207, doi:10.1029/2003GL019100, 2004.

23

Investigation of horizontal structures at mesospheric altitudes using coherent radar imaging

S. Sommer, G. Stober, C. Schult, M. Zecha, and R. Latteck

Leibniz Institute of Atmospheric Physics at the Rostock University, Schloss-Str. 6, 18225 Kühlungsborn, Germany

Correspondence to: S. Sommer (sommer@iap-kborn.de)

Abstract. The Middle Atmosphere Alomar Radar System (MAARSY) in Northern Norway (69.30° N, 16.04° E) was used to perform interferometric observations of Polar Mesosperic Summer Echoes (PMSE) in June 2012. Coherent Radar Imaging (CRI) using Capon's method was applied allowing a high spatial resolution. The algorithm was validated by simulation and trajectories of meteor head echoes. Both data sets show a good correspondence with the algorithm. Using this algorithm, the aspect sensitivity of PMSE was analysed in a case study, making use of the capability of CRI to resolve the pattern within the beam volume. No correction of the beam pattern was made yet. It was found in this case study, that no large variations in the scattering width and the scattering center occured apart from a very short period of time at the upper edge of the PMSE.

1 Introduction

Polar Mesospheric Summer Echoes (PMSE) are a phenomenon in the middle atmosphere at polar latitudes, where VHF radar waves are backscattered at mesospheric heights (e.g. Balsley et al., 1979; Hoppe et al., 1988). The overall mechanism of the formation of these PMSE is understood and widely accepted. 'Mesospheric neutral air turbulence in combination with a significantly reduced electron diffusivity due to the presence of heavy charged ice aerosol particles (radii $\sim$ 5–50 nm) leads to the creation of structures at spatial scales significantly smaller than the inner scale of the neutral gas turbulent velocity field itself' (Rapp and Lübken, 2004) and therefore scales of the radar's Bragg wavelength. The inner structure of PMSE is still unknown but a field of recent interest (Chilson et al., 2002; Chen et al., 2008). The aspect sensitivity of PMSE is an indicator of the scattering mechanism and describes the backscattered power in dependence on the off-zenith angle of the radar beam. If the power

decreases fast with increasing off-zenith angle, the aspect sensitivity is high and it is unlikely an homogenous scattering mechanism and vice versa (Röttger and Vincent, 1978; Hocking et al., 1986). It is essential to understand the aspect sensitivity to use PMSE as a tracer of atmospheric dynamics (Stober et al., 2012). In order to resolve the still unknown inner structures of PMSE, the Middle Atmosphere Alomar Radar System (MAARSY) on the Norwegian Island Andøya started its operation in 2010. MAARSY has a beam width of 3.6° at 3 dB and beam steering capabilities with off-zenith angles up to 30° without generating severe grating lobes. Furthermore the MAARSY antenna can be divided into several sub-arrays to perform interferometric analysis. These capabilities can be used to apply coherent radar imaging (CRI) and to get a deeper insight into the inner structures of PMSE. Coherent Radar Imaging was introduced and established by Kudeki and Sürücü (1991) and Woodman (1997) in atmospheric physics and used by e.g. Yu et al. (2001); Chilson et al. (2002); Chen et al. (2004, 2008) to investigate PMSE.

2 Experimental setup and calculation

2.1 Experimental setup

MAARSY employs an active phased array antenna system located at the island Andøya in Northern Norway (69.30° N, 16.04° E). It operates at 53.5 MHz and consists of 433 Yagi antennas, arranged in a circular shape with a diameter of 90 m. It has a half power beam width of 3.6° and a peak power of about 800 kW. Pulse-to-pulse beam steering is possible to an off-zenith angle up to 30° (Latteck et al., 2012). For receiving purposes in this experiment, the array was divided in seven sub-arrays (Fig. 1), each consisting of 49 antennas. The receiving channels for the seven sub-arrays have been phase-calibrated as described in Chau et al. (2013). On 21 June 2012, MAARSY was operated in an interferometric mode,

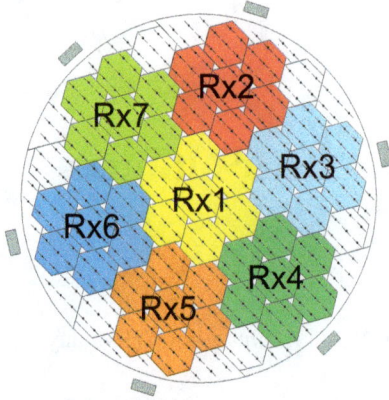

Fig. 1. Sketch of MAARSY. The array is divided into 55 sub-arrays of hexagonal shape. Seven of contiguous sub-arrays can be combined to a larger structure called an anemone and the seven anemones are colour-coded in the figure above. The boxes around the antenna field are the equipment buildings.

allowing the application of CRI. The sampling range was 75–111 km. Using a pulse length of 1 μs, the resulting range resolution is 150 m. The pulse repetition frequency was 1250 Hz, a 16 bit complementary coded pulse was transmitted and the received data was stored in 256 points complex time series for each sub-array.

2.2 Used method

The data was analysed by using Capon's method of CRI. Capon's method was found to be better than Fourier's method and faster than the Maximum Entropy method (Yu et al., 2000). The imaging method is based on the cross spectral function between the different receivers i and j and denoted by R_{ij}. Following Palmer et al. (1998), one can derive the cross correlation matrix in the time domain as

$$\mathbf{R}(t) = \begin{bmatrix} R_{11}(t) \ldots R_{1n}(t) \\ R_{21}(t) \ldots R_{2n}(t) \\ \vdots \qquad \vdots \\ R_{n1}(t) \ldots R_{nn}(t) \end{bmatrix} \tag{1}$$

The brightness as an indication for the backscattered power of Capon's method is given by

$$\mathbf{B}_c(t, \boldsymbol{k}) = \frac{1}{e^\dagger \mathbf{R}^{-1} e} \tag{2}$$

with

$$e = \begin{bmatrix} e^{j\boldsymbol{k} \cdot \boldsymbol{D}_1} \\ e^{j\boldsymbol{k} \cdot \boldsymbol{D}_2} \\ e^{j\boldsymbol{k} \cdot \boldsymbol{D}_3} \\ \vdots \\ e^{j\boldsymbol{k} \cdot \boldsymbol{D}_n} \end{bmatrix} \tag{3}$$

where $\boldsymbol{k}$ denotes the wavenumber vector and $\boldsymbol{D}_i$ the distance vector of the receiver i with respect to the origin (Palmer

et al., 1998), t the time, $e^\dagger$ represents the conjugate transpose of e and $\mathbf{R}^{-1}$ the inverse of $\mathbf{R}$.

The algorithm was applied after coherent integration of a time series containing 256 data points. The brightness was mapped on a rectangular grid with a meridional and zonal range of $-4°$ to $4°$ or $-6°$ to $6°$ for the simulation, and $-8°$ to $8°$ for the meteor head echo. The step width is 0.1°. As an initial validation and report of CRI with the MAARSY radar, however, we do not consider the beam weighting effect on the brightness distribution. The beam weighting effect may not be ignored for some circumstances (e.g. Chen et al., 2011; Chen and Furumoto, 2011, 2013).

3 Verification

3.1 Simulation

Assuming a point target, the backscattered signal can be simulated by a monochromatic wave, which can be written for one sub-array as

$$s_i(t) = A \cdot e^{j(2\pi f t + \varphi_i)} + N(t) \tag{4}$$

where A is the amplitude, f the transmitted radar frequency, t the time and φ the phase. To simulate internal and external disturbances, normally distributed random noise with the standard deviation σ and the amplitude N is added for the real and imaginary part independently. This was done for every sub-array for a single target and after that different signal locations were superimposed by adding the different signals of different targets. These signals were used to validate the CRI algorithm. For this simulation, the MAARSY configuration was used, with $f = 53.5$ MHz and the centers of two adjoining anemones 28 m apart. The amplitude was set to $A = 1$. Furthermore, Gaussian noise with $\sigma = 0.1$ was added . The results for three point targets are shown in Fig. 2 with different off-zenith angles. The left figure shows an off-zenith angle of 1°. The center of the brightness lies between the scatterers and the single centers can not be separated from each other. The triangular shape is preserved. The right figure shows a simulation with off-zenith angles of 2.5°. All three targets are separated as centers of brightness. Still, some signal is received from the zenith, where no scatterer is located. A more complex distribution of point scatterers is shown in Fig. 3. In contrast to Yu et al. (2000), we assume different centers of scattering. The scatterers are arranged in a sine-shaped way. When we decrease the signal to noise ratio (SNR), the less the shape of the point scatterers is preserved.

The SNR was calculated as $\text{SNR} = \left(A / \sqrt{2}\sigma \right)^2$. Hence, the distribution of the brightness depends not only on the distribution of the scatterers but also on the noise level of the observations. The addition of noise for point scatterer in a simulation is necessary, otherwise R becomes a singular matrix for point scatterers, which cannot be inverted. The ambiguity due to multiple phase cycles becomes a problem for large

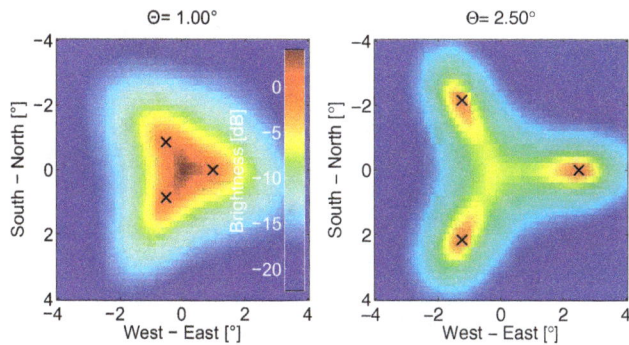

Fig. 2. Simulation of signals with known positions. The three x mark the position of the signal from point targets. Left: Three signals with off-zenith angles of 1° are superimposed, resulting in one center with the highest brightness in the middle. Right: Three signals with off zenith angles of 2.5° are superimposed, resulting in three separated centers of brightness.

off-zenith angles. The multiple phase cycles are introduced due to the given baseline length. In Fig. 3, the ambiguity is shown by black lines and determined by the baselines of the used sub-arrays. Even with no noise added, the pattern cannot be reproduced perfectly by the algorithm.

3.2 Verification using meteor head echoes

To verify the algorithm not only by simulated but real data, meteor head echoes were used. These echoes result from the plasma around the meteor when it enters the atmosphere. This is a good target due to the small size of the meteors (Westman et al., 2004). An event with two meteors entering the atmosphere with almost perpendicular trajectories were used to validate the CRI algorithm. The trajectories of the meteors are represented in Fig. 4. For this experiment, a pulse length of $49\,\mu s$ was used. The range was oversampled with 900 m. For the calculation of the brightness, a set containing 7 s of data were coherently integrated. The result is shown in Fig. 5 where the brightness distribution for five different heights are shown. The brightness is spread not only horizontally but vertically resulting from the oversampling, but the center of the brightness matches the trajectories. The algorithm is capable of resolving complex structures in real data.

4 Aspect sensitivity using CRI

Having validated our CRI algorithm, we performed a PMSE observation in summer 2012 to investigate the aspect sensitivity of this phenomenon for an arbitrarily choosen time. See Sect. 2.1 for the experimental setup. Following Chilson et al. (2002), the brightness distribution in the zenith beam can be regarded as an indicator of the aspect sensitivity during periods when the PMSE are uniform within the beam volume which corresponds to horizontal scales of approximately

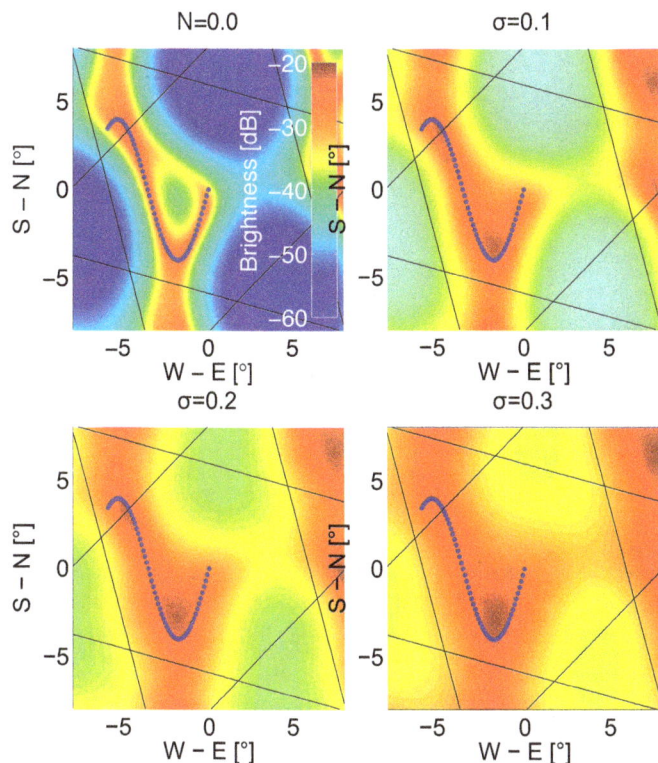

Fig. 3. Simulation of signals with known positions. The dots mark the position of the signal generated by point targets. The black lines indicate the ambiguity limit due to the antenna configuration. Upper left: No noise added reproduce the pattern of the scatterers. Upper right: A SNR of 50 preserves the shape of the scatterer distribution. Lower left: A SNR of 12.5 leads to a broader distribution of brightness. Lower right: A SNR of 5.6 leads to a smeared distribution of brightness.

5.7 km at a height of 90 km. Otherwise, the maximum of the brightness must be considered as the mean angle of arrival. The same approach was made for MAARSY. The difference made is that Chilson et al. (2002) applied a two-dimensional Gaussian fit, while we fitted the brightness meridional and zonal with a one-dimensional Gaussian fit of the form

$$f(x) = a \cdot \exp\left(-\frac{1}{2}\left(\frac{x-\mu}{\sigma}\right)^2\right) \tag{5}$$

where a is the amplitude, μ the deviation from the zenith, in zonal direction denoted by x, or meridional direction, denoted by y, respectively. σ denotes the meridional or zonal width. The CRI algorithm was applied to a time series of nine hours on 21 June 2012 for a height range from 80–89 km and the brightness was fitted using a Gaussian fit. To avoid the analysis of noise, only data with a maximum brightness $B > 75\,dB$ was included in the analysis. The results for the width and the deviation of the maximum from the zenith of the fit are shown in Fig. 6.

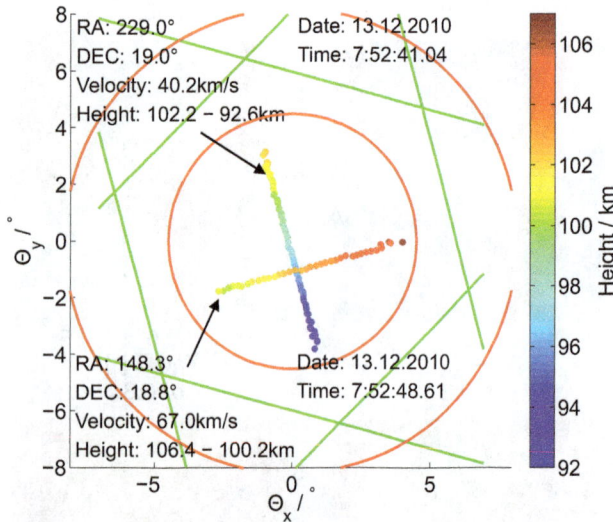

Fig. 4. Two meteors in the zenith beam of MAARSY. These trajectories were calculated by using interferometry. Both meteors enter the atmosphere 7 s apart and have almost perpendicular trajectories.

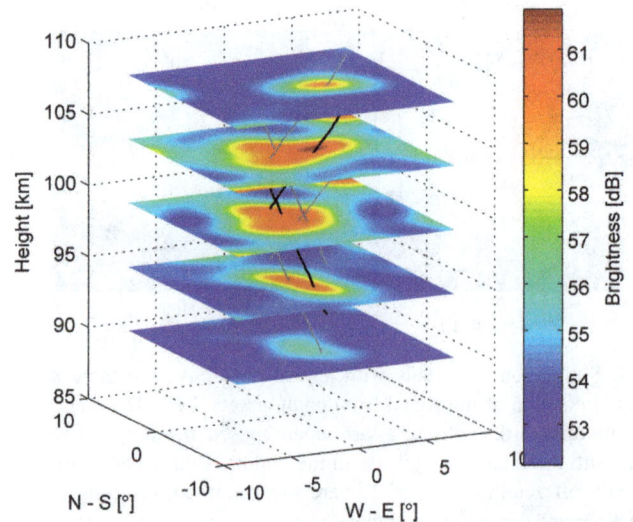

Fig. 5. Meteor trajectories of two meteors. The black lines indicate the center of the meteor trail, the light grey lines indicate the upper and lower limits of the meteor trail due to the used pulse length of $49\,\mu s$. The coloured rectangles represent the brightness in selected height gates. The trajectories and the brightness correspond well.

The top Fig. 6a shows the range-time power plot of the used time series to give an overview over the PMSE activity during this time. Figures 6b and 6c show the deviation of the fit's maximum in meridional and zonal direction, respectively. Figures 6d and 6e show the width of the fit in meridional and zonal directions. The width of the Gaussian function varies between $\sigma = 1°$ to $\sigma = 3°$. This corresponds to a full width at half maximum (FWHM) of FWHM = 2.4° to FWHM = 7°. Compared with the radar beam width of $FWHM_{beam} = 3.6°$ we used, the received pattern varies between a broadening of the beam pattern and a smaller received pattern, maybe due to aspect sensitivity. Most returned signals had a width of $\sigma < 2.2°$. Layers described by Chilson et al. (2002) are not found in this case study, except at about 13:00 UTC, where large deviation over 4° were found at the upper edge of the PMSE, but no layer shaped broadening in the returned width was found in this time series. It appears that in the lower part of the PMSE the aspect sensitivity is slightly increased compared to the upper part of the PMSE. This is similar to that observed by Chilson et al. (2002).

To inspect the altitudinal variation of aspect sensitivity in more detail, the PMSE was divided into three different parts for each time, where the upper and lower edge are defined by the two lowest and highest ranges of the PMSE and a part between those regions. The variations of the deviations and width for different parts of the PMSE are shown in Fig. 7. The left side shows the deviations of the maximum from the zenith while the right side shows the width of the Gaussian fit. In order to get an overview over the distribution of the scattering center, the expectation values of μ_x, μ_y, σ_x and σ_x are shown in Table 1. The maximum of the brightness is mostly centered around the zenith with no large variations over the different parts of the PMSE. The slight negative de-

viation of the scattering centers can result from the PMSE, which might be gently tilted or, although a phase calibration has been made, due to small phase instabilities between the receivers; we need more observations and data to verify this in the future. The width of the Gaussian fit does not vary much over the three different parts of the PMSE, although a slight decrease of aspect sensitivity with height was found in the width of the Gaussian fit in the zonal direction. On the other hand, in the width in meridional direction, no significant change in width was found. The decreasing aspect sensitivity in upper parts of the PMSE was also found by Zecha et al. (2001), who used the Alomar SOUSY VHF radar. Also, Hoppe et al. (1990) measured the aspect sensitivity with the SOUSY VHF radar and they found values of aspect sensitivity between 2° and 10° and a tendency of decreasing aspect sensitivity with height.

5   Summary and conclusions

A CRI algorithm using Capon's method was applied on PMSE measurements conducted with MAARSY and the algorithm was validated by both synthetic data and meteor head echoes. It was found that the CRI algorithm corresponds well with both the simulation and the trajectories of the meteor head echoes.

After that, the algorithm was used to analyse the aspect sensitivity of PMSE by applying a Gaussian fit to the meridional and zonal slices of the brightness. The brightness distribution can be regarded as an indicator for the aspect sensitivity when we assume, that the PMSE are uniform within the

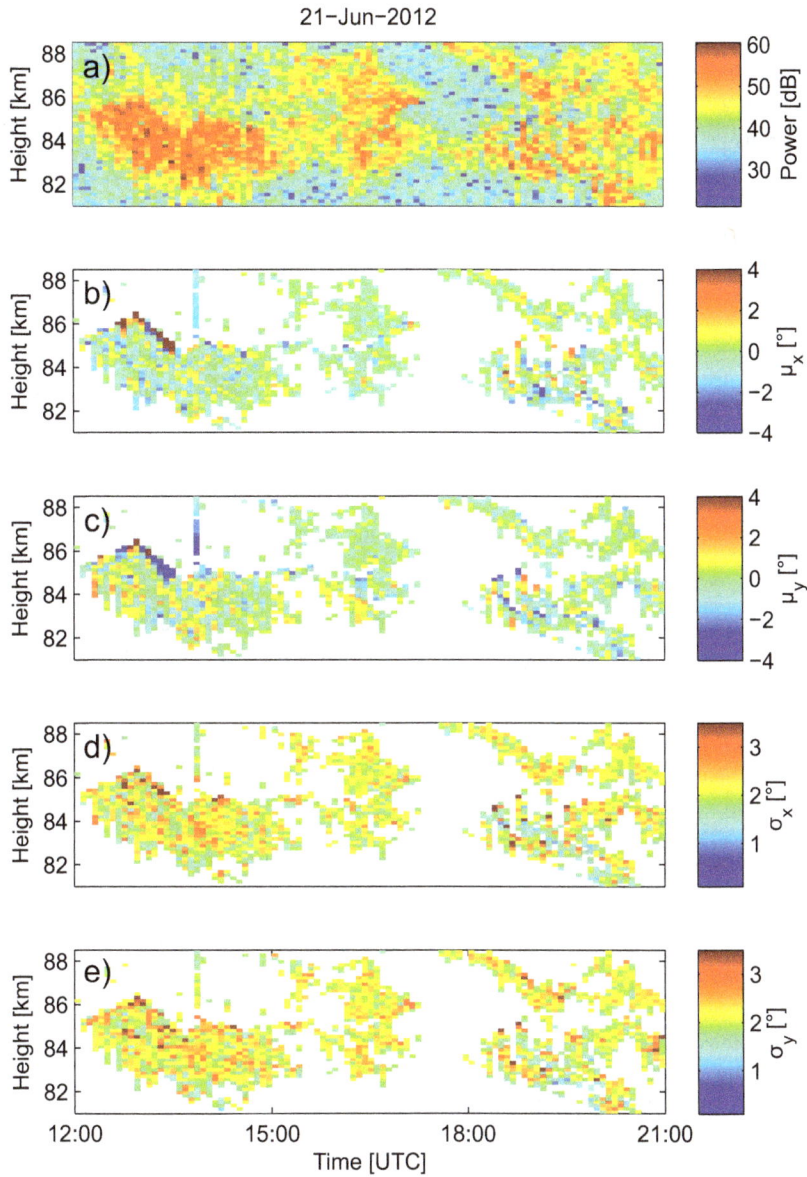

Fig. 6. Range-time pseudocolor plot of: (**a**) Power plot of the time series. (**b**) Deviation of the maximal brightness from the zenith in meridional direction. (**c**) Deviation of the maximal brightness from the zenith in zonal direction. (**d**) Width of the Gaussian fit in meridional direction. (**e**) Width of Gaussian fit in zonal direction.

radar beam. Is the PMSE not uniform within the radar beam and the pattern is too small to be resolved by the algorithm, it can only be seen as a mean angle of arrival.

In the dataset used, no layer shaped structures were found, but as a solitary event, in the upper boundary of the PMSE a large deviation from the zenith in both meridional and zonal direction was observed around 13:00 UTC.

Statistical estimates show that the scattering centers differ slightly from the zenith. The slight negative bias could be related to tilted structures in the echo layer, or results from small errors in the phase calibration of the receiving channels. On the other hand, statistical estimates show that the

Table 1. Expectation values of μ_x, μ_y, σ_x and σ_x. The error is the σ standard deviation. The derivation of the scatter centers from the zenith is around zero with a slight negative bias and the width of the brightness distribution is about $2°$.

Part of PMSE	μ_x [°]	μ_y [°]	σ_x [°]	σ_y [°]
Upper edge	-0.1 ± 1.3	-0.4 ± 1.4	2.2 ± 0.6	2.3 ± 0.7
Part between	-0.1 ± 1.0	-0.2 ± 1.0	2.1 ± 0.5	2.2 ± 0.5
Lower edge	-0.2 ± 0.8	-0.4 ± 1.4	2.0 ± 0.4	2.3 ± 0.7

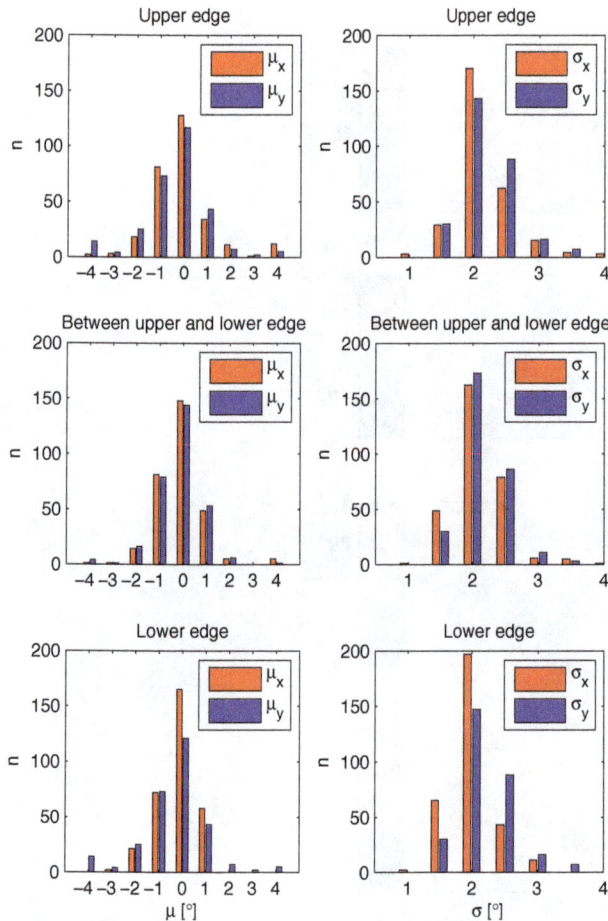

Fig. 7. Bar plots of the deviations from the zenith (left) and the width of the Gaussian fit (right) for different parts of the PMSE. The PMSE was divided in three parts, where the upper edge are the two uppermost heights and the two lowest heights are the lower edge. The part between both regions are pooled as one part. The histograms are fitted with a Gaussian fit (solid line) to get the mean values of μ_x, μ_y, σ_x, and σ_y in the different parts of the PMSE.

aspect sensitivity, indicated by the width of the brightness distribution, did not vary significantly from the lower to upper parts of the PMSE layer. Only a slight decrease in aspect sensitivity with height was found along the zonal cut.

Further data sets will be analysed in the future to demonstrate the capabilities of CRI with the MAARSY radar, and the beam weighting effects may be taken into account for some circumstances.

Acknowledgements. We like to thank to the Andya Rocket Range for the support while building and operating the MAARSY radar and the IAP personel for keeping the radar running. The radar development was build under grant 01 LP 0802A of Bundesmisterium für Bildung und Forschung. We thank the reviewers for the helpful comments on this paper. Topical Editor Matthias Förster thanks Jenn-Shyong Chen and an anonymous reviewer for their help in evaluating this paper.

References

Balsley, B., Ecklund, W., Carter, D., and Johnston, P.: Poker Flat MST Radar, 1st Results, Geophys. Res. Lett., 6, 921–924, 1979.

Chau, J. L., Renkwitz, T., Stober, G., and Latteck, R.: MAARSY multiple receiver phase calibration using radio sources, JASTP, doi:10.1016/j.jastp.2013.04.004, 2013.

Chen, J.-S. and Furumoto, J.: A novel approach to mitigation of radar beam weighting effect on coherent radar imaging using VHF atmospheric radar, IEEE Trans. Geosci. Remote Sens., 49, 3059–3070, doi:10.1109/TGRS.2011.2119374, 2011.

Chen, J.-S. and Furumoto, J.: Measurement of atmospheric aspect sensitivity using coherent radar imaging after mitigation of radar beam weighting effect, J. Atmos. Oceanic Technol., 30, 245–259, doi:10.1175/JTECH-D-12-00007.1, 2013.

Chen, J.-S., Hoffmann, P., Zecha, M., and Röttger, J.: On the relationship between aspect sensitivity, wave activity, and multiple scattering centers of mesosphere summer echoes: a case study using coherent radar imaging, Annales Geophysicae, 22, 807–817, 2004.

Chen, J.-S., Hoffmann, P., Zecha, M., and Hsieh, C.-H.: Coherent radar imaging of mesosphere summer echoes: Influence of radar beam pattern and tilted structures on atmospheric echo center, Radio Sci., 43, RS1002, 2008.

Chen, J.-S., Chen, C.-H., and Furu, J.: Radar beam- and range-weighting effects on three-dimensional radar imaging for the atmosphere, Radio Sci., 46, RS6014, 2011.

Chilson, P. B., Yu, T.-Y., Palmer, R. D., and Kirkwood, S.: Aspect sensitivity measurements of polar mesosphere summer echoes using coherent radar imaging, Annales Geophysicae, 20, 213–223, 2002.

Hocking, W., Ruster, R., and Czechowsky, P.: Absolute reflectivities and aspect sensitivities of VHF radiowave scatterers measured with the SOUSY radar, J. Atmos. Terr. Phys., 48, 131–144, doi:10.1016/0021-9169(86)90077-2, 1986.

Hoppe, U., Hall, C., and Röttger, J.: 1st observations of summer polar mesospheric backscatter with a 224 MHZ radar, Geophys. Res. Lett., 15, 28–31, 1988.

Hoppe, U.-P., Fritts, D., Reid, I., Czechowsky, P., Hall, C., and Hansen, T.: Multiple-frequency studies of the high-latitude summer mesosphere : implications for scattering processes, Journal of Atmospheric and Terrestrial Physics, 52, 907–926, doi:10.1016/0021-9169(90)90024-H, http://www.sciencedirect.com/science/article/pii/002191699090024H, 1990.

Kudeki, E. and Sürücü, F.: Radar interferometric imaging of field-aligned plasma irregularities in the equatorial electrojet, Geophys. Res. Lett, 18, 41–44, doi:10.1029/90GL02603, 1991.

Latteck, R., Singer, W., Rapp, M., Vandepeer, B., Renkwitz, T., Zecha, M., and Stober, G.: MAARSY: The new MST radar on Andoya-System description and first results, Radio Sci., 47, 2012.

Palmer, R., Gopalam, S., Yu, T., and Fukao, S.: Coherent radar imaging using Capon's method, Radio Sci., 33, 1585–1598, 1998.

Rapp, M. and Lübken, F.-J.: Polar mesosphere summer echoes (PMSE): review of observations and current understanding, Atmos. Chem. Phys., 4, 2601–2633, 2004, http://www.atmos-chem-phys.net/4/2601/2004/.

Röttger, J. and Vincent, R.: VHF radar studies of tropospheric velocities and irregularities using spaced antenna techniques, Geophys. Res. Lett., 5, 917–920, 1978.

Stober, G., Latteck, R., Rapp, M., Singer, W., and Zecha, M.: MAARSY-The new MST radar on Andøya: First results of spaced antenna and Doppler measurements of atmospheric winds in the troposphere and mesosphere using a partial array, ARS, 291–298, doi:10.5194/ars-10-291-2012, 2012.

Westman, A., Wannberg, G., and Pellinen-Wannberg, A.: Meteor head echo altitude distributions and the height cutoff effect studied with the EISCAT HPLA UHF and VHF radars, Ann. Geophys., 22, 1575–1584, 2004,
http://www.ann-geophys.net/22/1575/2004/.

Woodman, R.: Coherent radar imaging: Signal processing and statistical properties, Radio Sci., 32, 2373–2391, 1997.

Yu, T., Palmer, R., and Hysell, D.: A simulation study of coherent radar imaging, Radio Sci., 35, 1129–1141, 2000.

Yu, T.-Y., Palmer, R. D., and Chilson, P.: An investigation of scattering mechanisms and dynamics on PMSE using coherent radar imaging, J. Atmos. Solar-Terres. Phys., 63, 1797–1810, 2001.

Zecha, M., Röttger, J., Singer, W., Hoffmann, P., and Keuer, D.: Scattering properties of PMSE irregularities and refinement of velocity estimates, JASTP, 63, 201–214, 2001.

24

Occurrence frequencies of polar mesosphere summer echoes observed at 69° N during a full solar cycle

R. Latteck and J. Bremer

Leibniz Institute of Atmospheric Physics at the Rostock University, Schloss-Str. 6, 18225 Kühlungsborn, Germany

Correspondence to: R. Latteck (latteck@iap-kborn.de)

Abstract. Polar mesosphere summer echoes (PMSE) are strong enhancements of received signal power at very high radar frequencies occurring at altitudes between about 80 and 95 km at polar latitudes during summer. PMSE are caused by inhomogeneities in the electron density of the radar Bragg scale within the plasma of the cold summer mesopause region in the presence of negatively charged ice particles. Thus the occurrence of PMSE contains information about mesospheric temperature and water vapour content but also depends on the ionisation due to solar wave radiation and precipitating high energetic particles. Continuous and homogeneous observations of PMSE have been done on the North-Norwegian island Andøya (69.3° N, 16.0° E) from 1999 until 2008 using the ALWIN VHF radar at 53.5 MHz. In 2009 the Leibniz-Institute of Atmospheric Physics in Kühlungsborn, Germany (IAP) started the installation of the Middle Atmosphere Alomar Radar System (MAARSY) at the same location. The observation of mesospheric echoes could be continued in spring 2010 starting with an initial stage of expansion of MAARSY and is carried out with the completed installation of the radar since May 2011. Since both the ALWIN radar and MAARSY are calibrated, the received echo strength of PMSE from 14 yr of mesospheric observations could be converted to absolute signal power. Occurrence frequencies based on different common thresholds of PMSE echo strength were used for investigations of the solar and geomagnetic control of the PMSE as well as of possible long-term changes. The PMSE are positively correlated with the solar Lyman α radiation and the geomagnetic activity. The occurrence frequencies of the PMSE show slightly positive trends but with marginal significance levels.

1 Introduction

The phenomenon of strong radar echoes from the mesopause region during summer is well known from VHF radar observations at frequencies between ~ 2 MHz and ~ 1 GHz at polar and middle latitudes for more than 30 yr. These so called Polar Mesosphere Summer Echoes (PMSE) are caused by inhomogeneities in the electron density of a size comparable to the radar Bragg scale (about 3 m at 50 MHz radar frequency) in the presence of negatively charged aerosol particles. At mesospheric heights under normal condition irregularities in the electron density distribution in the order of about 3 m are smoothed out by molecular diffusion (Lübken et al., 2002). Small scale structures of e.g. charged ice particles created by turbulent advection leads to similar structures in the distribution of free electrons and positive ions due to multipolar coupling between all charged species (Rapp et al., 2008). These small scale structures in the electron gas may well exist a considerable time after molecular diffusion has already destroyed any structures in the neutral gas distribution (Rapp and Lübken, 2003). First mesospheric summer echoes were observed at the end of the 1970s at mid-latitudes above the Harz mountains in Germany (Czechowsky et al., 1979) before the first echoes of this type were seen at polar latitudes above Poker Flat (Ecklund and Balsley, 1981). A detailed review about the early observations of PMSE can be found in Cho and Röttger (1997) and an overview on the current understanding of this phenomenon has been published by Rapp and Lübken (2004).

This paper gives an overview about the long-term observations of PMSE obtained with two VHF radars at the Norwegian island Andøya (69.3° N, 16.0° E) from 1999 until 2012. In contrast to earlier investigations by Bremer et al. (2006) and Bremer et al. (2009) the dependency of PMSE occurrence rates on solar and geomagnetic activity is discussed on

Table 1. Basic radar and experiment parameters relevant for volume reflectivity determination.

radar period	ALWIN 1998–2008	ALWIN64 2009	MAARSY 2010	MAARSY 2011/2012
P_t	36 kW	36 kW	250 kW	736 kW
Tx ant.	144	6	147	433
G_t	28.3 dBi	15.6 dBi	29.0 dBi	33.5 dBi
θ	6°	14.1°	6°	3.6°
Rx ant.	24	64	7	7
G_r	20.6 dBi	20.1 dBi	15.5 dBi	15.5 dBi
e	0.58	0.58	0.54	0.54
τ	300 m	300 m	210 m	210 m
$\rightarrow c_{sys}$	2.5×10^{-8}	2.3×10^{-7}	1.3×10^{-8}	4.3×10^{-9}
m	32	64	32	32
n	16	8	8	8
g_r	101 dB	107 dB	101 dB	101 dB

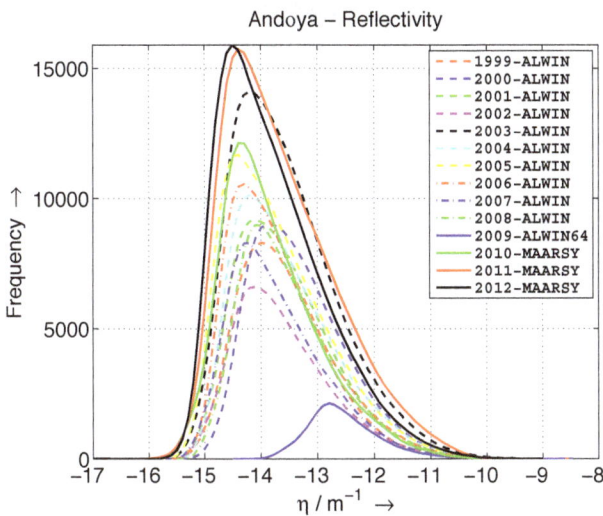

Fig. 1. Annual distributions of PMSE volume reflectivity as observed with ALWIN and MAARSY on Andøya between 1999 and 2012.

the basis of radar volume reflectivity which allows the comparison of observations from different radar systems.

2 Observation of polar mesosphere summer echoes at Andøya from 1999 until 2012

2.1 Radars used for PMSE observations

The observation of PMSE at Andøya started in the early 1990s using the mobile SOUSY radar (Czechowsky et al., 1984) and since 1994 the ALOMAR SOUSY radar (Singer et al., 1995). The latter was replaced in 1998 by the ALWIN radar (Latteck et al., 1999) which allowed unattended and remote controlled operation. After 10 yr of nearly continuous operation the ALWIN radar was switched off in September 2008 to be replaced by MAARSY (Latteck et al., 2010, 2012b) the more powerful and flexible Middle Atmosphere Alomar Radar System. Parts of the ALWIN antenna array and the container housing the transmitter and receiving units were moved approximately 100 m westward of the old radar site to be used for PMSE observation during the construction of the MAARSY antenna array in 2009. This interim solution called ALWIN64 (Latteck et al., 2010, 2012b) used six of the newly designed MAARSY antennas for transmission and 64 of the old ALWIN Yagi antennas for reception. The successive installation of MAARSY started in September 2009 upon completion of the new antenna array. The radar control and data acquisition hardware as well as 217 transceiver modules were installed in spring of 2010. A second stage of expansion to 343 transceiver modules was brought into service in November 2010 and the system was finally upgraded to 433 transceiver modules in May 2011 (Latteck et al., 2012a).

In this paper we concentrate on the period of continuous PMSE observation at Andøya starting in summer 1999 until autumn 2012, using ALWIN and MAARSY. Both radars were run at 53.5 MHz but most of the other technical parameters as e.g. the peak power P_t, the gains for the transmitting (G_t) and receiving (G_r) antennas, the antenna beam width θ and the loss factor e as well as operation parameters as e.g. the used effective pulse width τ and the number of used code elements m or the number of coherent integrations n were different. Table 1 lists the most important parameters of the used radars during the various periods of operation.

2.2 Comparability of radar observations

In order to compare PMSE occurrence rates the received echo power was converted into radar volume reflectivity which is a system independent parameter in contrast to, e.g., relative signal strength or signal-to-noise ratio, since its calculation considers the individual radar characteristics and experiment configurations. Radar volume reflectivity η is defined as the power which would be scattered if all powers were scattered isotropically with a power density equal to that of the backscattered radiation, per unit volume and per unit incident power density (Hocking, 1985). It can be expressed as

$$\eta = \frac{P_r \, 128 \, \pi^2 \, 2 \ln(2) \, r^2}{P_t \, G_t \, G_r \lambda^2 \, e \, \theta_{[1/2]}^2 \, c \, \tau} \tag{1}$$

where r is the range to the scatterers, G_t and G_r are the gain of the transmitting and receiving antenna respectively, $\theta_{[1/2]}$ is the half power half-width of the transmitting antenna beam, λ is the radar wavelength, e is the system efficiency containing mainly the losses of the antenna feeding system, P_t is the transmitted peak power, P_r is the received signal power, c is the speed of light, and τ is the effective pulse width (Hocking and Röttger, 1997). The factor $2 \ln(2)$ is a correction term related to the non-uniform antenna gain

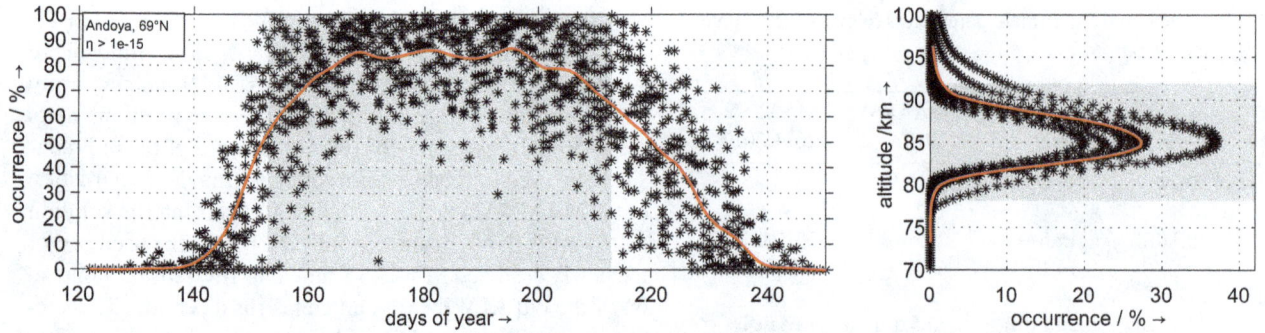

Fig. 2. Mean seasonal variation (left, thick red line) and mean height variation (right, thick red line) of PMSE occurrence rates at Andøya based on VHF-radar observations using ALWIN and MAARSY between 1999 and 2012. The star-symbols indicate the corresponding occurrence rates for the individual years. The gray shaded period and height range were used for the determination of mean PMSE occurrences for further investigation.

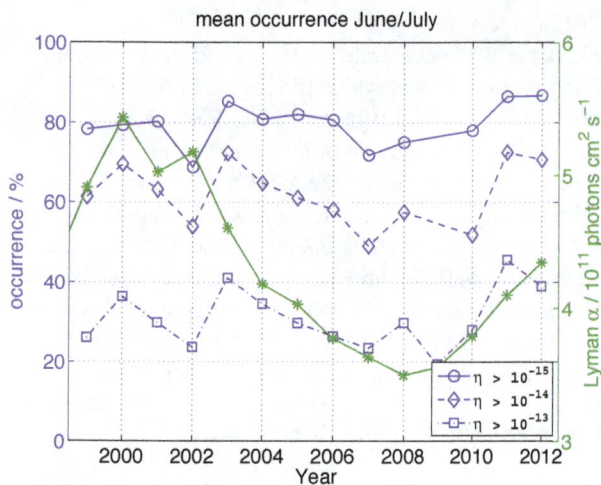

Fig. 3. Mean occurrence rates of PMSE for the months June and July from 1999 until 2012 derived from radar volume reflectivities $\eta \geq 10^{-15}$ (blue circles), $\eta \geq 10^{-14}$ (blue diamonds) and $\eta \geq 10^{-13}$ (blue squares) based on VHF-radar observations using ALWIN, ALWIN64 and MAARSY. The green stars represent the mean Lyman α values for June/July.

over the half-power beam-width (Probert-Jones, 1962; Skolnik, 1990). All system dependent parameters of Eq. (1) can be combined into a system factor c_{sys} as shown in Table 1 for the different periods and radar configurations. Hence the radar reflectivity η depends only on the range to the scatterers r and the absolute value of the received signal power P_r

$$\eta = P_r \cdot c_{sys} \cdot r^2 \qquad (2)$$

The correct determination of P_r requires the calibration of the receiving path of radar system as e.g. described in Latteck et al. (2008).

The determining of PMSE is based on 5 min averages of radar volume reflectivity. A PMSE event was defined as a radar reflectivity enhancement above the detection limit, but

for a minimum duration of 20 min (i.e. 4 consecutive 5-min averages) in one range gate (Latteck et al., 2008).

Figure 1 shows the annual distribution of PMSE volume reflectivity observed with ALWIN, ALWIN64 and MAARSY on Andøya between 1999 and 2012. The left borders of the distributions illustrate the differences in sensitivity of the operational modes of the radars, most convincingly shown by the solid blue line representing the ALWIN64 observations in 2009. The differences in echo detection sensitivity is also represented in the absolute values of the peaks of the distribution of the various years. This parameter is also predominantly affected by the annual variation of PMSE occurrence caused by geophysical variations, as can clearly be seen by comparing e.g. the dashed curves representing ALWIN results only.

2.3   Long term changes in PMSE occurrence rates

A qualitative comparison of the PMSE occurrence rates of the various years required the consideration of even marginal small differences in echo detecting sensitivity due to system and operational differences of the radars. Therefore a minimum value $\eta_{min} = 10^{-15}$ m^{-1} was defined and occurrence rates were calculated for PMSE greater than this threshold for all the year between 1999 and 2012. The left plot in Fig. 2 illustrates the results as daily values (black stars) of the individual years and a polynomial fit (red line) through the corresponding daily averages. The characteristic of the derived mean seasonal variation of PMSE is similar to the results found earlier by Bremer et al. (2009). Here, however, the start of the season is 4 days earlier near 16 May.

For the further study of long term changes in PMSE occurrence rates the possible influence of ionisation caused by solar and geomagnetic activity have been investigated following the procedure used by Bremer et al. (2009). Seasonal mean values of PMSE for the time period from 1 June until 31 July and the height range between 78.5 and 92 km (gray shaded areas in Fig. 2) have been calculated for every year and compared with corresponding mean values of the solar Lyman α radiation and of the geomagnetic Ap index. The Lyman α radiation is the dominant ionisation source (ionisation of nitric oxide) in the undisturbed ionospheric D region whereas the geomagnetic Ap index may be an indicator for precipitating high energetic particle flux (Bremer et al., 2009). Since the echo detection sensitivity for ALWIN64 in 2009 was extremely lower compared to the other years, two more sets of PMSE occurrence rates for the individual years were derived using thresholds of $\eta_{min} = 10^{-14}$ m^{-1} and $\eta_{min} = 10^{-13}$ m^{-1}. The latter allowed to include the 2009 observations in the study. Figure 3 shows the corresponding PMSE occurrence rates OR_{-15}, OR_{-14} and OR_{-13} derived for the three thresholds $\eta \geq 10^{-15}$ m^{-1} (blue circles), $\eta \geq 10^{-14}$ m^{-1} (blue diamonds) and $\eta \geq 10^{-13}$ m^{-1} (blue squares) respectively. The green stars are the mean Lyman α values for June/July representing mean solar activity during the last solar cycle. All three occurrence rate series show a similar behavior with slight variations but on different levels and indicate no obvious dependence on the solar cycle.

Qualitative results of the dependency of three PMSE occurrence rates OR_{-15}, OR_{-14} and OR_{-13} on the solar Lyman α radiation as well as on the geomagnetic Ap index are shown in the left, middle and right plots of the upper and middle panels of Fig. 5, respectively. The correlation between Lyman α and the PMSE occurrence rates OR_{-15} representing most echoes is slightly negative, whereas a positive correlation with Lyman α is given for OR_{-14} (middle panel) and OR_{-13} (right panel). A general positive correlation exists in the comparison of all three PMSE series with Ap, indicating an increasing PMSE occurrence with increasing ionisation level caused by enhanced geomagnetic activity.

The influence of both parameters on the PMSE occurrence has been removed by a twofold regression analysis

$$OR' = a + b \cdot Ly\,\alpha + c \cdot Ap \qquad (3)$$

following the procedure described in Bremer et al. (2009) in order to investigate possible long-term variations of PMSE. The differences $\Delta OR_{-15}, \Delta OR_{-14}$ and ΔOR_{-13} between the original observed occurrence rates OR_{-15}, OR_{-14} and OR_{-13} and the adjusted occurrences rates OR' are shown in the three plots of the lower panel of Fig. 5, respectively. An estimated linear trend line shows a slightly positive trend of $0.39 \pm 0.67\%/a, 0.52 \pm 0.83\%/a$ and $0.64 \pm 0.78\%/a$ for $\Delta OR_{-15}, \Delta OR_{-14}$ and ΔOR_{-13}, respectively. The error values of the different trends have been estimated for a significance level of 95%. As can be seen for all trends that the correct significance levels are smaller than 95% (77%, 80%, 90%).

3 Discussion

The present study was conducted to update earlier investigations (Bremer et al., 2006, 2009) with data series obtained

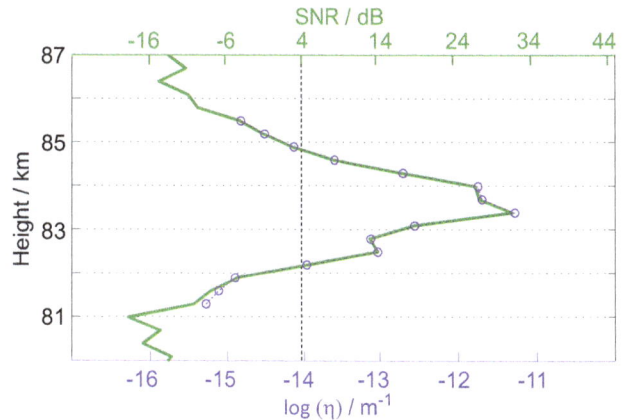

Fig. 4. Height profile (10 min average) of a PMSE obtained on 15 July 2008 at 12:00 UT using ALWIN shown as signal-to-noise ratio (green line) as well as volume reflectivity (blue dashed line and circles). The vertical black dashed line shows the connection between the threshold as used in Bremer et al. (2009) for ALWIN data and the corresponding value in volume reflectivity.

at the same location but with the more powerful and flexible Middle Atmosphere Alomar Radar System (MAARSY). In contrast to these earlier studies the PMSE occurrence rates were derived on the basis of radar volume reflectivity using a common threshold instead of using several SNR-thresholds adjusted to the system parameters. This allows a more flexible comparison of observations from different radar systems as once the received echo power is converted into reflectivity various thresholds can easily be chosen in order to consider the different PMSE detection limits of the various radars based on differences in system parameters and experiment configurations as shown in Fig. 1.

Three common thresholds of volume reflectivity OR_{-15}, OR_{-14} and OR_{-13} have been used to investigate the dependence of PMSE occurrence on solar and geomagnetic activity. The results are presented in the upper and middle panels of Fig. 5. The correlation between Lyman α and PMSE occurrence rates OR_{-15} (Fig. 5, left top plot) representing all detected echoes is slightly negative whereas a positive correlation with Lyman α is given for OR_{-14} (Fig. 5, middle top plot) and OR_{-13} (Fig. 5, right top plot) both representing well pronounced echoes only. The latter also eliminate daily variations on PMSE occurrence due to daily variation of background noise. Furthermore OR_{-14} is very comparable with the SNR-threshold as used in Bremer et al. (2009) for ALWIN data as shown in Fig. 4. The results indicate a generally low dependence of PMSE occurrence on solar activity and also confirms the conclusion by Bremer et al. (2009) that various processes caused by solar radiation, reduce the influence of increasing electron density due to increasing ionisation which actually should increase PMSE.

The correlation found between the PMSE occurrence series and the geomagnetic activity represented by Ap was

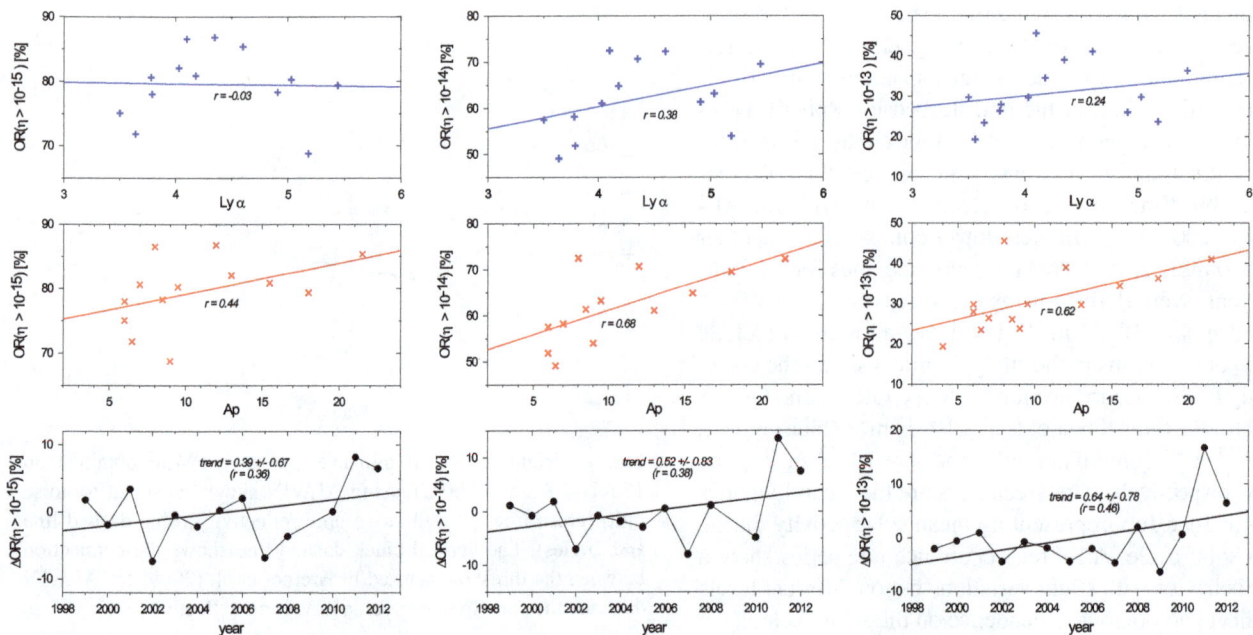

Fig. 5. Dependence of the PMSE occurrence rates OR_{-15} (left panels), OR_{-14} (middle panels) and OR_{-13} (right panels) on the solar Lyman α radiation (upper panels) and on geomagnetic Ap index (middle panels) as well as long-term variation of the PMSE occurrence rates after elimination of their solar and geomagnetically caused parts (lower panels).

positive for OR_{-15} (Fig. 5, left middle plot), OR_{-14} (Fig. 5, center plot) as well as for OR_{-13} (Fig. 5, right middle plot). This is an indication for a direct influence of precipitating fluxes of high energetic particles to the increase of electron densities and therefore the increase of PMSE occurrence. The PMSE trends in the lower part of Fig. 5 are for all three investigated thresholds of volume reflectivity positive with significance levels between about 77–90 %.

4 Summary and conclusion

Polar mesospheric summer echoes from 14 yr of continuous VHF radar observations at Andøya (69° N) have been investigated for solar and geomagnetic control as well as for possible long-term changes. The investigation is based on occurrence rates derived from radar volume reflectivity of the received echoes considering the differences in technical performance and experimental operation of the radars ALWIN and MAARSY used at the same site during the long period of observation. The given differences in detection sensitivity of the used radars were taken into account by using various common thresholds of PMSE echo strength for the determination of the occurrence frequencies.

The correlation of PMSE occurrence with the solar Lyman α radiation depends on the applied threshold for the detection of PMSE. A positive correlation is only seen if well pronounced echoes are taken into account. Nevertheless the results confirm the conclusion by Bremer et al. (2009) that various processes caused by solar radiation reduce the influ-

ence of increasing electron density due to increasing ionisation which should actually increase PMSE.

A significant correlation between the PMSE occurrence rate and the geomagnetic activity represented by Ap was found independently of the used PMSE detection thresholds, confirming the direct influence of precipitating fluxes of high energetic particles to the increase of electron densities and therefore the increase of PMSE.

After eliminating the influences of solar and geomagnetic activity, the PMSE occurrence frequencies show slightly positive trends but with small significance levels. The latter is probably caused by the short period of observation of 14 yr.

Future activities will be directed to the integration of PMSE data from earlier years obtained with the ALOMAR SOUSY radar between 1994 and 1997. Since ALWIN data can be directly converted from SNR into volume reflectivity and vice versa as illustrated in Fig. 4, the ALWIN data set could be used to adjust the older (SOUSY) and newer (MAARSY) data to a common threshold for long term investigation of PMSE occurrence rates. This will improve the overall data availability and should also improve the significance of the trends.

References

Bremer, J., Hoffmann, P., Höffner, J., Latteck, R., Singer, W., Zecha, M., and Zeller, O.: Long-term changes of mesospheric summer echoes at polar and middle latitudes, J. Atmos. Solar Terr. Phys., 68, 1940–1951, doi:10.1016/j.jastp.2006.02.012, 2006.

Bremer, J., Hoffmann, P., Latteck, R., Singer, W., and Zecha, M.: Long-term changes of (polar) mesosphere summer echoes, J. Atmos. Solar-Terr. Phys., 71, 1571–1576, doi:10.1016/jastp.2009.03.010, 2009.

Cho, J. Y. N. and Röttger, J.: An updated review of polar mesosphere summer echoes: Observation, theory, and their relationship to noctilucent clouds and subvisible aerosols, J. Geophys. Res., 102, 2001–2020, 1997.

Czechowsky, P., Rüster, R., and Schmidt, G.: Variations of mesospheric structures in different seasons, Geophys. Res. Lett., 6, 459–462, 1979.

Czechowsky, P., Schmidt, G., and Rüster, R.: The mobile SOUSY Doppler radar: Techical design and first results., Radio Sci., 19, 441–450, 1984.

Ecklund, W. L. and Balsley, B. B.: Long-term observations of the Arctic mesosphere with the MST radar at Poker Flat, Alaska, J. Geophys. Res., 86, 7775–7780, 1981.

Hocking, W. K.: Measurements of turbulent energy dissipation rates in the middle atmosphere by radar techniques: A review, Radio Sci., 20, 1403–1422, 1985.

Hocking, W. K. and Röttger, J.: Studies of polar mesosphere summer echoes over EISCAT using calibrated signal strengths and statistical parameters, Radio Sci., 32, 1425–1444, 1997.

Latteck, R., Singer, W., and Bardey, H.: The ALWIN MST radar – Technical design and performances, in: Proceedings of the 14th ESA Symposium on European Rocket and Balloon Programmes and Related Research, Potsdam, Germany (ESA SP–437), edited by Kaldeich-Schürmann, B., 179–184, 1999.

Latteck, R., Singer, W., Morris, R. J., Hocking, W. K., Murphy, D. J., Holdsworth, D. A., and Swarnalingam, N.: Similarities and differences in polar mesosphere summer echoes observed in the Arctic and Antarctica, Ann. Geophys., 26, 2795–2806, doi:10.5194/angeo-26-2795-2008, 2008.

Latteck, R., Singer, W., Rapp, M., and Renkwitz, T.: MAARSY – The new MST radar on Andøya/ Norway, Adv. Radio Sci., 8, 219–224, doi:10.5194/ars-8-219-2010, 2010.

Latteck, R., Singer, W., Rapp, M., Renkwitz, T., and Stober, G.: Horizontally resolved structures of radar backscatter from polar mesospheric layers, Adv. Radio Sci., 10, doi:10.5194/ars-10-1-2012, 2012a.

Latteck, R., Singer, W., Rapp, M., Vandepeer, B., Renkwitz, T., Zecha, M., and Stober, G.: MAARSY – The new MST radar on Andøya: System description and first results, Radio Sci., 47, RS1006, doi:10.1029/2011RS004775, 2012b.

Lübken, F.-J., Rapp, M., and Hoffmann, P.: Neutral air turbulence and temperatures in the vicinity of polar mesosphere summer echoes, J. Geophys. Res., 107, 4273, doi:10.1029/2001JD000915, 2002.

Probert-Jones, J. R.: The Radar Equation in Meteorology, Q. J. Roy. Meteorol. Soc., 88, 485–495, 1962.

Rapp, M. and Lübken, F.-J.: On the nature of PMSE: Electron diffusion in the vicinity of charged particles revisited, J. Geophys. Res., 108, 8437, doi:10.1029/2002JD002857, 2003.

Rapp, M. and Lübken, F.-J.: Polar mesosphere summer echoes (PMSE): Review of observations and current understanding, Atmos. Chem. Phys., 4, 2601–2633, doi:10.5194/acp-4-2601-2004, 2004.

Rapp, M., Strelnikova, I., Latteck, R., Hoffmann, P., Hoppe, U.-P., Häggström, I., and Rietveld, M. T.: Polar mesosphere summer echoes (PMSE) studied at Bragg wavelengths of 2.8 m, 67 cm, and 16 cm, J. Atmos. Solar-Terr. Phys., 70, 947–961, doi:10.1016/j.jastp.2007.11.005, 2008.

Singer, W., Keuer, D., Hoffmann, P., Czechowsky, P., and Schmidt, G.: The ALOMAR SOUSY radar: Technical design and further developments, in: Proceedings of the 12th ESA Symposium on European Rocket and Balloon Programmes and Related Research, Lillehammer, Norway (ESA SP–370), 409–415, 1995.

Skolnik, M.: Radar handbook, McGraw-Hill, 1990.

25

Some anomalies of mesosphere/lower thermosphere parameters during the recent solar minimum

Ch. Jacobi[1]**, P. Hoffmann**[1]**, M. Placke**[2]**, and G. Stober**[2]

[1]Institute for Meteorology, University of Leipzig, Stephanstr. 3, 04103 Leipzig, Germany
[2]Leibniz Institute of Atmospheric Physics at the Rostock University, Schlossstraße 6, 18225 Kühlungsborn, Germany

Abstract. The recent solar minimum has been characterized by an anomalous strong decrease of thermospheric density since 2005. Here we analyze anomalies of mesosphere/lower thermosphere parameters possibly connected with this effect. In particular, nighttime mean LF reflection heights measured at Collm, Germany, show a very strong decrease after 2005, indicating a density decrease. This decrease is also visible in mean meteor heights measured with VHF meteor radar at Collm. This density decrease is accompanied by an increase of gravity wave (GW) amplitudes in the upper mesosphere and a decrease in the lower thermosphere. On the decadal scale, GWs are negatively correlated with the background zonal wind, but this correlation is modulated in the course of the solar cycle, indicating the combined effect of GW filtering and density decrease.

1   Introduction

It is widely known that solar variability influences the atmosphere (Gray et al., 2010), e.g., the dynamics of the middle and upper atmosphere. In particular, search for an effect of the 11-year solar Schwabe cycle has been undertaken, for example, to explain part of the observed variability of mesosphere and lower thermosphere (MLT), which can be studied by radars. Indeed, indication for a solar effect has been found in MLT radar wind time series over Central Europe (Jacobi and Kürschner, 2006; Keuer et al., 2007). This effect is more pronounced in summer than in winter owing to the more disturbed middle atmosphere during winter, and essentially consists in a stronger mesospheric jet during solar maximum. In the summer lower thermosphere, above the mean wind reversal, this results in weaker westerly winds during solar maximum, i.e. a negative correlation between mean wind and solar flux. This effect is mainly caused by the reaction of the stratosphere and mesosphere on solar flux

Correspondence to: Ch. Jacobi
(jacobi@uni-leipzig.de)

changes, and consequently decreases with altitude (Keuer et al., 2007).

The lower thermosphere wind reversal is owing to momentum deposition of gravity waves (GWs) in the upper mesosphere. According to linear theory, GWs with wind speed amplitudes exceeding the intrinsic phase speed, i.e., the difference between phase speed and mean wind speed, break and transfer part of their momentum to the background flow. This means that a GW amplitude remains at the breaking level, equaling the intrinsic phase speed, which is called a saturated GW. During summer/winter, strong mesospheric easterlies/westerlies prevent westward/eastward travelling GWs from propagating to the upper mesosphere, because their intrinsic phase speeds become small or even zero (a so called critical line). This means that in summer/winter only eastward/westward travelling GWs remain in the MLT region. Since these waves are still saturated, during summer/winter a stronger mesospheric easterly/westerly jet is expected to be connected with larger GW amplitudes. Consequently, in the upper mesosphere a positive correlation between solar flux and GW amplitudes is expected. Jacobi et al. (2006) have reported such a connection using GW proxy from Collm E-region drift measurements.

Solar cycles can differ from one to another. Especially, the recent solar minimum has been extremely extended and extraordinarily deep. Consequently, it led to extreme upper atmosphere reactions, in particular a decrease of thermospheric density (Emmert et al., 2010; Solomon et al., 2010) which exceeds the expectations that would have been based on conventional solar indices like the sunspot number or F10.7.

Lower ionospheric electron density reacts on the solar cycle, which leads to an 11-year modulation of radio wave reflection heights (e.g., Bremer and Berger, 2002; Bremer, 2005). These have been observed, e.g., by Kürschner and Jacobi (2003) who found that the Collm LF reflection heights (177 kHz, distance to transmitter about 160 km) are about 2 km lower during solar maximum than during solar minimum. This result has confirmed earlier findings (Entzian, 1967). Owing to the increased ionisation during solar maximum, however, this effect is superposed by thermal shrinking of the mesosphere during solar minimum, since the middle atmosphere has a solar cycle signal of about 2 K difference between maximum and minimum (Keckhut et al., 1995). This thermal shrinking is usually overcompensated by increased ionisation. We are interested, whether the recent extreme solar minimum has led to anomalous signatures either in MLT wind or density. We focus on the summer MLT, which is not that much influenced by stratospheric planetary wave activity.

2   Measurements

2.1   Collm LF lower ionospheric drifts, reflection heights, and GW estimates

At Collm Observatory, MLT winds have been obtained by D1 LF radio wind measurements from 1959–2008, using the ionospherically reflected sky wave of three commercial radio transmitters. The data are combined to half-hourly zonal and meridional mean wind values. The virtual reflection heights have been estimated since late 1982 using measured travel time differences between the separately received ionospherically reflected sky wave and the ground wave (Kürschner et al., 1987). More details of the Collm LF system are given in Jacobi (2011).

Since the LF reference height changes in the course of the day, a continuous time series at some fixed height is not available and consequently GW spectra cannot be calculated. However, using the method presented by Gavrilov et al. (2001a, b), horizontal wind fluctuations can be obtained which may serve as GW proxy. In brief, the method uses differences between two consecutive half-hourly means of the horizontal wind, if the mean reflection height does not change more than 1 km between these time intervals. Calculation of such differences combined with previous half-hourly averaging of the data is equivalent to a numerical filter passing harmonics with periods of 0.7–3 h with a maximum at about 1 h (Gavrilov et al., 2001a). Then the squared differences are averaged for 10 km height ranges during the time interval under consideration. Jacobi et al. (2006) has used this method to analyze the Collm dataset from 1984–2003. They found an 11-year solar cycle with larger GW amplitudes during solar maximum, but their dataset did not include the recent solar minimum.

2.2   Collm meteor radar

At Collm Observatory (51.3° N, 13° E), a SKiYMET meteor radar (MR) is operated on 36.2 MHz to measure horizontal winds in the 80–100 km height range, meteor rates and heights, and further meteor parameters since August 2004. The radar and the hourly wind detection are described in Jacobi (2011). Monthly mean wind parameters are obtained from one month of half-hourly winds applying a multiple regression analysis including the mean wind and tidal compo-

nents. Based upon 2-hourly means, GW variances and fluxes are obtained according to Hocking (2005) by projecting the 2-hourly mean GW fluxes in a 3 km height gate to the radial direction of the respective meteor, and minimizing the difference between projected and measured meteor trail drift variances. Details can be found in Placke et al. (2011). The data are averaged to obtain 3-monthly means for each height gate under consideration.

2.3   GW potential energy from SABER

The SABER instrument on the TIMED satellites (Russell et al., 1999; Mertens et al., 2001, 2004) scans the atmosphere from about 52° of one hemisphere to 83° of the other. This latitude range is reversed by a yaw manoeuvre every 60 days. Due to the sun-synchronous orbital geometry the spacecraft passes the equator always at the same local time (12:00 LT) on the day side. Each single temperature profile, having a vertical resolution of 0.5 km, is high-pass filtered to analyze waves with vertical wavelength of up to 6 km. These filtered data represent deviations T' of temperature from the background. Their average specific potential energy in a 10 km vertical window is obtained through

$$E_{\mathrm{p}} = \frac{1}{2}\left(\frac{g}{N}\right)^2 \overline{\left(\frac{T'}{\overline{T}}\right)^2},\tag{1}$$

with the bar denoting averaging over the respective 10 km window, and N and g being the buoyancy frequency and the acceleration due to gravity, respectively. The vertical profile of GW activity is obtained through shifting the data window by 1 km step and repeating the procedure. We use seasonal means at 45° N of zonally averaged potential energy here. The method of potential energy determination has frequently been applied to GPS radio occultations (e.g., Fröhlich et al., 2007). Note that limb scanning of the atmosphere by SABER only reveals certain parts of the GW spectrum due to the integration along the line of sight (Preusse et al., 2006). Another limitation is made by the chosen vertical filter, which allows only GWs of short wavelengths to be studied.

3   GW proxy decadal variability

Jacobi et al. (2006) had shown that there is a solar cycle influence on GW activity as measured by LF over Collm. Figure 1 presents the summer (JJA) mean time series of GW variance $\zeta'^2 = u'^2 + v'^2$ at 100 km virtual height, which represents approximately 91 km real height (Jacobi, 2011). The data are an update from Jacobi et al. (2006). The 13-monthly mean sunspot number is added. Clearly, there is a solar cycle in the GW variance, and from visual inspection a decreasing long-term trend is also visible. We thus added, as a red solid line, a least squares fit of a linear trend superposed by a solar cycle

$$\zeta'^2 = a + b \times yr + c \times R,\tag{2}$$

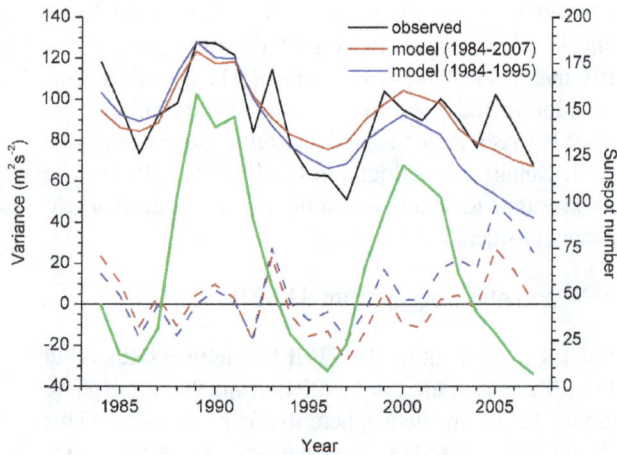

Fig. 1. JJA mean LF GW proxy, and fit including linear trend and a solar cycle according to Eq. (2) at 100 km virtual height (approx. 91 km real height). The fit was performed both using the complete dataset 1984–2007 (red curve) and using part of the dataset until 1995 (blue curve). In the lower part the respective residuals are given as dashed lines. The sunspot number is also added as green line.

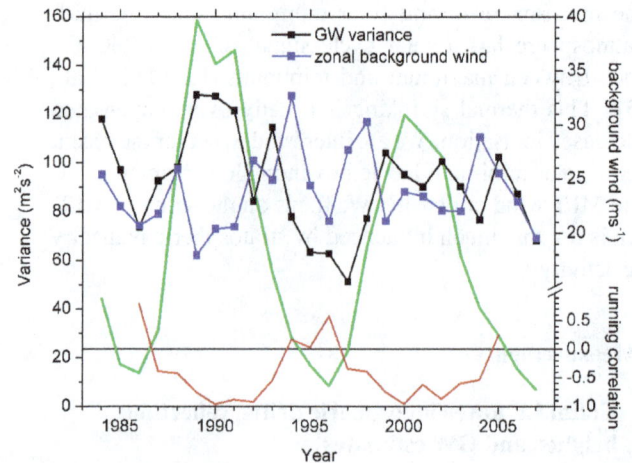

Fig. 2. JJA mean LF GW proxy (black) and zonal mean wind (blue) at 100 km virtual height. In the lower part of the figure, running correlation coefficients between GW proxy and mean wind are added. The sunspot number is added as green line.

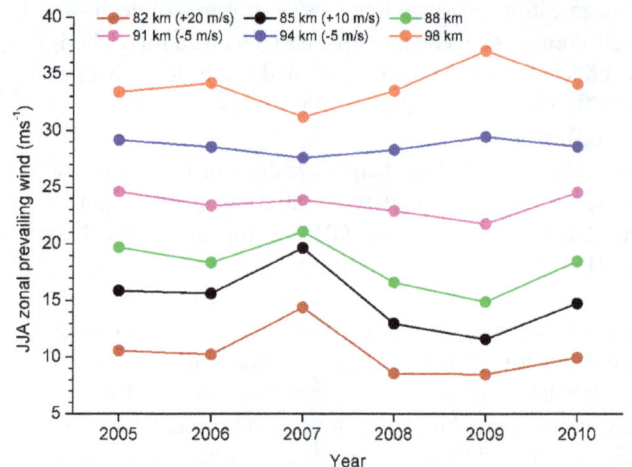

Fig. 3. Collm June–August mean zonal prevailing winds measured by meteor radar.

with R being the 13-monthly smoothed relative sunspot number. We are only interested in long-term and qualitative connections, so employing other widely used solar proxies as F10.7 are not superior to use of R here. As can be seen in the lower part of the figure, the residuals are not normally distributed, and generally the model is not valid during the 1996 as well as the recent solar minimum. Using the same model, but including only data until 1995 (blue line) reveals that the time interval until the early 2000s is well represented by that model, and after 2004 there is a drastic change. We therefore conclude that there is a possible change in dynamical regime since about 2004.

In the case of saturated GWs, linear theory predicts amplitudes proportional to the intrinsic phase speed. Consequently, since GW phase speeds must be positive (eastward) in the summer MLT owing to the filtering effect of the stratospheric and mesospheric easterlies, a negative correlation is expected between the background wind and the GW amplitudes. In Fig. 2 we present GW variance together with the background mean zonal wind at 100 km virtual height. Note that the background wind is simply the mean of the zonal wind averaged over those times when GW amplitudes have been calculated, and thus may deviate from the prevailing wind. There is an overall anticorrelation between GW variances and zonal winds, as expected from linear theory. However, during solar minimum the correlation reverses. The running correlation (Kodera, 1993) between GW proxy and mean zonal wind is added in the lower part of Fig. 2. Due to the shortness of the time series, only 5 data points are used for each calculation. A clear solar cycle modulation is visible. The running correlation is correlated with the

sunspot number time series with a correlation coefficient of $r = -0.80$. It is also remarkable that this modulation takes place during each solar minimum since 1986. Note also that the increase of GW variances in 2005, when solar flux already decreases and decreasing GW variances are expected, has its counterpart in a peak in 1993. We may conclude that there is obviously a different regime of mean wind-GW coupling during each solar minimum, which is, however, more emphasized during the recent minimum.

LF height measurements at Collm have been terminated in late 2007, so that the solar minimum is not completely covered by them. To analyze winds and waves during the minimum, in Fig. 3 MR summer mean zonal prevailing winds at 6 height gates are presented. Clearly, interannual variability of winds in the upper and lower height gates is opposite.

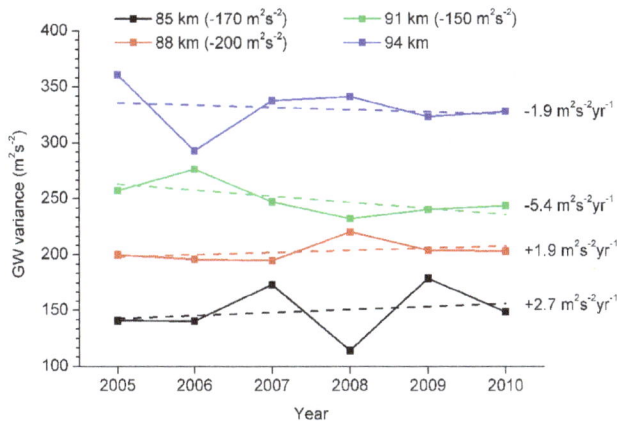

Fig. 4. Horizontal wind variance calculated within 2-h intervals using Collm MR wind measurements for 4 height gates.

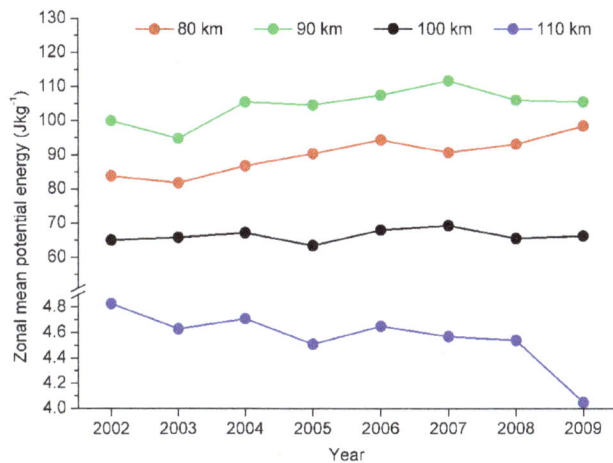

Fig. 5. JJA mean potential energy at 45° N from SABER temperature profiles. Data are averages over a 10 km vertical window, and means over all longitudes.

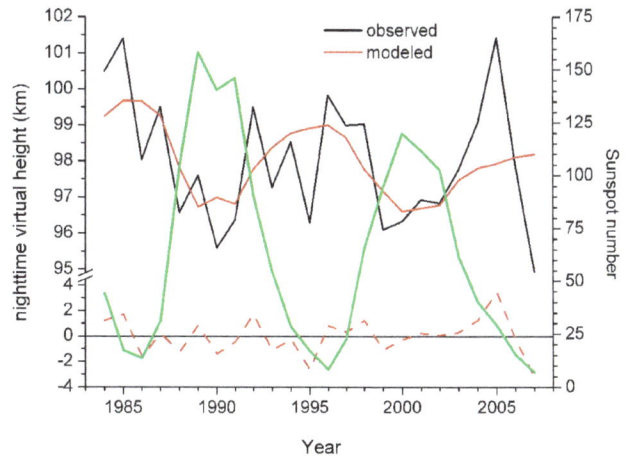

Fig. 6. Collm LF virtual nighttime (22:00–02:00 LT) reflection heights. A linear fit according to Eq. (2) is added as red line, as well as the residuals (red dashed line). The sunspot number is added as green line.

energy, while Figs. 1–4 represent point measurements and non-zonal structures are likely to exist in the MLT. Note that the above mentioned trends are only valid for heights above 90 km, while for the lower height gates the winds and GW amplitudes behave in an opposite manner.

4 LF virtual heights and mean meteor altitudes

The solar modulation of correlation between zonal wind and GW amplitudes suggests that the decrease of reference heights during solar minimum of wind systems may play a role. Collm LF virtual nighttime (22:00–02:00 LT) reflection heights are shown in Fig. 6. A multiple linear fit after Eq. (2), but analyzing virtual height instead of variance, is added as red line, as well as the residuals (red dashed line). The sunspot number is added as green line. Note that real heights and height differences are much smaller than virtual height differences. Thus, the strong decrease of LF heights after 2005, for example, represents a real height decrease of about 2 km only. From the residuals in Fig. 6 one can see that the recent minimum is outstanding. Generally, there is an ionization driven solar cycle of reflection heights such that these are lower during solar maximum than during solar minimum. Thus, the strong decrease after 2005 is unexpected. However, a similar variability has already been observed after 1993 during the last solar minimum, although with much smaller amplitudes, LF reflection height variability is influenced by changes in ionization and mesospheric shrinking. According to Fig. 6 this would mean that during the recent solar minimum thermal shrinking has overcompensated the ionization effect.

Note that the reference height changes shown in Fig. 6 do not show the real variations of a line of constant density,

This is explained by GW acceleration and filtering in the mesosphere. In the case of strong/weak mesospheric easterlies, GW amplitudes are large/small, which then lead to strong/weak vertical wind shear. Figure 3 shows that above 91 km winds are decreasing until 2007, which is qualitatively consistent with the decrease of the LF winds during 2005–2007. This also agrees with the zonal LF wind decrease after 1994 (Fig. 2). Note that after 2007, MR winds in the uppermost height gats are increasing again, which would be consistent with the increasing LF winds after 1996 in Fig. 2.

The LF zonal wind decrease during the first part of the solar minimum is accompanied by GW proxy amplitude decrease. MR GW analyses of the recent minimum (Fig. 4) again show this effect qualitatively. The same is the case for SABER potential energy (Fig. 5), although here only the GW amplitudes above ~100 km decrease with time. This may be due to the fact that we present zonal mean potential

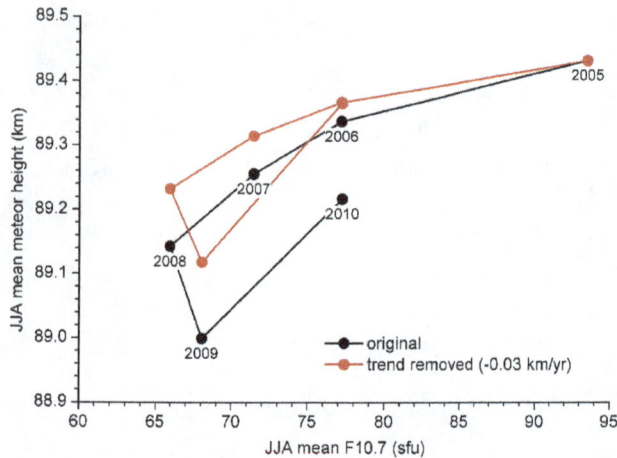

Fig. 7. JJA mean meteor heights over Collm vs. F10.7 solar flux. The red curve represents meteor heights after removing a linear trend of $-30\,myr^{-1}$.

for example. A better proxy is the mean height of meteors because these, constant meteor parameters as mass and velocity assumed, burn at a height that is determined by the density distribution. Figure 7 shows June–August mean meteor heights from 2005–2010, plotted against the solar flux. As expected, the meteor heights increase with solar activity. Note, however, that the mean meteor heights during 2008 and 2010, at the same level of solar flux, are different. This is partly due to the fact that the decrease is superposed by the long-term changes of mesospheric density. Bremer and Peters (2008) found a long-term decrease of $-30\,myr^{-1}$ from LF reflection heights (and excluding the solar cycle). Subtracting this from the measured meteor heights in Fig. 7 (red line) shows that then the 2008 and 2010 heights, at the same solar flux, have exactly the same values when long-term cooling of the middle atmosphere is taken into account.

There remains some sort of hysteresis, so that the density increases about one year later than the solar flux does. Similar delay has been found in other parameters, too. A delay of one year of noctilucent cloud occurrence and solar activity has been reported by DeLand et al. (2006) and Bremer et al. (2009). Ortiz de Adler and Elias (2008) showed a similar hysteresis in ionospheric foF2 data. Jacobi et al. (2008) showed that MLT planetary wave activity lags the solar cycle by 1–3 years.

5   Discussion and conclusions

Linear theory predicts that GW amplitudes are proportional to the intrinsic phase speed and thus, in the case of a given GW with specified phase speed, by the zonal wind itself. Therefore a general negative correlation between mean wind and GW amplitudes is expected in the upper mesosphere. Figure 2 indeed shows such a connection for solar maxi-

mum. During solar maximum, mesospheric eastward winds are stronger, leading to a negative correlation between solar flux and mean wind there (e.g., Keuer et al., 2007). However, stronger winds and consequently larger GW amplitudes lead to a stronger GW drag, stronger wind shear, and therefore a decreasing solar cycle effect on the mean wind with height, which reverses at some altitude. Such a tendency is visible in the data shown by Keuer et al. (2007) and leads to the reversed interannual variability of the mean winds at lower and upper MR height gates seen in Fig. 3.

The positive correlation between GW amplitudes and zonal winds during solar minimum is unexpected at first glance, but may be explained by a downward shift of the wind systems owing to thermal shrinking of the mesosphere. In such a case, at fixed geometric altitude the reference altitude with respect to the wind system is shifted upward, and therefore the increasing mean wind with decreasing solar flux (which is typical for the mesosphere) changes to a decreasing mean wind with decreasing solar flux, which is typical for the lower thermosphere. This may explain the positive correlation during solar minimum, while there is a negative correlation during the other years, then simply in accordance with linear theory.

The outlined connection for solar minimum cannot be explained by linear theory, at least for the lower thermosphere, because this would, in the case of saturated GWs, always predict negative correlation between GW amplitudes and mean wind at any height. The outlined process also does not explain the change of mean wind tendencies after 2007. Comparison of Figs. 4 and 5 shows that after 2007 decreasing/increasing mean winds are connected with increasing/decreasing GW amplitudes at lower/upper height gates, as expected by linear theory. Thus, there is probably both an effect of GW-mean wind interaction and thermal shrinking, leading to different effects on the mean wind, but for specific years one cannot conclusively predict which one is dominating.

Whether or not the above mentioned coupling processes really work requires more detailed analyses, including more satellite analyses, further radars, and numerical modeling. However, the observations of MLT GWs, mean winds, and reference heights already suggest that there is a height shift during solar minimum which may influence vertical coupling between mesosphere and lower thermosphere. The recent solar minimum represents an extreme case, but the fundamental variability, as shown by the LF measurements, was not qualitatively (although quantitatively) different from the last solar minimum, at least as far as the LF measurements at Collm are concerned. In the thermosphere, however, density decrease during the recent minimum was extreme. Thus, there are still open questions concerning solar variability and its effect on the MLT.

Acknowledgements. This study was supported by Deutsche Forschungsgemeinschaft under JA 836/22-1. Topical Editor Matthias Förster thanks Edward Kazimirovsky and Suvarna Fadnavis for their help in evaluating this paper.

References

Bremer, J.: Detection of long-term trends in the mesosphere/lower thermosphere from ground-based radio propagation measurements, Adv. Space Res., 35, 1398–1404, 2005.

Bremer, J. and Berger, U.: Mesospheric temperature trends derived from ground-based LF phase-height observations at mid-latitudes: comparison with model simulations, J. Atmos. Solar-Terr. Phys., 64, 805–816, 2002.

Bremer, J. and Peters, D.: Influence of stratospheric ozone changes on long-term trends in the meso- and lower thermosphere, J. Atmos. Solar-Terr. Phys., 70, 1473–1481, 2008.

Bremer, J., Hoffmann, P., Latteck, R., Singer, W., and Zecha, M.: Long-term changes of (polar) mesosphere summer echoes, J. Atmos. Solar-Terr. Phys., 71, 1571–1576, 2009.

DeLand, M. T., Shettle, E. P., Thomas, G. E., and Olivero, J. J.: A quarter-century of satellite polar mesospheric cloud observations, J. Atmos. Solar-Terr. Phys., 68, 9–29, 2006.

Emmert, J. T., Lean, J. L., and Picone, J. M.: Record-low thermospheric density during the 2008 solar minimum, Geophys. Res. Lett., 37, L12102, doi:10.1029/2010GL043671, 2010.

Entzian, G.: Der Sonnenfleckenzyklus in der Elektronenkonzentration der D-Region, Kleinheubacher Ber., 12, 309–313, 1967.

Fröhlich, K., Schmidt, T., Ern, M., Preusse, P., de la Torre, A., Wickert, J., and Jacobi, Ch.: The global distribution of gravity wave energy in the lower stratosphere derived from GPS data and gravity wave modelling: Attempt and challenges, J. Atmos. Solar-Terr. Phys., 69, 2238–2248, 2007.

Gavrilov, N. M., Jacobi, Ch., and Kürschner, D.: Climatology of ionospheric drift perturbations at Collm, Germany, Adv. Space Res., 27, 1779–1784, 2001a.

Gavrilov, N. M., Jacobi, Ch., and Kürschner, D.: Short-period variations of ionospheric drifts at Collm and their connection with the dynamics of the lower and middle atmosphere, Phys. Chem. Earth, 26, 459–464, 2001b.

Gray, L. J., Beer, J., Geller, M., Haigh, J. D., Lockwood, M., Matthes, K., Cubasch, U., Fleitmann, D., Harrison, G., Hood, L., Luterbacher, J., Meehl, G. A., Shindell, D., van Geel, B., and White, W.: Solar influences on climate, Rev. Geophys., 48, RG4001, doi:10.1029/2009RG000282, 2010.

Hocking, W. K.: A new approach to momentum flux determinations using SKiYMET meteor radars, Ann. Geophys., 23, 2433–2439, doi:10.5194/angeo-23-2433-2005, 2005.

Jacobi, Ch.: Meteor radar measurements of mean winds and tides over Collm (51.3° N, 13° E) – comparison with LF drift measurements 2005–2007, Adv. Radio Sci., this issue, 2011.

Jacobi, Ch. and Kürschner, D.: Long-term trends of MLT region winds over Central Europe, Phys. Chem. Earth, 31, 16–21, 2006.

Jacobi, Ch., Gavrilov, N. M., Kürschner, D., and Fröhlich, K.: Gravity wave climatology and trends in the mesosphere/lower thermosphere region deduced from low-frequency drift measurements 1984–2003 (52.1° N, 13.2° E), J. Atmos. Solar-Terr. Phys., 68, 1913–1923, 2006.

Jacobi, Ch., Hoffmann, P., and Kürschner, D.: Trends in MLT region winds and planetary waves, Collm (52° N, 15° E), Ann. Geophys., 26, 1221–1232, doi:10.5194/angeo-26-1221-2008, 2008.

Keckhut, P., Hauchecorne, A., and Chanin, M. L.: Midlatitude long-term variability of the middle atmosphere: Trends and cyclic and episodic changes, J. Geophys. Res., 100, 18887–18897, 1995.

Keuer, D., Hoffmann, P., Singer, W., and Bremer, J.: Long-term variations of the mesospheric wind field at mid-latitudes, Ann. Geophys., 25, 1779–1790, doi:10.5194/angeo-25-1779-2007, 2007.

Kodera, K.: Quasi-decadal modulation of the influence of the equatorial quasi-biennial oscillation on the north polar stratospheric temperatures, J. Geophys. Res., 98, 7245–7250, 1993.

Kürschner, D. and Jacobi, Ch.: Quasi-biennial and decadal variability obtained from long-term measurements of nighttime radio wave reflection heights over central Europe, Adv. Space Res. 32, 1701–1706, doi:10.1016/S0273-1177(03)00773-2, 2003.

Kürschner, D., Schminder, R., Singer, W., and Bremer, J.: Ein neues Verfahren zur Realisierung absoluter Reflexionshöhenmessungen an Raumwellen amplitudenmodulierter Rundfunksender bei Schrägeinfall im Langwellenbereich als Hilfsmittel zur Ableitung von Windprofilen in der oberen Mesopausenregion, Z. Meteorol., 37, 322–332, 1987.

Mertens, C. J., Mlynczak, M. G., Lopez-Puertas, M., Wintersteiner, P. P., Picard, R. H., Winick, J. R., Gordley, L. L., and Russell III, J. M.: Retrieval of mesospheric and lower thermospheric kinetic temperature from measurements of CO_2 15 μm earth limb emission under non-LTE conditions, Geophys. Res. Lett., 28, 1391–1394, 2001.

Mertens, C. J., Schmidlin, F. J., Goldberg, R. A., Remsberg, E. E., Pesnell, W. D., Russell III, J. M., Mlynczak, M. G., López-Puertas, M., Wintersteiner, P. P., Picard, R. H., Winick, J. R., and Gordley, L. L.: SABER observations of mesospheric temperatures and comparisons with falling sphere measurements taken during the 2002 summer MaCWAVE campaign, Geophys. Res. Lett., 31, L03105, doi:10.1029/2003GL018605, 2004.

Ortiz de Adler, N. and Elias, A. G.: Latitudinal variation of *fo*F2 hysteresis of solar cycles 20, 21 and 22 and its application to the analysis of long-term trends, Ann. Geophys., 26, 1269–1273, doi:10.5194/angeo-26-1269-2008, 2008.

Placke, M., Stober, G., and Jacobi, Ch.: Gravity wave momentum fluxes in the MLT – Part I: Seasonal variation at Collm (51.3° N, 13.0° E). J. Atmos. Solar-Terr. Phys., in press, doi:10.1016/j.jastp.2010.07.012, 2011.

Preusse, P., Ern, M., Eckermann, S. D., Warner, C. D., Picard, R. H., Knieling, P., Krebsbach, M., Russell, J. M., Mlynczak, M. G., Mertens, C. J., and Riese, M.: Tropopause to mesopause gravity waves in August: measurement and modelling, J. Atmos. Solar-Terr. Phys., 68, 1730–1751, 2006.

Russell III, J. M., Mlynczak, M. G., Gordley, L. L., Tansock, J., and Esplin, R.: An overview of the SABER experiment and preliminary calibration results, in: Proceedings of the SPIE, 3756, 44th Annual Meeting, Denver, Colorado, 18–23 July, 277–288, 1999.

Solomon, S. C., Woods, T. N., Didkovsky, L. V., Emmert, J. T., and Qian, L.: Anomalously low solar extreme-ultraviolet irradiance and thermospheric density during solar minimum, Geophys. Res. Lett., 37, L16103, doi:10.1029/2010GL044468, 2010.

26

Distortion of meteor count rates due to cosmic radio noise and atmospheric particularities

G. Stober[1,*]**, C. Jacobi**[1]**, and D. Keuer**[2]

[1]University Leipzig, Institute for Meteorology, Stephanstr. 3, 04103 Leipzig, Germany
[2]University of Rostock, Leibniz-Institute of Atmospheric Physics, Schlossstr. 6, 18225 Kühlungsborn, Germany
[*]present address: University of Rostock, Leibniz-Institute of Atmospheric Physics, Schlossstr. 6, 18225 Kühlungsborn, Germany

Abstract. The determination of the meteoroid flux is still a scientifically challenging task. This paper focusses on the impact of extraterrestrial noise sources as well as atmospheric phenomena on the observation of specular meteor echoes. The effect of cosmic radio noise on the meteor detection process is estimated by computing the relative difference between radio loud and radio quiet areas and comparing the monthly averaged meteor flux for fixed signal-to-noise ratios or fixed electron line density measurements. Related to the cosmic radio noise is the influence of D-layer absorption or interference with sporadic E-layers, which can lead to apparent day-to-day variation of the meteor flux of 15–20%.

1 Introduction

During the last years all-sky SKiYMET meteor radars have become a frequently used sensor system to study the mesopause region. The diversity of radar experiments reaches from mesopause region wind (Jacobi et al., 2007, 2008) and temperature (Hocking et al., 1999, 2001; Singer et al., 2004; Stober et al., 2008) measurements to astrophysical observations on meteor fluxes (Campbell-Brown and Close, 2007), radiant position (Jones and Jones, 2006) or velocity determination (Baggaley, 2003; Hocking et al., 2004; Stober, 2009) of single meteors.

Recently, this kind of meteor radar is also used to estimate the meteoroid masses (Stober et al., 2009). The basis for this analysis is the calibration of the radar, which was performed according to Latteck et al. (2008). A calibrated radar enables one to determine the electron line density q in the specular

Correspondence to: G. Stober
(stober@iap-kborn.de)

point, which is a direct measure of the ablation rate dm/dt of the meteoroid:

$$q = \frac{\beta(v)}{\mu v^2} \frac{dm}{dt}. \tag{1}$$

Here β is the ionization efficiency, μ is the mass of an ablated atom and v is the velocity of the meteor. In combination with meteor ablation models and the determined meteor flux it is possible to estimate the deposited meteoric mass in the upper mesosphere and lower thermosphere (MLT) region.

The determination of the total meteor flux is a scientifically challenging task for two reasons. There is no sensor available to observe the complete spectra of meteoric particles reaching from faint radio meteors (μm-sized) to the optical bright fireballs/bolides (m-sized). The other aspect is that each method suffers from measurement limitations e.g. observation geometry or detection thresholds. The meteor radar measurements provide a meteor count rate according to the radar parameters (antenna gain, frequency, noise level, observation geometry). This meteor count rate can, in principle, be converted to a real meteor flux by considering these parameters. In particular, the observation geometry for specular meteors requires some computational effort. Campbell-Brown and Jones (2006) estimated the meteor flux for the six sporadic meteor sources considering the minimum detectable electron line density, the declination angle of the source radiant and the population index for sporadic meteoroids.

The minimum detectable electron line density is given by the background radio noise level, which defines the detection threshold for a meteor and also the minimum measurable meteoroid mass (Stober et al., 2009). The upper detection threshold is given by the requirement that only partial reflecting meteor trails (underdense meteors) can be used to determine the electron line density. If total reflection occurs

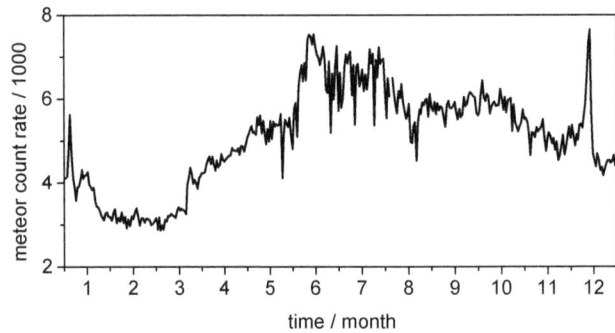

Fig. 1. August 2004 to October 2009 composite of the daily meteor count rate over Collm.

Table 1. Technical data and main parameters employed by the Collm meteor radar.

geographic position	51.3° N, 13° E
frequency	36.2 MHz
pulse repetition frequency	2144 Hz
transmitting power	6 kW
height range	70–110 km
range resolution	2 km
coherent integrations	4-point
pulse width	13 μs
angular resolution	2°
receiver bandwidth	50 kHz
transmitting antenna	3-element Yagi
receiving antenna	5-channel interfero-meter of 2-element Yagi

on the trail the mass ablation rate cannot be measured, due to a saturation of electron line density (overdense meteors).

Estimating the meteor flux from meteor radar observations of specular echoes therefore includes only meteors within the mass range given by the minimum and maximum detectable electron line density or meteoroid mass, respectively. However, the meteor flux can be extrapolated by using the population index r for given meteor population. This index describes the magnitude distribution or mass distribution, respectively, within a specific group of meteors with a power law. The sporadic meteor background can be characterized with a population index of $r = 2$. This parameter enables one to estimate the expected meteor count rate outside the detectable mass range and is therefore a suitable parameter to estimate the total meteor flux within a meteor population independent of selection effects due to the choice of radar parameters.

2 Collm meteor radar

The measurements presented in this paper were recorded with the Collm meteor radar (51.3° N, 13° E). The radar is a commercially produced all-sky SKiYMET radar, which detects ambipolar plasma trails of meteoroids entering the Earth's atmosphere. The radar measures the backscattered radio wave at the specular point of the rapidly diffusing trails. The interferometric design of the receiving antennas provides the position of the meteor in the sky. The radial drift velocity, measured through the receiver phase shift with time at each receiver, is suitable to observe MLT winds. The diffusion of the meteor trails can be used to estimate the mesopause region temperature through estimating the ambipolar diffusion coefficient. A detailed description of the system can be found in Hocking, 1999; Hocking et al., 2001, 2004.

The receiving antenna array consists of five two-element Yagis, which are aligned in an asymmetric cross with base lengths between the receivers of 2 and 2.5 wavelengths. The transmitter antenna is placed a few meters aside the receiver antenna array. The transmitter antenna is a three-element

Yagi. This leads to a characteristic dipole antenna pattern (Stober et al., 2009). Each antenna is connected to a receiver with a 70 m coaxial cable. The cable loss introduced through these cables is 1.61 dB. In Table 1 the radar settings and major parameters are summarized.

For each confirmed meteor event, 4 s of data are recorded, one second before the t_0-point of the meteor (defined as the time when the meteor head reaches specular condition) and 3 s after it (Hocking et al., 2001). The sky temperature is derived using only the first 400 pulses, i.e. before the t_0-point, of the in-phase and quadrature components to avoid any contamination from the meteor trail. A detailed description of the procedure can be found in Stober et al., 2009.

3 Meteor climatology

The meteor flux as measured by standard meteor radars is characterized by a higher meteor count rate during the summer and autumn months and a clear minimum in spring. This seasonal pattern is mainly caused by the different sporadic meteor sources (Campbell-Brown and Jones, 2006; Stober, 2009). However, a more detailed analysis of the meteor flux indicates a strong variability from day to day, which cannot be explained by the different meteor sources or the occurrence of meteor showers. This variability can, among others, be caused by external interferences, ionospheric D-layers or sporadic E-layers. In Fig. 1 the measured meteor flux over Collm is shown. The figure clearly indicates the position of the prominent meteor showers of the Quadrantids (begin of January) and the Geminids (mid December), as well as the general increased meteor flux during the summer months. However, the graph also shows that there is a strong variability during the summer, with day-to-day differences of up to 1000 meteors. In the next section we investigate probable reasons for these variations.

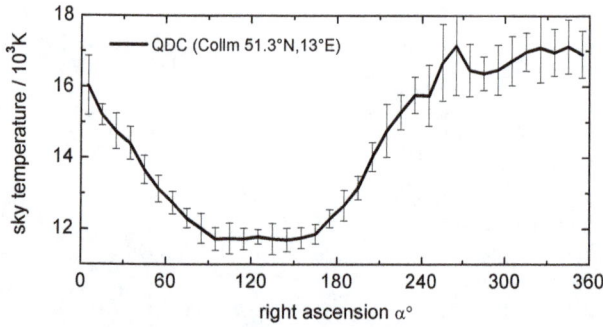

Fig. 2. QDC used for the determination of the absorption.

4   Background radio noise

Besides the variation of the background radio noise due to existing D-layers or sporadic E-layers, there is a natural variation of the radio signature due to the beam pointing direction. For wide beam meteor radars, this variation is mainly caused by our home galaxy. In Fig. 2 the resulting Quiet Day Curve (QDC) of the Collm meteor radar is shown. The procedure to determine this QDC is described in detail by Stober et al., 2009b. The figure illustrates the characteristic sidereal variation of the cosmic radio noise for a beam moving along the celestial sphere.

The variation during one sidereal day is in the order of 1.8–2 dB for the Collm meteor radar. This results in the fact that the meteor detection threshold shifts by approximately 2 dB to smaller meteoroids during times, when the beam points towards a region with a reduced cosmic radio intensity and vice versa. In Fig. 3 this effect is demonstrated by comparing the meteor count rate for a constant signal-to-noise ratio and a constant electron line density as detection threshold. The two curves represent one month averages of the daily meteor count rate over local time. The radar applies a constant trigger level of 5 dB above the noise floor, which represents a compromise between a high acceptance rate and noise spike rejections. However, for a precise estimate of the meteor flux the detectable mass range has to be kept constant and should not include any variation with time. The differences between the two lines in Fig. 3 result from the shift in the detection threshold caused by the sidereal variation of the cosmic noise level.

The impact of the QDC (2 dB variation) on meteor count rates is approximately 8% or $200 - 300$ meteors per day for a constant threshold of 8 dB. Assuming a threshold of only 6 dB would lead to a difference of 18% in the meteor count rate. The lower the signal-to-noise ratio, the smaller meteoroids can be detected. Hence, the estimation of the meteor flux can be significantly improved by using only meteors within a fixed electron line density interval, instead of counting all possible detections. The minimum detectable electron line density was taken as $q = 6.3 \cdot 10^{12} 1/\text{m}$ to com-

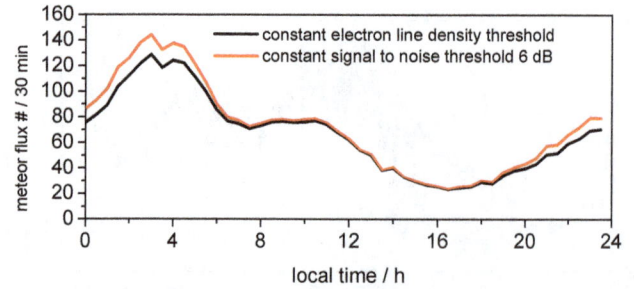

Fig. 3. Comparison of the impact of constant signal-to-noise-ratio and fixed electron line density filters on the determination of the meteor flux.

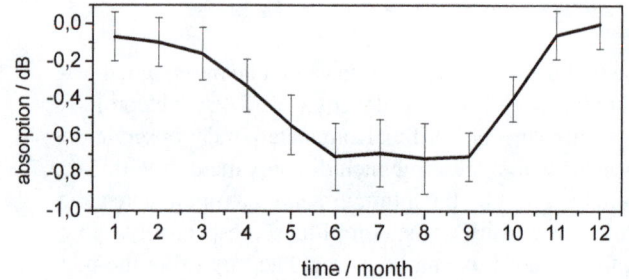

Fig. 4. Seasonal absorption measured with the Collm meteor radar.

pute the reference meteor flux. This effect is also strongly dependent on the geographic position of the meteor radar. On the southern hemisphere the QDC can reach a difference between radio loud and radio quiet areas of 8 dB.

5   Ionospheric distortions on the meteor count rate

The Collm meteor radar shows strong seasonal variation of the absorption of cosmic radio noise. One reason is the location of the radar site, which is surrounded by trees. In Fig. 4 the apparent energy loss in the course of the year is visualized. The energy loss due to the surrounding vegetation was estimated to $0.5 - 1$ dB by comparing the absorption curves from the meteor radar in Juliusruh with the Collm radar. The strongest absorption occurred during times with precipitation (see also Sect. 7).

The seasonal absorption consists of contributions from the radar environment and ionospheric phenomena like ionospheric D-layers or sporadic E-layers, which can be considered to explain the observed seasonal pattern. The absorption due to D-layers results in a similar effect like the diurnal variation of the cosmic radio background. The absorption was estimated by comparing a QDC with the radio noise measurements during the whole year. The result is shown in Fig. 5. The figure is based on a 4 h running mean absorption computed from the noise observation.

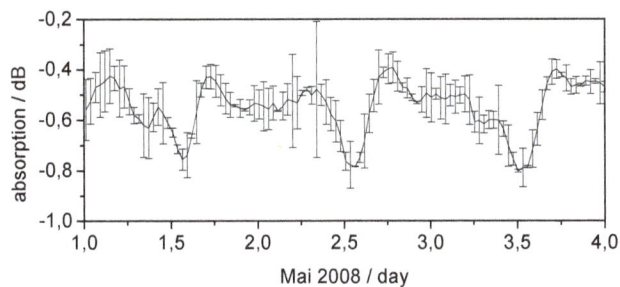

Fig. 5. Absorption computed from meteor radar noise observations.

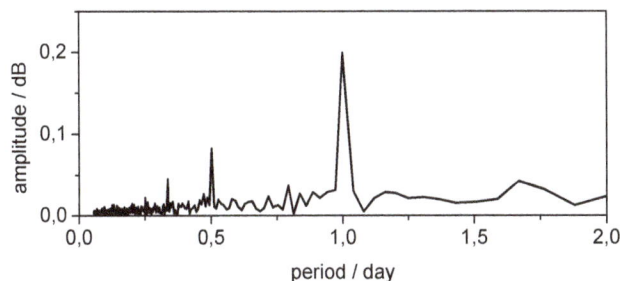

Fig. 6. Spectrum of the absorption observed in September 2008.

Figure 5 illustrates a relative absorption of 0.2–0.3 dB that occurs always around 12:00 LT. These values show a variation through the year with a minimum of 0.05–0.1 dB during the winter and the maximum of 0.3 dB in summer. Considering the error of the absorption measurements, which is computed to 0.2 dB leads to the conclusion that the absorption pattern seems not to be significant at first sight. The phase of this minimum remains stable throughout the year at 12:00 LT. A more detailed analysis reveals a minor minimum in the evening hours between 18:00–22:00 LT, which may belong to a sporadic E-layer. A Fourier transform was used to extract the characteristic frequencies from the time series. The spectral components are given in Fig. 6.

The spectrum shown in Fig. 6 indicates an apparent peak at 1 day period and further subharmonics. The calculated spectral amplitude consists of the integral absorption due to the D-layer and probable sporadic E-layers. However, the meteor radar can not deliver a constant absorption measurement due to a lot of external interferences. The derived absorption amplitude due to ionospheric phenomena is still much weaker than the absorption caused by tropospheric distortions. The seasonal course of the absorption needs to be considered in the estimation of meteor fluxes. Stober et al. (2009) demonstrated the use of a calibrated meteor radar to measure meteoroid masses from underdense trails observed in the specular point. The key element of this method is the accurate measurement of the electron line density. The seasonal absorption pattern is a particular problem for this analysis, because the radio waves have to pass the D-region

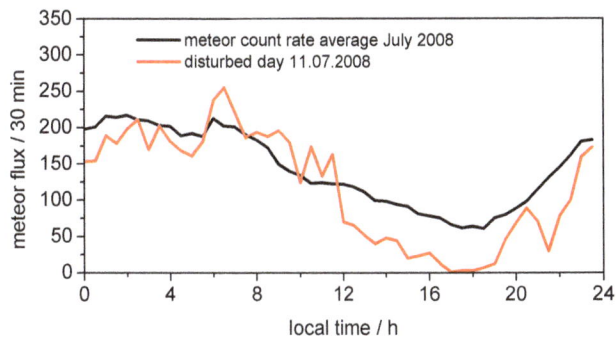

Fig. 7. Example of monthly mean meteor flux (black) compared to a disturbed day (red).

twice. The result is a signal amplitude loss of 1.4 dB during summer. Hence, the electron line density is biased by a factor of 2.

6 Tropospheric interferences

During summer one sometimes also recognizes a decrease of the meteor count rate in the afternoon or evening hours. Figure 7 visualizes a typical situation which happened in July 2008. The black line represents the monthly averaged meteor count rate over LT. The red line visualizes the meteor count rate for 11 07 2008. In Fig. 7 the meteor count rates at this day starts to decrease at 12:00 LT, which is caused by an approaching thunderstorm with extensive lightning activity. This results in an increased noise level. The radar starts to sample at a range of 6 km, which results in strong tropospheric echoes created on the strong gradients of this thunderstorm front. The second source that may lead to an almost complete breakdown of the meteor detection are lightnings. Their plasma channels have an almost fractal structure and are therefore suitable to be a specular reflector. Also, the plasma density reaches a sufficient level to partially reflect the transmitted radio waves. However, the event on 11 07 2008 represents an extreme heavy thunderstorm with almost 5 h of lightning and storm gust activity in the vicinity of the Collm radar. The storm front was also detected by the meteor radar in Juliusruh, which did record real bursts of probable meteor events in the count rate. These bursts were characterized by many signatures within a few milliseconds and covering several height gates. The typical Gaussian height distribution of the observed meteors around 90 km vanished during such a burst. Hence these spikes in the recorded data are caused by lightnings, which can also provide enough ionized plasma within a rapidly diffusing channel to result in measurable signal for the meteor radar.

7   Conclusions

In this paper we focus on the impact of radio noise variations and ionospheric phenomena on the estimation of the meteor flux using VHF radars. The results demonstrate that the daily variation of the QDC has to be considered for the determination of the meteor flux. In the case of the Collm meteor radar the effect changes the measured count rate by 8–10%. Using a fixed lower and upper electron line density threshold ensures that the measured meteor population fits into a then well defined mass range of the meteoroids.

Furthermore, we investigated the disturbances caused by ionospheric phenomena on the meteor detection process, which can be considered to be of minor importance for the meteor flux determination. However, particle precipitation or solar flares can lead to large ionospheric D-region electron line densities, which can lead to a significant absorption of the transmitted radio waves.

The biggest impact for the Collm meteor radar is the absorption caused by the properties of the radar site itself. This work points out that the electron line densities are biased towards a factor of 2 between summer and winter. This has to be considered also for the determination of corrected meteor fluxes based upon a fixed electron line density threshold. The second important distortion of the meteor flux during the summer was also identified. The event from 11 07 2008 could be clearly related to a heavy thunderstorm crossing the Collm station.

This work represents several aspects of likely distortions of the meteor detection, which need to be corrected for to obtain reliable estimates of the meteor flux. In combination with calibrated radars a well-determined meteor flux then may be used to estimate the meteoric mass deposit in the MLT region, which is still a challenging scientific task.

Acknowledgements. Special thanks to colleagues from the IAP in Kühlungsborn for advice, useful discussions and support during the calibration. The technical support and maintenance of the radar at Collm by F. Kaiser is acknowledged. Topical Editor M. Förster thanks E. D. Schmitter, J. Wickert, and an anonymous reviewer for their help in evaluating this paper.

References

Baggaley, W.: Radar Observations, in: Meteors in the Earths Atmosphere, edited by: Williams, I. P. and Murad, E., Cambridge University Press, 123–147, 2002.

Campbell-Brown, M. and Close, S.: Meteoroid structure from radar head echoes, Mon. Not. R. Astron. Soc., 382, 1309–1316, 2007.

Campbell-Brown, M. and Jones, J.: Annual variation of sporadic radar meteor rates, Mon. Not. R. Astron. Soc., 367, 709–716, 2006.

Hocking, W.: Temperatures using radar-meteor decay times, Geophys. Res. Lett., 26, 3297–3300, 1999.

Hocking, W., Fuller, B., and Vandepeer, B.: Real-time determination of meteor-related parameters utilizing modern digital technology, J. Atmos. Solar-Terr. Phys., 63, 155–169, 2001.

Hocking, W., Singer, W., Bremer, J., Mitchell, N., Batista, P., Clemesha, B., and Donner, M.: Meteor radar temperatures at multiple sites derived with SKiYMET radars and compared to OH, rocket and lidar measurements, J. Atmos. Solar-Terr. Phys., 66, 585–593, 2004.

Jacobi, C., Fröhlich, K., Viehweg, C., Stober, G., and Kürschner, D.: Midlatitude mesosphere/lower thermosphere meridional winds and temperatures measured with meteor radar, Adv. Space Res., 39, 1278–1283, 2007.

Jacobi, C., Stober, G., and Kürschner, D.: Connection between winter mesopause region temperatures and diurnal LF reflection height variations measured at Collm, Adv. Space Res., 41, 1428–1433, 2008.

Jones, J. and Jones, W.: Meteor radiant activity mapping using single-station radar observations, Mon. Not. R. Astron. Soc., 367, 1050–1056, 2006.

Latteck, R., Singer, W., Morris, R. J., Hocking, W. K., Murphy, D. J., Holdsworth, D. A., and Swarnalingam, N.: Similarities and differences in polar mesosphere summer echoes observed in the Arctic and Antarctica, Ann. Geophys., 26, 2795–2806, 2008, http://www.ann-geophys.net/26/2795/2008/.

Singer, W., von Zahn, U., and Weiß, J.: Diurnal and annual variations of meteor rates at the arctic circle, Atmos. Chem. Phys., 4, 1355–1363, 2004, http://www.atmos-chem-phys.net/4/1355/2004/.

Stober, G.: Astrophysical Studies on Meteors using a SKiYMET All-Sky Meteor Radar, Ph.D. thesis, Institute for Meteorology, University Leipzig, 2009.

Stober, G., Jacobi, C., Fröhlich, K., and Oberheide, J.: Meteor radar temperatures over Collm (51.3° N 13° E), Adv. Space Res., 42, 1253–1258, 2008.

Stober, G., Singer, W., and Jacobi, C.: Meteoroid mass determination from specular observed underdense trails, J. Atmos. Solar-Terr. Phys., submitted 2009.

Stober, G., Singer, W., Jacobi, C., and Hoffmann, P.: Cosmic Noise Observations using a mid-latitude Meteor Radar, J. Atmos. Solar-Terr. Phys., submitted 2009.

27

Meteor heights during the recent solar minimum

Ch. Jacobi

Institute for Meteorology, University of Leipzig, Stephanstr. 3, 04103 Leipzig, Germany

Correspondence to: Ch. Jacobi (jacobi@uni-leipzig.de)

Abstract. Average meteor heights have been continuously observed using a SKiYMET VHF radar at Collm (51.3° N, 13.0° E) since late summer of 2004. Initially, the daily mean meteor height was about 89.4 km. Since that time, average meteor heights have decreased. This is consistent with earlier results on middle atmosphere temperature change from the literature and from earlier results of low-frequency reflection height changes measured at Kühlungsborn and Collm. During the recent solar minimum 2008/2009 the meteor heights further decreased. Linear fitting of a trend and a solar cycle to the heights reveals a linear decrease of about $-56\,\mathrm{m\,year^{-1}}$ and a solar cycle effect of $+450\,\mathrm{m}$ per 100 sfu. Assuming that meteor heights, on a long-term average, approximately refer to a level of constant pressure, this decrease can be converted to a mean middle atmosphere linear temperature decrease of $-0.23\,\mathrm{K\,year^{-1}}$ and a solar cycle effect of $+1.8\,\mathrm{K}$ per 100 sfu during the last decade, which is in the range of observed trends reported in the literature.

1 Introduction

The increase of greenhouse gas concentration leads to warming of the troposphere, but cooling of the middle and upper atmosphere (Beig et al., 2003; Lastovicka et al., 2006). This long-term cooling of the stratosphere and mesosphere has been shown by model calculations and has been observed during the last decades also (see, e.g., Beig et al., 2003; Beig, 2011a). It should lead to a decrease of layers of constant pressure in the middle atmosphere (e.g., Lübken et al., 2013). This decrease has been registered using LF phase heights (Bremer and Berger, 2002; Bremer and Peters, 2008), and it was found that the height of radio wave reflections decreases according to mesospheric cooling. For virtual heights of about 95 km obtained during the time interval 1983–2001, Kürschner and Jacobi (2003) found a de-

creasing trend as well. At E region altitudes, also a decrease of the reference heights with time has been reported (e.g., Hall et al., 2011), at least for the majority of analysed time series (Bremer, 2008). On the decadal time scale, the 11-year solar cycle is one of the most important sources of middle atmosphere variability (e.g., Li et al., 2011; Beig, 2011b). From observations, long-term cooling and solar effect in the mesosphere is found to be dependent on altitude; as an approximate order of magnitude a trend of $-2\,\mathrm{K\,decade^{-1}}$ and a solar effect of $+1\,\mathrm{K}\,100\,\mathrm{sfu^{-1}}$ has been found. It has to be taken into account that these analyses usually do not include the recent strong solar minimum.

A widely used method to observe mesopause region (80–100 km) atmospheric parameters are VHF meteor radar (MR) measurements. MR can measure meteor altitudes, Doppler winds, diffusion coefficients, and further parameters (e.g., Hocking et al., 2001; Stober et al., 2012). The observed meteor altitudes depend on the meteor parameters like velocity, entrance angle, meteor mass and composition. In addition, the atmospheric density plays an important role so that meteors on an average reach lower altitudes if the density decreases (e.g., Stober et al., 2012). In the case of progressive cooling of the middle atmosphere, one may therefore expect long-term decrease of mean meteor altitudes. Actually, this may affect long-term trend analyses of winds and tidal parameters, as has been discussed by Merzlyakov and Portnyagin (1999) and Jacobi et al. (2005), for instance.

In the following, 9 years of meteor heights are used to analyse possible height changes during the time interval 2004–2013. Although the available dataset is much too short to reliably derive long-term trends, tendencies can be detected to support the hypothesis of decreasing meteor altitudes and thus middle atmosphere cooling with time.

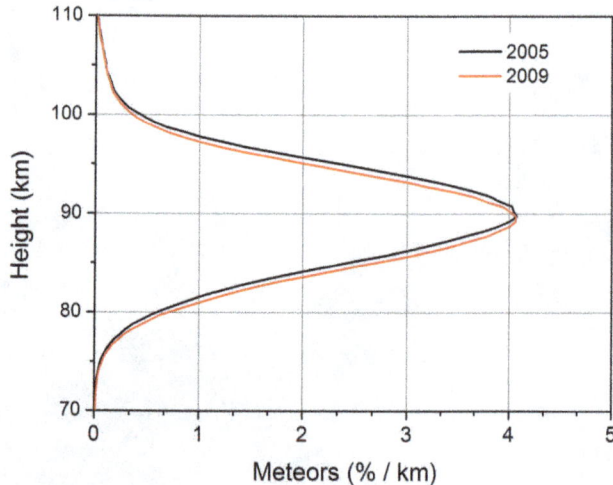

Figure 1. Histograms of meteor heights in 2005 (black) and 2009 (red), given in % per 1 km height interval. In total, 1.97 and 2.37 million meteors have been registered in 2005 and 2009, respectively.

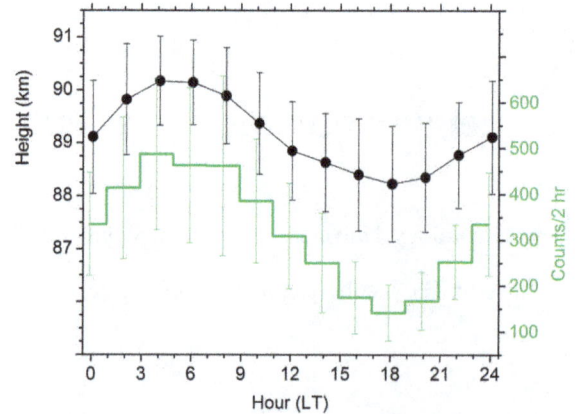

Figure 2. 9 year mean diurnal cycle of 2-hourly mean meteor heights (black) and number of meteors per 2 h rime interval (green). Data used are from September 2004 through August 2013. Error bars show standard deviations calculated using the 2 h means.

2 Collm meteor radar measurements

At Collm, Germany (51.3° N, 13.0° E), a SKiYMET all sky MR is operated at 36.2 MHz since summer 2004. The antenna system consists of one 3-element Yagi transmitting antenna and five 2-element Yagi receiving antennas, forming an interferometer. Peak power is 6 kW. Pulse repetition frequency is 2144 Hz, but effectively only 536 Hz due to 4-point coherent integration. The sampling resolution is 1.87 ms. The angular and range resolutions are ∼ 2° and 2 km, respectively. The pulse width is 13 μs, the receiver bandwidth is 50 kHz.

We consider zenith angles between 0° and 70°, and distances of up to 400 km from the transmitter. The height distribution of all individual meteors during two different years is given in Fig. 1. For the present analysis, meteor count rates and mean heights are taken every 2 h. The long-term mean diurnal cycles of mean meteor heights and count rates are shown in Fig. 2. In the following, daily and monthly mean meteor heights are calculated as arithmetic averages of the 2-hourly means. Note that these values are lower than the peak heights shown in Fig. 1, owing to the uneven distribution of meteor count rates in the course of a day.

3 Results

Monthly mean meteor altitudes, calculated from the 2-hourly mean altitudes are shown in Fig. 3. The red line is a smoothed curve obtained after applying a 6 pt FFT filter. On can see that from the beginning of the registrations until 2009 the heights have decreased by about 500 m. Since then they have remained approximately constant. Minimum altitudes are registered in 2009. During that year, solar activity had reached

its minimum in the course of the extended 23/24 solar minimum (e.g., Nikutowski et al., 2011). Since then solar activity has increased again, although the current solar cycle 24 is weaker than previous ones. Meteor heights, however, did not increase again substantially. This is possibly due to the fact that after 2009 a possible increase owing to the increasing solar activity and connected warming of the middle atmosphere is just compensated by a long-term decreasing trend.

To discriminate a possible long-term tendency from a solar cycle effect, annual mean altitudes z were calculated from the monthly means. The time series is shown in Fig. 4 as a black line with squares. Since the measurements have started in summer 2004, annual means are here calculated as averages over the time interval from September through August. A least-squares fit was applied:

$$z = z_0 + a \times t + b \times F, \qquad (1)$$

with t as the time in years and F as the F10.7 solar flux index in sfu. The obtained coefficients are $a = -56$ m year^{-1} and $b = +450$ m per 100 sfu. The annual mean F10.7 values are added in Fig. 4 as a green line. The meteor height time series after subtracting the solar cycle is given as a red line, and the linear trend curve is added. The obtained linear trend is on the order of magnitude of that reported by Bremer and Peters (2008) but smaller than the one obtained from LF heights at Collm (Kürschner and Jacobi, 2003). However, the latter have used virtual heights that overestimate the real heights and therefore the trends as well. The height time series after removing the long-term tendency is shown as blue line in Fig. 4. Clearly, the minimum altitude is found in 2009 during the deep solar minimum.

The annual mean heights plotted vs. the F10.7 solar flux are shown in Fig. 5. Clearly, during the increasing part of solar cycle 24 the meteor heights are lower than during earlier

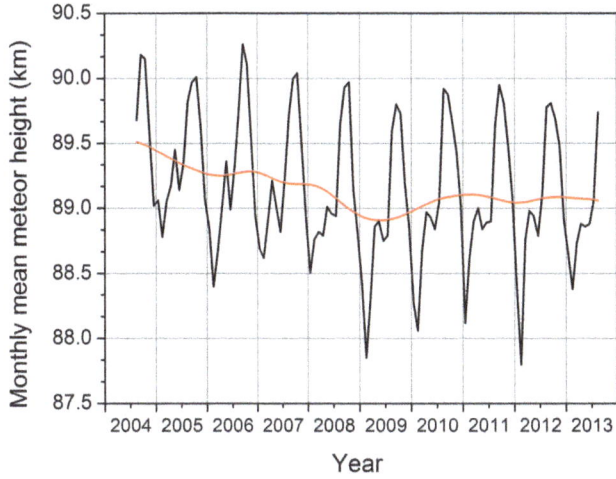

Figure 3. Monthly mean meteor heights, calculated from 2-hourly mean heights. The smoothed curve after applying a 6 pt FFT filter is given in red.

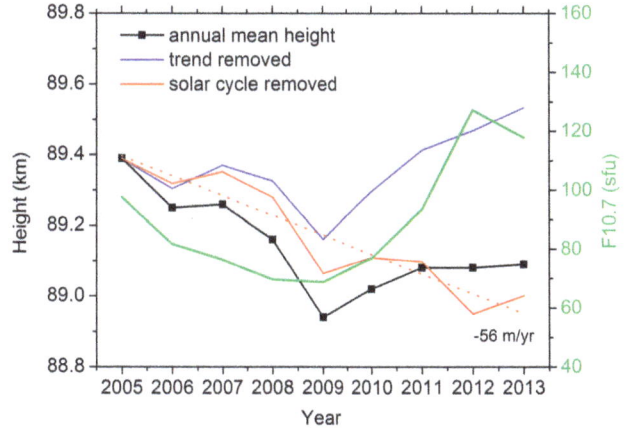

Figure 4. Annual mean meteor heights (black curve). The annual mean F10.7 values are added as green line. The blue and red lines represent annual means after removing the decadal trend or solar effect, respectively. The dotted red line is a linear fit to the annual means after removing the solar cycle. The annual means refer to the time interval from September of the previous year through August.

years, i.e. during the decreasing part of solar cycle 23. After removing the long-term tendency (blue line with squares in Fig. 5) the two parts of the height time series, i.e., during the last part of solar cycle 23 and the first part of solar cycle 24 show a good correspondence. One may also note that the difference between 2005 and 2009 amounts to about -450 m.

4 Middle atmosphere temperature trends

Bremer and Berger (2002) had estimated mean middle atmosphere cooling rates from pressure level height decrease by a registered height difference Δz. Assuming that at the beginning and end of a registration the pressure values p_b and p_e are the same

$$p_b(z) = p_e(z + \Delta z), \qquad (2)$$

one may use the hydrostatic equation and ideal gas law

$$p_0 \exp \left\{ -\frac{g}{R} \int_{z_0}^{z} \frac{dz}{T(z)} \right\} = p_0 \exp \left\{ -\frac{g}{R} \int_{z_0}^{z+\Delta z} \frac{dz}{T(z) + \Delta T} \right\}, (3)$$

with R as the gas constant and g as acceleration due to gravity, taken as constant here. The temperature profile $T(z)$ is taken from the CIRA 1986 climatology (Fleming et al., 1990). Assuming a constant temperature change ΔT with height and no height change at some lower boundary z_0, one obtains

$$\int_{z_0}^{z} \frac{dz}{T(z)} = \int_{z_0}^{z+\Delta z} \frac{dz}{T(z) + \Delta T}, \qquad (4)$$

which can be solved for ΔT. Note that Bremer and Berger (2002), taking the temperature change height distribution

from literature results, were able to obtain a vertical profile of temperature change. Since, however, this requires external input this is not done here. Here, $z_0 = 20$ km is assumed, and the resulting mean middle atmosphere temperature tendencies are -2.3 K decade^{-1} for the long-term tendency and $+1.8$ K per 100 sfu for the solar cycle effect. According to a t test, the trends are significant at the 95 % or 90 % level for the long-term decrease and the solar cycle effect, respectively. The results do not much depend on the temperature profile $T(z)$ chosen. This is, because the change in temperature is small compared with the mean temperature itself, and differences between climatologies (CIRA, MSIS, ...) are of the order of few percent. For example, using NRLMSISE-00 predictions (Picone et al., 2002) delivers the same result (at an accuracy of 0.1 K), and even in the extreme case of temperatures taken as constant with height the trends do not change by more than 10 %.

5 Discussion and conclusions

From the analysis of 9 years of meteor altitudes one may find a long-term decreasing tendency and a possible solar cycle effect. The trend values are within the range of reported middle atmosphere temperature trends (Li et al., 2011; Beig, 2011a) and solar cycle effects (Beig, 20011b). However, most values of observed trends reported in the literature are based on time series that do not include the deep solar minimum 23/24 and therefore numbers may differ in detail.

In contrast to the reported decreasing trends, model calculations by Berger and Lübken (2011) and Lübken et al. (2013) show positive temperature trends in summer over Germany since the middle 1990s. They attribute this increase to the increase of stratospheric ozone since then. This is,

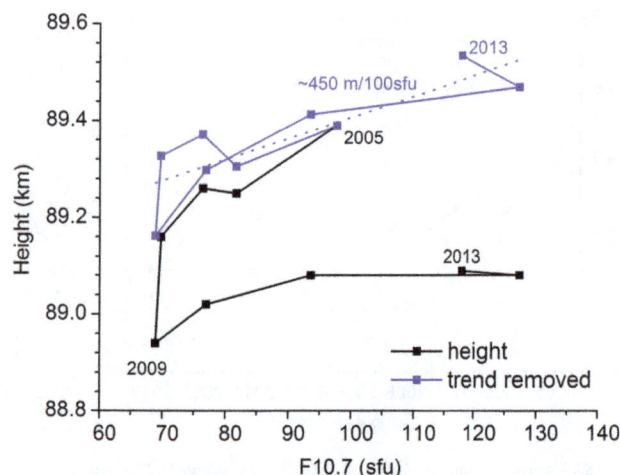

Figure 5. Annual mean meteor heights vs. annual mean F10.7 solar radio fluxes (black curve). The blue values represent annual means after removing the decadal trend. The dotted blue line is a linear fit to the annual means after removing the trend. The indicated years denote the time interval from September of the previous year through August.

to a certain degree, consistent with analyses of Jacobi and Kürschner (2007), who had used an updated dataset from Kürschner and Jacobi (2003), and who found that during the early 2000s the LF reflection heights did not decrease. However, the LF measurements at Collm have been interrupted in 2007, so that these results cannot be compared with the MR heights. Offermann et al. (2010), on the other hand, showed negative mesopause region temperature trends derived from OH* temperatures over Wuppertal, however, they also reported that these trends in summer are smaller compared to the winter ones. Hall et al. (2012) also reported decreasing temperatures in the high latitude mesopause region during the last decade. One may conclude that there are observations that indicate a continuing cooling of the middle atmosphere, but there is still debate on this point.

The here analyzed time series is very short and therefore no definite quantitative conclusions on real trends and in particular on the relative role of solar and long-term trend effects can be drawn. In addition, results from such a short time series cannot be extrapolated, even if they are significant in a purely statistical sense. On the other hand, the obtained tendencies for the mean altitudes at least qualitatively confirm most tendencies reported in the literature, which shows that MR heights may be used as an indicator for middle atmosphere temperature trends.

Acknowledgements. F10.7 solar indices have been provided by NGDC through ftp://ftp.ngdc.noaa.gov/STP/SOLAR_DATA/. CIRA temperatures have been provided by BADC through http://www.badc.nerc.ac.uk/data/cira. NRLMSISE-00 data has been provided by CCMC, GSFC, NASA on http://ccmc.gsfc.nasa.gov/modelweb/.

Edited by: M. Förster
Reviewed by: two anonymous referees

References

Beig, G.: Long-term trends in the temperature of the mesosphere/lower thermosphere region: 1. Anthropogenic influences, J. Geophys. Res., 116, A00H11, doi:10.1029/2011JA016646, 2011a.

Beig, G.: Long term trends in the temperature of the mesosphere/lower thermosphere region: 2. Solar response, J. Geophys. Res., 116, A00H12, doi:10.1029/2011JA016766, 2011b.

Beig, G., Keckhut, P., Lowe, R. P., Roble, R. G., Mlynczak, M. G., Scheer, J., Fomichev, V. I., Offermann, D., French, W. J. R., Shepherd, M. G., Semenov, A. I., Remsberg, E. E., She, C. Y., Lübken, F. J., Bremer, J., Clemesha, B .R., Stegman, J., Sigernes, F., and Fadnavis, S.: Review of mesospheric temperature trends, Rev. Geophys., 41, 1015, doi:10.1029/2002RG000121, 2003.

Berger, U. and Lübken, F.-J.: Mesospheric temperature trends at midlatitudes in summer. Geophys. Res. Lett., 38, L22804, doi:10.1029/2011GL049528, 2011.

Bremer, J.: Long-term trends in the ionospheric E and F1 regions, Ann. Geophys., 26, 1189–1197, 2008, 2008.

Bremer, J. and Berger, U.: Mesospheric temperature trends derived from ground-based LF phase-height observations at midlatitudes: comparison with model simulations, J. Atmos. Sol.-Terr. Phy., 64, 805–816, 2002.

Bremer, J. and Peters, D.: Influence of stratospheric ozone changes on long-term trends in the meso- and lower thermosphere, J. Atmos. Sol.-Terr. Phys., 70, 1430–1440, doi:10.1016/j.jastp.2008.03.024, 2008.

Fleming, E. L., Chandra, S., Barnett, J. J., and Corney, M.: Zonal mean temperature, pressure, zonal wind and geopotential height as function of latitude, Adv. Space Res., 10, 11–59, 1990.

Hall, C. M., Rypdal, K., and Rypdal, M.: The E region at 69° N, 19° E: Trends, significances, and detectability, J. Geophys. Res., 116, A05309, doi:10.1029/2011JA016431, 2011.

Hall, C. M., Dyrland, M. E., Tsutsumi, M., and Mulligan, F. J.: Temperature trends at 90 km over Svalbard, Norway (78° N 16° E), seen in one decade of meteor radar observations, J. Geophys. Res., 117, D08104, doi:10.1029/2011JD017028, 2012.

Hocking, W. K., Fuller, B., and Vandepeer, B.: Real-time determination of meteor-related parameters utilizing modern digital technology, J. Atmos. Sol.-Terr. Phy., 63, 155–169, 2001.

Jacobi, Ch. and Kürschner, D.: Possible climate change response of the mesosphere/lower thermosphere region, Proceedings of the International Symposium "Atmospheric Physics: Science and Education", St. Petersburg, 11.-13.9.2007, 33–37, 2007.

Jacobi, Ch., Portnyagin, Yu. I., Merzlyakov, E. G., Solovjova, T. V., Makarov, N. A., and Kürschner, D.: A long-term comparison of mesopause region wind measurements over Eastern and Central Europe, J. Atmos. Sol.-Terr. Phy., 67, 227–240, 2005.

Kürschner, D. and Jacobi, Ch.: Quasi-biennial and decadal variability obtained from long-term measurements of nighttime radio wave reflection heights over central Europe, Adv. Space Res., 32, 1701–1706, doi:10.1016/S0273-1177(03)90465-0, 2003.

Lastovicka, J., Akmaev, R. A., Beig, G., Bremer, J., and Emmert, J. T.: Global Change in the Upper Atmosphere, Science, 314, 1253–1254, doi:10.1126/science.1135134, 2006.

Li, T., Leblanc, T., McDermid, I. S., Keckhut, P., Hauchecorne, A., and Dou, X.: Middle atmosphere temperature trend and solar cycle revealed by long-term Rayleigh lidar observations, J. Geophys. Res., 116, D00P05, doi:10.1029/2010JD015275, 2011.

Lübken, F.-J., Berger, U., and Baumgarten, G.: Temperature trends in the midlatitude summer mesosphere, J. Geophys. Res., 118, 13347–13360, doi:10.1002/2013JD020576, 2013.

Merzlyakov, E. G. and Portnyagin, Y. I.: Long-term changes in the parameters of winds in the midlatitude lower thermosphere (90–100 km), Izvestiya, Atmos. Ocean. Phys., 35, 428–493, 1999.

Nikutowski, B., Brunner, R., Erhardt, Ch., Knecht, St., and Schmidtke, G.: Distinct EUV minimum of the solar irradiance (16–40 nm) observed by SolACES spectrometers onboard the International Space Station (ISS) in August/September 2009, Adv. Space Res., 48, 899–903, 2011.

Offermann, D., Hoffmann, P., Knieling, P., Koppmann, R., Oberheide, J., and Steinbrecht, W.: Long-term trends and solar cycle variations of mesospheric temperature and dynamics, J. Geophys. Res., 115, D18127, doi:10.1029/2009JD013363, 2010.

Picone, J. M., Hedin, A. E. Drob, D. P., and Aikin, A. C.: NRLMSISE-00 empirical model of the atmosphere: Statistical comparisons and scientific issues, J. Geophys. Res., 107, 1468, doi:10.1029/2002JA009430, 2002.

Stober, G., Jacobi, Ch., Matthias, V., Hoffmann, P., and Gerding, M.: Neutral air density variations during strong planetary wave activity in the mesopause region derived from meteor radar observations, J. Atmos. Sol.-Terr. Phy., 74, 55–63, doi:10.1016/j.jastp.2011.10.007, 2012.

Horizontally resolved structures of radar backscatter from polar mesospheric layers

R. Latteck, W. Singer, M. Rapp, T. Renkwitz, and G. Stober

Leibniz Institute of Atmospheric Physics at the Rostock University, Schloss-Str. 6, 18225 Kühlungsborn, Germany

Correspondence to: R. Latteck (latteck@iap-kborn.de)

Abstract. The Leibniz-Institute of Atmospheric Physics in Kühlungsborn, Germany (IAP) installed a new powerful VHF radar on the North-Norwegian island Andøya (69.30° N, 16.04° E) from 2009 to 2011. The new Middle Atmosphere Alomar Radar System (MAARSY) replaces the existing ALWIN radar which has been in continuous operation on Andøya for more than 10 yr. MAARSY is a monostatic radar operated at 53.5 MHz with an active phased array antenna consisting of 433 Yagi antennas each connected to its own transceiver with independent control of frequency, phase and power of the transmitted signal. This arrangement provides a very high flexibility of beam forming and beam steering. It allows classical beam swinging operation as well as experiments with simultaneous multiple beams and the use of modern interferometric applications for improved studies of the Arctic atmosphere from the troposphere up to the lower thermosphere with high spatial-temporal resolution. The installation of the antenna was completed in August 2009. An initial expansion stage of 196 transceiver modules was installed in spring 2010, upgraded to 343 transceiver modules in December 2010 and the installation of the radar was completed in spring 2011. Beside standard observations of tropospheric winds and Polar Mesosphere Summer Echoes, multi-beam experiments using up to 91 beams quasi-simultaneously in the mesosphere have been carried out using the different expansion stages of the system during campaigns in 2010 and 2011. These results provided a first insight into the horizontal variability of Polar Mesosphere Summer and Winter Echoes in an area of about 80 km by 80 km with time resolutions between 3 and 9 min.

1 Introduction

The phenomenon of strong radar echoes from the mesopause region during summer is well known from VHF radar observations at polar and middle latitudes for more than 30 yr. These so-called Polar Mesosphere Summer Echoes (PMSE) are caused by inhomogeneities in the electron density of a size comparable to the radar Bragg scale in presence of negatively charged aerosol particles. An overview on the current understanding of this phenomenon has been published by Rapp and Lübken (2004).

Similar echoes can be detected by VHF radars during the winter months in the mid mesosphere from ∼55–85 km altitude. Due to their similarity in observation method and location to PMSE these echoes were named polar mesosphere winter echoes (PMWE) (Czechowsky et al., 1979; Ecklund and Balsley, 1981; Kirkwood et al., 2002). PMWE are much less frequent than their summer counterparts (Zeller et al., 2006) and the underlying physics is currently debated (Lübken et al., 2006; Kirkwood et al., 2006) and subject of current active research.

In this study we present the first 3-dimensionally resolved observations of PMSE and PMWE obtained during the construction period of the Middle Atmosphere Alomar Radar system (MAARSY). The design and functionality of both the initial and second stage of extension of MAARSY used for this observation are described.

2 System description

MAARSY is a monostatic radar located on the North-Norwegian island Andøya (69.30° N, 16.04° E). The radar is designed for atmospheric studies from the troposphere to the lower thermosphere, especially for the investigation of horizontal structures of PMSE. The general idea of MAARSY has been presented by Latteck et al. (2010).

The system is composed of an active phased antenna consisting of 433 array elements and an identical number of transceiver modules. The operational radar frequency is 53.5 MHz and the maximum peak power is approximately 800 kW. The 433 3-element linear polarized Yagi antennas

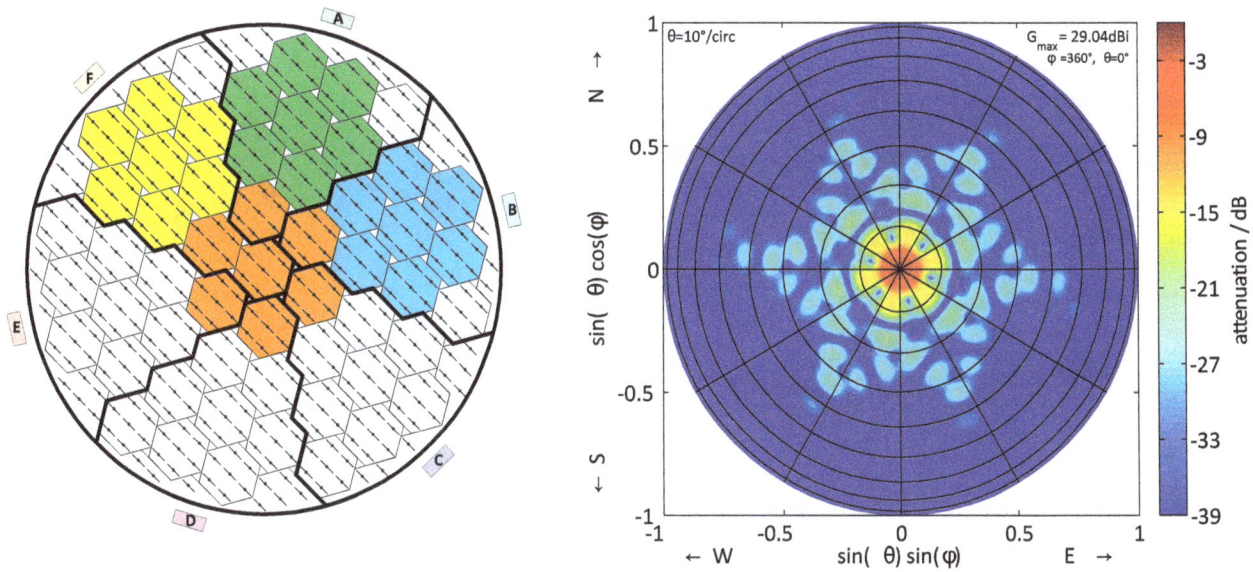

Fig. 1. Initial expansion stage of MAARSY as installed in May 2010. Antenna subgroups connected to transceiver modules are highlighted in color in the left panel. Three antenna structures (green, blue, orange) were used to form a central symmetric antenna beam with a half power beam width of 6° for classical Doppler beam steering. A computed radiation pattern for a beam pointing to 0° (zenith) is shown in the right panel. The receiving signals related to the individual hexagon antenna structures marked in yellow were connected to 7 receivers and used for spaced antenna experiments in the mesosphere.

arranged in an equilateral triangle grid structure with a spacing of 4 m forms a nearly circular antenna array with a diameter of approximately 90 m corresponding to an aperture of ~6300 m². The symmetric antenna radiation pattern of the array has a beam width of 3.6° (full width at half power), a directive gain of 33.5 dBi and an almost symmetric first sidelobe with more than 17 dB suppression with respect to the main lobe.

Each antenna of the array is connected to its own transceiver module individually controllable in frequency, phase, and gain settings on a pulse-to-pulse basis. This arrangement allows very high flexibility of beam forming and beam steering with a symmetric radar beam and arbitrary beam pointing directions.

On reception the antenna array can be divided in 61 subarrays, 55 of them are symmetrical structures (hexagon) consisting of 7 antennas each, 6 subgroups consisting of 8 antennas each are located at the perimeter of the array. 7 adjacent hexagonal structures as e.g. indicated by the colored areas in the left panel of Fig. 4 can be further combined to 7 so called "anemone" antenna structures. Each receiving signal of a hexagonal or anemone antenna sub-structure can be selected as an input of a 16 channel data acquisition unit. This allows a wide range of receiving arrangements with different antenna configurations for interferometric or multi-receiver applications. Additionally, separately located receiving antennas used for e.g. interferometer observations of meteors or for boundary layer observations can be switched to the data acquisition via an antenna interface unit.

A detailed technical description of MAARSY is given by Latteck et al. (2012).

3   3-D resolved structures of mesospheric echoes obtained during the construction period of MAARSY

The construction of the MAARSY antenna array and the infrastructure for radar control and communication was completed in August 2009. The radar control hardware for synchronization, triggering and communication needed in each of the six equipment buildings, the master control system, an interim design of the combining unit and the data acquisition hardware housed in the main building was installed during winter 2009/2010. The radar operation started with an initial expansion of 196 transceiver modules in May 2010. A second stage of expansion to 343 transceiver modules was brought into service in November 2010. First results using the two stages of extension during campaigns in summer and December 2010 are presented here to demonstrate the performance and functionality of the new radar during the construction period.

3.1   Results using the initial stage of expansion

3.1.1   Design of the initial stage of expansion

The installed transceiver modules of the initial expansion stage were connected to antenna subgroups as highlighted in color in Fig. 1. Three anemone antenna structures as

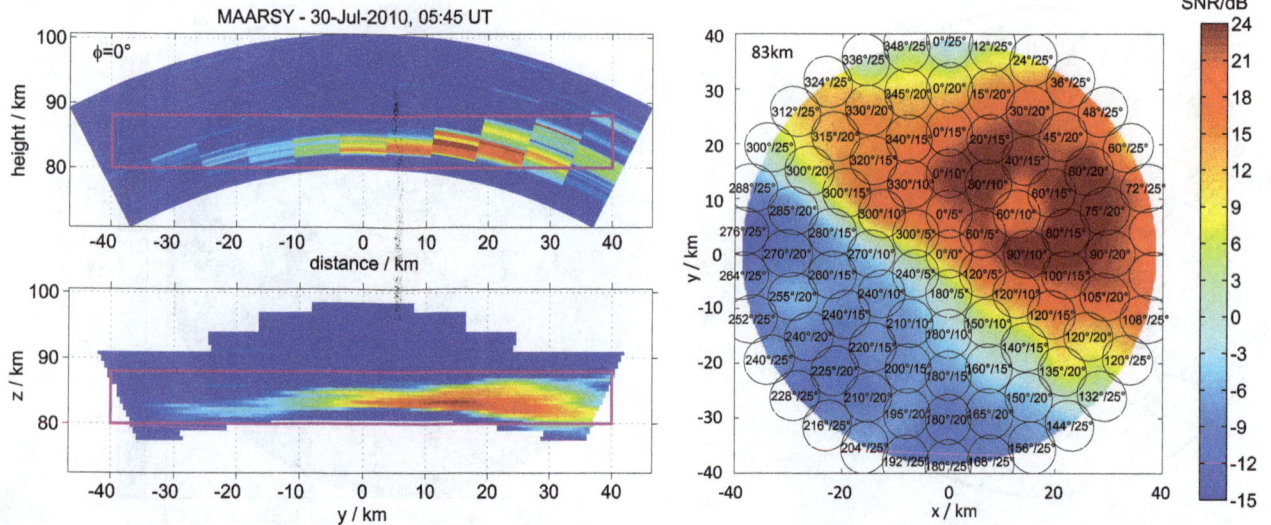

Fig. 2. Vertical and horizontal structure of a PMSE observed with 91 beams quasi-simultaneously on 30 July 2010. The upper left plot shows the SNR determined from observations along the North-South axis ($\varphi = 0°$) using 10 beams changing off-zenith angle after 4 coherent integrations. The lower left panel depict the same situation but interpolated to a grid with 500 m horizontal and vertical resolution. The right plot shows a slice at 83 km height through the 3-dimensional interpolated data obtained from all 91 beams. The tagged ovals overlayed on the contour plot mark the areas at 83 km altitude illuminated by the beams.

Table 1. Allocation of MAARSY antenna structures to the inputs of the signal processing unit as used with the initial and second stage of expansion. Single capital letters indicate anemone structures as e.g. A = "anemone A", "F-xx" stands for a selected hexagon of anemone F. "Met-xx" and "But-xx" are signals from separate located antennas as from the meteor antennas and the Butler array (Renkwitz et al., 2011).

signal processing unit inputs	initial stage of expansion	second stage of expansion
01	Met-01	A
02	Met-02	B
03	Met-03	C
04	Met-04	D
05	Met-05	E
06	A	F
07	B	M
08	M	A+B+C+D+E+F+M
09	F-01	But-01
10	F-02	But-02
11	F-03	But-03
12	F-04	But-04
13	F-05	But-05
14	F-06	But-06
15	F-07	But-07
16	A+B+M	But-08

marked in green, cyan and orange (A, B, and M) representing 147 antennas connected to transceiver modules were used to form a central symmetric antenna beam with a half power

beam width of 6° for transmission at an effective peak power of ~218 kW. On reception the combined IF signals of the anemone structures A, B and M were split and one half of the three split signals were combined and connected to signal processing channel 16 to be used in Doppler beam swinging (DBS) mode. The other half of the split anemone signals were led to three individual signal processing channels and used as spaced antennas (SA) in tropospheric experiments. Additionally to these three SA channels the seven IF signals belonging to the hexagons of equipment building F as marked in yellow in Fig. 1 were connected to further seven signal processing channels to be used as spaced antennas for mesospheric experiments. The allocation of the antenna structures to the signal processing channels is summarized in Table 1.

The data acquisition unit used with the 2010 partial expansion stages of MAARSY was capable to sample 8 dual channels only. Hence a multiplexer was used to switch the 16 dual outputs of the base band down converters to the 8 dual inputs of the digitizers.

3.1.2 Experiment description

MAARSY was designed for middle atmosphere studies, especially the investigation of horizontal structures of PMSE. In order to scan the atmosphere with a number of radar beams, the transceivers can store up to 50 parameter settings for e.g. phase offsets for beam pointing, and the parameters can be changed with the PRF. Due to limitations in the maximum data rate of the acquisition system used with the partial installations in 2010, it was not possible to operate a single experiment configured with 50 different beam directions and full coverage of the PMSE altitudes in summer 2010. Hence a sequence of experiments each configured for a reduced number of beam directions was used to perform first horizontal scans of PMSE in a quasi-simultaneous mode.

The maximum data rate and a sufficient coverage of altitude range around 85 km resulted in a experiment configuration as summarized in Table 2. Nine identical experiments with 10 different oblique beam directions each were configured. The vertical beam position was added to one experiment. The overall 91 beam positions are tagged to ovals indicating the illuminated areas at 83 km and overlayed on a contour plot of SNR in the right panel of Fig. 2. The beam position was changed after every 4 coherent integrations during the experiments runtime of 30 s and 33 s, respectively. The whole sequence of 9 experiments illuminated an area of about 80 km diameter at an altitude range between 77 km and 88 km. The scanning experiments were followed by a standard mesospheric experiment running for 35 s and covering a wider range in vertical direction with a larger number of coherent integrations. The total overall sequence runtime, corresponding to the time resolution of the scan, was 503 s due to additional time required for the reconfiguration of the hardware between the experiments.

3.1.3 3-D resolved structures of PMSE

A first mesospheric scan experiment using the described experiment sequence was conducted from 30 July until 4 August in 2010. The height-distance-intensity plot in the upper left panel of Fig. 2 shows an example of signal-to-noise ratio (SNR) obtained with with 11 beams along the North-South direction on 30 July 2010. The color coded areas represent vertical slices of the radar pulse volumes illuminated by the 11 beams with 300 m radial resolution. The range information of the profiles were first converted to altitude using the nominal off-zenith angle of the corresponding beams, then interpolated to a vertical grid with a resolution of 100 m and the resulting altitude values were converted back to ranges. The new polar coordinates of the interpolated data were then converted to cartesian coordinates and finally interpolated to a 2-dimensional horizontal grid with a equal resolution of 100 m. The lower left panel of Fig. 2 shows the resulting 2-dimensional vertical slice of SNR corresponding to the example shown in the upper left panel. The pink frames in both plots mark the area which is zoomed in and presented in the upper right panel of Fig. 3.

The horizontal slice through the PMSE at 83 km depicted in the right panel of Fig. 2 shows a sharp decrease of SNR of about 30 dB along a 120° to 300° axis. A stack of horizontal slices at different altitudes and a zoomed vertical intersection of the same example all with 100 m horizontal and vertical resolution are presented in Fig. 3. The 3-dimensional resolved PMSE shows a maximum or core located north-

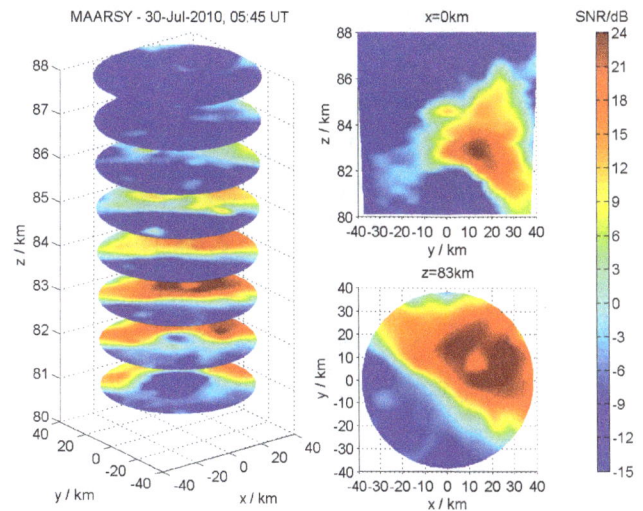

Fig. 3. 3-D structure of a PMSE quasi-simultaneously observed with 91 beams on 30 July 2010.

Table 2. Parameters of MAARSY experiments used for horizontal scanning of PMSE in August 2010 and PMWE in December 2010. The values in brackets correspond to the one experiment configured for 11 beam directions.

Period	08/2010	12/2010
Pulse repetition frequency	1400 Hz	1250 Hz
No. of coherent integrations	4	4
Wave form (pulse coding)	8 bit coco	8 bit coco
Sub-pulse length	2 μs	2 μs
Pulse length	16 μs	16 μs
→ Inter pulse period	714.9 μs	800 μs
→ Effective PRF	700 Hz	
→ Duty cycle	2.24 %	2 %
Sampling start range	77.1 km	65.1 km
Sampling end range	97.5 km	85.5 km
Sampling resolution	300 m	300 m
→ range gates	69	77
No. of data points p. exp.	20 480 (22 528)	14 336
No. of beam directions	10 (11)	7
→ No. of data pts. p. beam	512	1024
→ Time resolution Δt	57.1 ms (62.9 ms)	22.4 ms
→ Nyquist frequency	8.75 Hz (7.95 Hz)	22.32 Hz
→ Experiment runtime	29.3 s (32.2 s)	22.9 s
No. of experiments	9	4

eastward of the radar location. This indicates that PMSE or the underlying structures and processes reveal a pronounced spatial variability that asks for in-depth studies in the future.

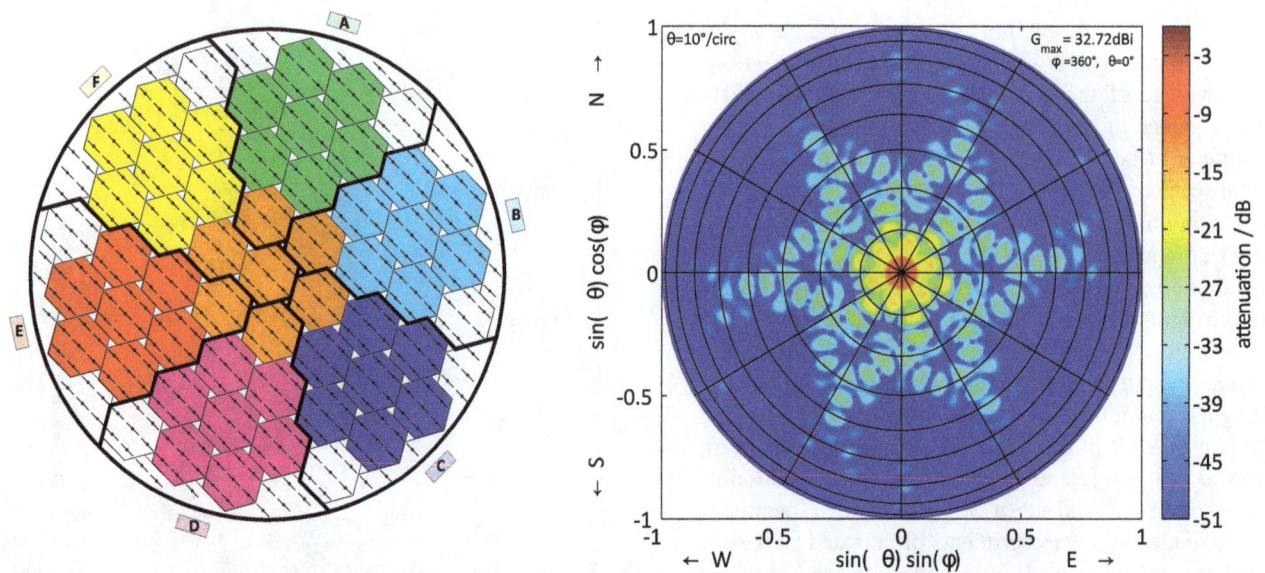

Fig. 4. Second expansion stage of MAARSY as installed in November 2010. 343 antennas of the colored hexagon structures as shown in the left panel were connected to transceiver modules and formed a radar beam with a half power beam width of 4° for Doppler beam operation. The computed radiation pattern for a beam pointing to 0° (zenith) using these seven anemones is shown in the right panel. The combined IF signals related to the seven anemone antenna structures (green, cyan, blue, magenta, red, yellow and orange) were connected to seven receivers and used for spaced antenna experiments in the troposphere.

3.2   Results using the second stage of expansion

3.2.1   Design of the second stage of expansion

After completion of the 2010 summer campaign for PMSE observation further transceiver modules were installed in November 2010. The second stage of expansion contained 343 transceivers connected to antennas as marked by colored hexagon structures in Fig. 4. The symmetrical setup formed a radar beam with a half power beam width of $\sim 4°$ for transmission. On reception the combined signals related to the seven anemone antenna structures A, B, C, D, E, F and M as marked in green, cyan, blue, magenta, red, yellow and orange, respectively, were split and connected to seven signal processing channels as shown in Table 1. The total combined receiving signal of all seven anemone antenna structures was led to signal processing channel 8. The upper eight inputs 9…16 of the signal processing units were connected via AIU to eight output ports of the 16 channel Butler matrix as described in detail in Renkwitz et al. (2011).

3.2.2   3-D resolved structures of polar mesosphere winter echoes

The new allocation of eight combined receiving signals corresponding to seven anemone structures and the full array to eight signal processing inputs allowed a non-multiplexing operation of MAARSY since December 2010. Since PMWE are a rare phenomenon with much shorter occurrence rate than PMSE a sequence of four experiments with six different oblique beam directions and a vertical beam was used

in a sequence in order to reduce the sequence runtime for a quasi-simultaneous 3-D-scan. The detailed experiment configuration is summarized in Table 2. The four scanning experiments pointing to 25 different beam directions in total ran one after another with 23 s runtime each. The scan was followed by a standard mesospheric monitor experiment with 27 s runtime using a vertical beam only but a larger number of coherent integrations and range gates. The overall sequence runtime was 189 s due to the additionally needed times for experiment reconfiguration. The reduced number of beam directions (25) compared to the PMSE scanning experiment (91, see Sect. 3.1.3) reduced the illuminated area to about 40 km diameter in the altitude range of interest between 65 km and 82 km.

Since PMWE are also weaker in absolute reflectivity than PMSE the small number of coherent integrations used in the scanning experiments makes it difficult to detect echoes. The event from 31 December 2010 at around 19:00 UT was strong enough to backscatter signals from all transmitted directions down to 15° off-zenith. Figure 5 shows horizontally resolved structures of a PMWE at different heights. The maximum scaling of the color coded SNR is lower by 21 dB compared to the plots in Fig. 3 demonstrating the much weaker strength of PMWE compared to PMSE. The vertical slice along the east-west direction shows a layered structure with a thickness of 2–3 km tilted by approximately 30°. The horizontal slices through the PMWE depict sharp gradients of SNR indicating a spatially localized process behind. A detailed discussion of this very first 3-D-observation of a PMWE is given in Rapp et al. (2011).

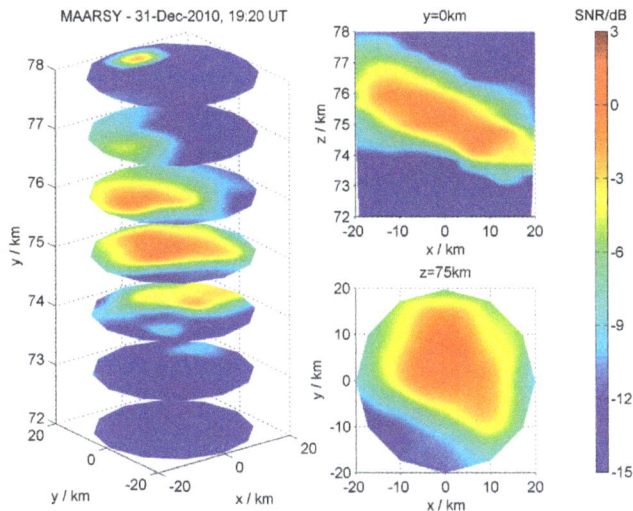

Fig. 5. 3-D structure of a PMWE quasi-simultaneously observed with 25 beams on 31 December.

4 Summary

The present paper describes the design and functionality of the two major stages of expansion of the Middle Atmosphere Alomar Radar System (MAARSY) during the installation in 2010. After the construction of the antenna array in summer 2009 and the installation of an initial expansion stage of 196 transceiver modules the radar operation started in May 2010. The second stage of expansion to 343 transceiver modules was brought into service in November 2010.

The major objective for building MAARSY, the investigation of horizontal structures of Polar Mesosphere Echoes, was tested with the partial systems using multi-beam experiments with up to 91 beams quasi-simultaneously during campaigns in August and December 2010. The presented first examples of horizontally resolved structures of Polar Mesosphere Summer and Winter Echoes demonstrate the performance of MAARSY and give a first insight into the three dimensional variability of PMSE and PMWE.

The final extension of MAARSY to 433 transceiver modules has recently been completed in May 2011.

Acknowledgements. The authors would like to thank IAP personnel who worked hard with the installation of MAARSY in particular Jörg Trautner, Thomas Barth, Jens Wedrich, Norbert Engler, Dieter Keuer, Marius Zecha, Hans-Jürgen Heckl, Torsten Köpnick, Manja Placke, and Qiang Li, as well as the students Ding Tao, Gunnar Keuer, Christian Schernus, Sophie Latteck, Danilo Hauch and Richard Hünerjäger. We also thank S. Fukao, T. Sato and M. Yamamoto for their suggestions and discussions in the early stage of planing the radar. We are indebted to the staff of the Andøya Rocket Range for their permanent support. The radar development was supported by grant 01 LP 0802A of Bundesmisterium für Bildung und Forschung. Topical Editor Matthias Förster thanks Michael Rietveld and Martin Friedrich for their help in evaluating this paper.

References

Czechowsky, P., Rüster, R., and Schmidt, G.: Variations of mesospheric structures in different seasons, Geophys. Res. Lett., 6, 459–462, 1979.

Ecklund, W. L. and Balsley, B. B.: Long-term observations of the Arctic mesosphere with the MST radar at Poker Flat, Alaska, J. Geophys. Res., 86, 7775–7780, 1981.

Kirkwood, S., Barabash, V., Belova, E., Nilsson, H., Rao, T. N., Stebel, K., Osepian, A., and Chilson, P. B.: Polar Mesosphere Winter Echoes during Solar Proton Events, Adv. Polar Upper Atmos. Res., 16, 111–125, 2002.

Kirkwood, S., Chilson, P., Belova, E., Dalin, P., Häggström, I., Rietveld, M., and Singer, W.: Infrasound – the cause of strong Polar Mesosphere Winter Echoes?, Ann. Geophys., 24, 475–491, doi:10.5194/angeo-24-475-2006, 2006.

Latteck, R., Singer, W., Rapp, M., and Renkwitz, T.: MAARSY – the new MST radar on Andøya/Norway, Adv. Radio Sci., 8, 219–224, doi:10.5194/ars-8-219-2010, 2010.

Latteck, R., Singer, W., Rapp, M., Vandepeer, B., Renkwitz, T., Zecha, M., and Stober, G.: MAARSY - The new MST radar on Andøya: System description and first results, Radio Sci., 47, RS1006, doi:10.1029/2011RS004775, 2012.

Lübken, F.-J., Strelnikov, B., Rapp, M., Singer, W., Latteck, R., Brattli, A., Hoppe, U.-P., and Friedrich, M.: The thermal and dynamical state of the atmosphere during polar mesosphere winter echoes, Atmos. Chem. Phys., 6, 13–24, doi:10.5194/acp-6-13-2006, 2006.

Rapp, M. and Lübken, F.-J.: Polar mesosphere summer echoes (PMSE): Review of observations and current understanding, Atmos. Chem. Phys., 4, 2601–2633, doi:10.5194/acp-4-2601-2004, 2004.

Rapp, M., Latteck, R., Stober, G., and Singer, W.: First 3-dimensional observations of polar mesosphere winter echoes: resolving space-time ambiguity, J. Geophys. Res., 116, A11307, doi:10.1029/2011JA016858, 2011.

Renkwitz, T., Singer, W., Latteck, R., and Rapp, M.: Multi beam observations of cosmic radio noise using a VHF radar with beam forming by a Butler matrix, Adv. Radio Sci., 9, 349–357, doi:10.5194/ars-9-349-2011, 2011.

Zeller, O., Zecha, M., Bremer, J., Latteck, R., and Singer, W.: Mean characteristics of mesosphere winter echoes at mid- and high-latitudes, J. Atmos. Sol.-Terr. Phy., 68, 1087–1104, doi:10.1016/j.jastp.2006.02.015, 2006.

29

Data analysis of low frequency transmitter signals received at a midlatitude site with regard to planetary wave activity

E. D. Schmitter

University of Applied Sciences Osnabrueck, 49076 Osnabrueck, Germany

Correspondence to: E. D. Schmitter (e.d.schmitter@hs-osnabrueck.de)

Abstract. More than 2 yr of continuously recorded signal amplitude data from the MSK transmitters NRK/TFK (37.5 kHz, Iceland) and NSY (45.9 kHz, Sicily) received at (52° N 8° E) in the time range from August 2009 to September 2011 are analyzed with regard to planetary wave activity. Wavelet analysis of the day/night amplitude ratio reveals clear evidence of quasi 16 day periods mainly during winter time as well as traces of 5 and 10 day periods on both paths. The amplitude ratio is well correlated to the typical stratospheric (10 hPa) seasonal temperature profile – more clearly to be seen on the northern path. The results are in line and an extension of manifold research with regard of ionospheric absorption phenomena caused by atmospheric wave activity. Continuous monitoring of transmitters in the 40 kHz frequency range proved as an inexpensive tool for investigating mesospheric response to forcing from below.

1 Introduction

Remote monitoring of lower ionosphere conditions using low and very low transmitters signal propagation is a well known method for several decades. Especially MSK (Minimum Shift Keying) transmitters prove as useful in this respect because of their constant amplitude emissions. For several years we have analysed the signal amplitude variations of 2 transmitters, NRK/TFK (37.5 kHz , 63.9° N 22.5° W, L = 5, Iceland) and NSY (45.9 kHz , 37.13° N 14.44° E, L = 1.4, Sicily) received a midlatitude site (52° N 8° E). The NW path to Iceland is particularly suited to study lower ionosphere forcing from above (especially auroral particle precipitation, Schmitter, 2010), whereas both paths yield useful information about forcing processes from below. In Schmitter (2011) we presented data analysis and propagation modeling of the southern path to Sicily. In this paper we extend the analysis with respect to forcing from below to the northern path and use wavelet analysis for period identification. A lot of research has been done with respect to the coupling of atmospheric phenomena (tidal, gravity and planetary waves) to the ionosphere and an important remote sensing tool in this respect is lower ionospheric absorption of radio wave propagation between ground based transmitters and receivers (Schwentek, 1974; Fraser, 1977; Lastovicka et al., 1994b; Lastovicka, 1994). Our work is based on measurements using 40 kHz transmitters near to the very low frequency range (3–30 kHz) at a distance of around 2000 km. Propagation at this frequency is sensitive to the conditions in the lowest part of the ionosphere (around 65 to 85 km) whereas most of the cited work uses long and medium frequencies (some hundred kHz to a few MHz) penetrating further into the ionosphere. The essential parameters of the lower ionosphere, i.e. electron density and collision frequency, are modulated by atmospheric waves ascending from tropospheric and stratospheric heights to the mesosphere (Lauter et al., 1984; Lastovicka et al., 1994a).

Planetary waves are disturbances having zonal wavelengths of the scale of the earth's radius. Excited mainly by orographic and land-sea temperature contrasts a broad spectrum of planetary waves is generated. The cavity properties of the atmosphere then allow for mainly westward travelling normal modes with periods near to 2, 5, 10 and 16 days to propagate into the stratosphere and mesosphere. For a thorough introduction into the topic see Volland (1988). By coupling to the ionosphere they can effect radio wave propagation through modulation of the electron density and neutral density profiles. The latter profile controls the electron collision frequency. Also their variability is suggested to be a signature of possible climatic changes (Pogoreltsev et al., 2009). In following we shortly describe the used wavelet data analysis method and afterwards we discuss results of the last two years of monitoring (August 2009 to September 2011).

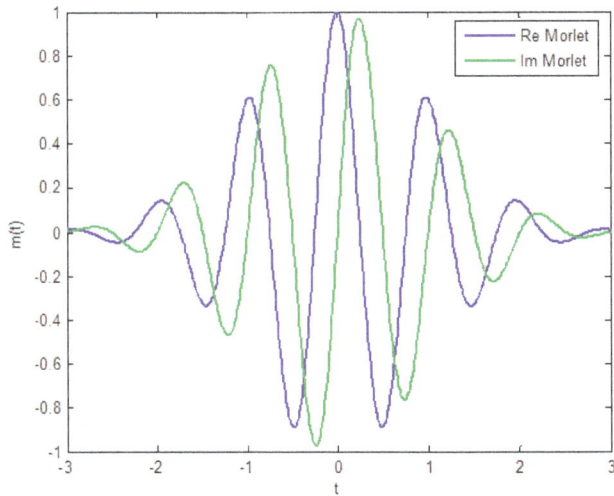

Fig. 1. Morlet wavelet: wave packet with real and imaginary part. The std. deviation of the Gaussian envelope function represents the time domain scaling period s. Its Fourier transform is a real Gaussian acting like a bandpass with center at $f = 1/s$ in the frequency domain.

2 Data analysis

For period identification we use a wavelet transformation W of the data $y(t)$

$$W[y](s,t) = \frac{1}{s} \int_{-\infty}^{+\infty} y(z)\, m(\frac{1}{s}(z-t)) \mathrm{d}z$$

of the Morlet type (Goupillaud et al., 1984) with: $m(t) = e^{-t^2/2}\, e^{2\pi i t}$. In the time domain the scaled Morlet $m(t/s) = e^{-t^2/2s^2}\, e^{2\pi i t/s}$ is a complex valued wave packet (Fig. 1) with width $2s$ (more precisely: s is the standard deviation of the Gaussian envelope function). In the frequency domain its Fourier transform F acts like a band pass filtering out the scaling period s: $F[m(t/s)](\omega) = e^{-(\omega-2\pi)^2 s^2/2}$ with a maximum at $\omega = 2\pi f = 2\pi/s$ or $f = 1/s$ and a full width at half maximum of $\Delta f = 0.375/s$. The ripple spacings in the Morlet periodograms shown in this paper reflect the period s indicated on the vertical axis, respectively.

The upper part of Fig. 2 displays the zonally averaged temperature at the 10 mbar level (about 30 km height) between 65 and 25 degree northern latitude (data from ftp.cpc.ncep. noaa.gov) and its wavelet decomposition. Both radio propagation paths are in this latitude range.

Figure 2, lower part, and Fig. 3, upper part, show the signal amplitude ratio noon/midnight (2 h averages) and wavelet decompositions for the NRK and NSY propagation paths to 52° N 8° E. A ratio of 0 dB indicates equal average amplitudes at noon and midnight, whereas negative values mainly result from reduced noon amplitudes and indicate increased day over absorption. We see strong absorption events with large variability during northern winter time (October to April) and significantly reduced absorption and variability during summer (May to July). These features are more prominently to be seen on the northern propagation path (NRK, Iceland to 52° N 8° E) but also evident on the southern path (NSY, Sicily to 52° N 8° E). During summer time the NRK (northern path) absorption is low and exhibits few variation and correlates well with the zonally averaged 10 mb temperature course. The NSY (southern path) exhibits significant absorption variation also during summer time and the difference of the averaged day/night signal ratio between winter and summer is less pronounced than with the northern path. On the northern path we additionally see a low variation level around −5 dB during mid of August to end of September. With both radio signals as well as the stratosphere temperature we see common maxima around June and July with low variation and periods of high variability during winter. We relate the variability to planetary wave activity: the wavelet decompositions of the radio signal time series clearly show 12 to 17 days periods during winter time on both paths, more intensely during winter 2009/2010 than during winter 2010/2011. In fact there are also traces of periods near 2, 5 and 10 days in both radio wavelet decompositions, however not showing up in the cross correlation of the two radio time series, in contrast to the near 16 day period. So with some caution we suggest that normal planetary wave modes below 16 days can be also identified in the propagation data, however not in timely coincidence. The commonality of the near 16 day periodicities is highlighted by the wavelet cross correlation between the NRK and NSY time series shown in Fig. 4 (top). Plotted is the absolute value of the product of the two single wavelet periodograms. The 7 day periods are caused by weekly drop outs of the transmitters. In the middle of Fig. 4 we see the product of the two radio data wavelet periodogramms and the zonal temperature periodogramm (3-fold cross correlation). Prominent are the near 16 day period correlations during winter 2009/2011 between the two propagation paths as well as with the temperature data. Less significant but nevertheless clear are the near 13 and 18 day period correlations during winter 2010/2011.

We have to check how far solar/geomagnetic activity interferes with the conception that the signal absorption variability is a fingerprint of planetary wave activity: Fig. 3 (bottom) displays the the geomagnetic disturbance index Ap and its wavelet decomposition for the same time range. The 27 day synodic solar rotation and its first harmonic (13.5 day period) are clearly visible and pronounced during times of increased geomagnetic activity (starting significantly around March 2010 with the new solar cycle) but is not in line with the winter time maxima of low frequency signal ratio periodicities. There is no significant cross correlation between the Ap index and radio propagation periodograms up to the 27 day range (Fig. 4, bottom). However there are clearly common periods with all data in the 27–30 day range. Also noticeable with the Ap wavelet plot are the 5 and 9 day periods. The 9-day periodicities are related to the recurrence

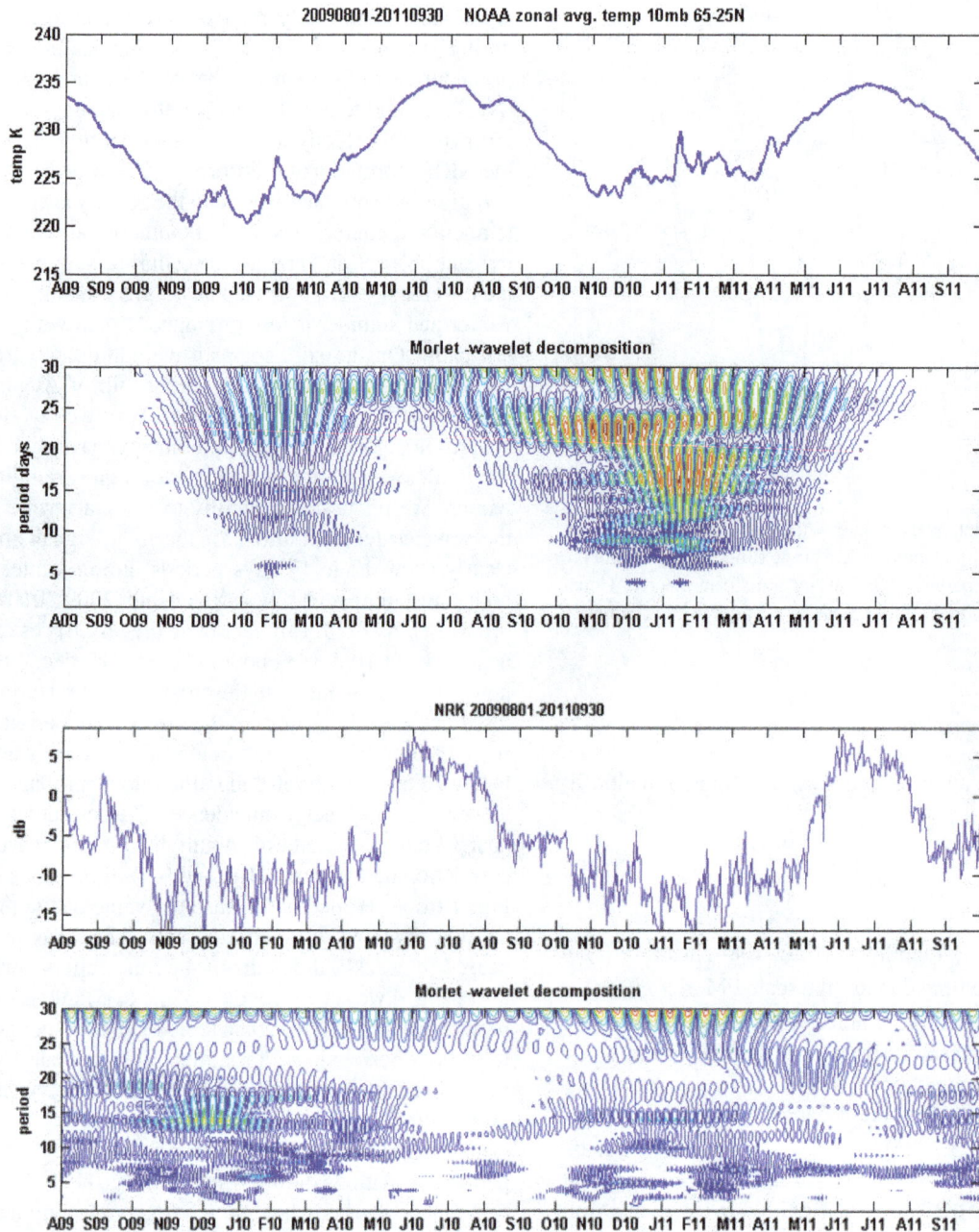

Fig. 2. Time range 1 August 2009 to 30 September 2011, top: 10 hPa (about 30 km height) zonal mean temperatures for the latitude range 65 to 25 degree north (data from ftp.cpc.ncep.noaa.gov) and the wavelet decomposition. Below: signal amplitude ratio noon/midnight (2 h averages) and wavelet decomposition for the NRK (Iceland) propagation path to 52° N 8° E.

of fast solar wind streams due to solar coronal holes which are distributed roughly 120 degrees apart in longitude modulating geomagnetic activity and also thermospheric densities (e.g. Lei at al., 2008; Mukhtarov et al., 2010; Emery et al., 2011, we thank one of the referees for pointing us to this result). The 5 day periods around March/April and August/September in 2010 and 2011 roughly coincide with the equinoxes, an effect which is related to the well known

semiannual variation of geomagnetic activity with equinoctial maxima (McIntosh, 1959).

3 Summary and conclusions

Wavelet analysis – especially using Morlet wave packets – is well suited to localize the intensity of periodicities not only in frequency (period scale) but also in time. Cross correlation

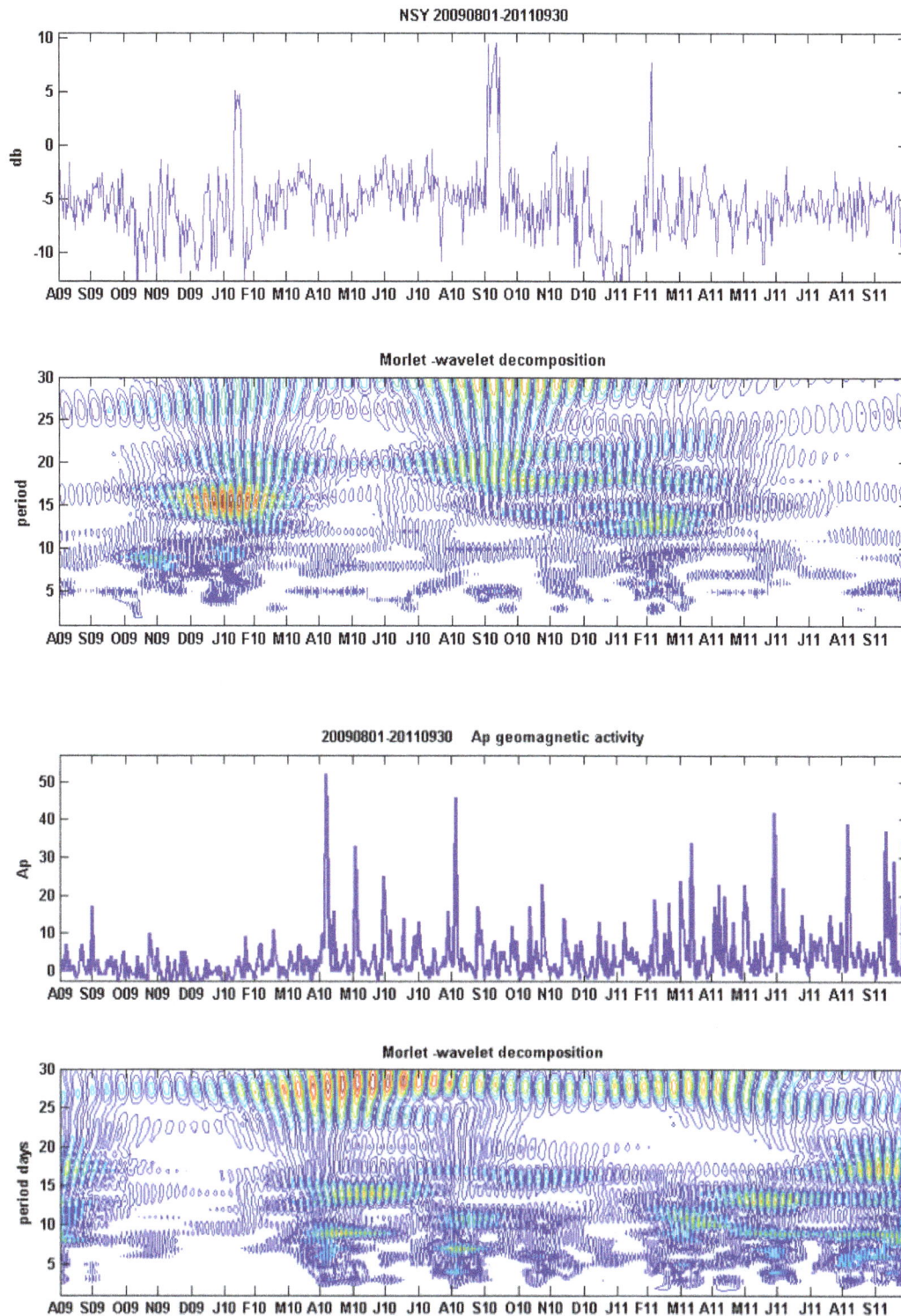

Fig. 3. Time range 1 August 2009 to 30 September 2011, top: signal amplitude ratio noon/midnight (2 h averages) and wavelet decomposition for the NSY (Sicily) propagation path to 52° N 8° E. Bottom: time series of the geomagnetic Ap index and its wavelet decomposition. The 27 day synodic sun rotation period and its first harmonic (13.5 days) are prominent during times of higher geomagnetic activity. We also note 5 and 9 day periods with the Ap index, see text.

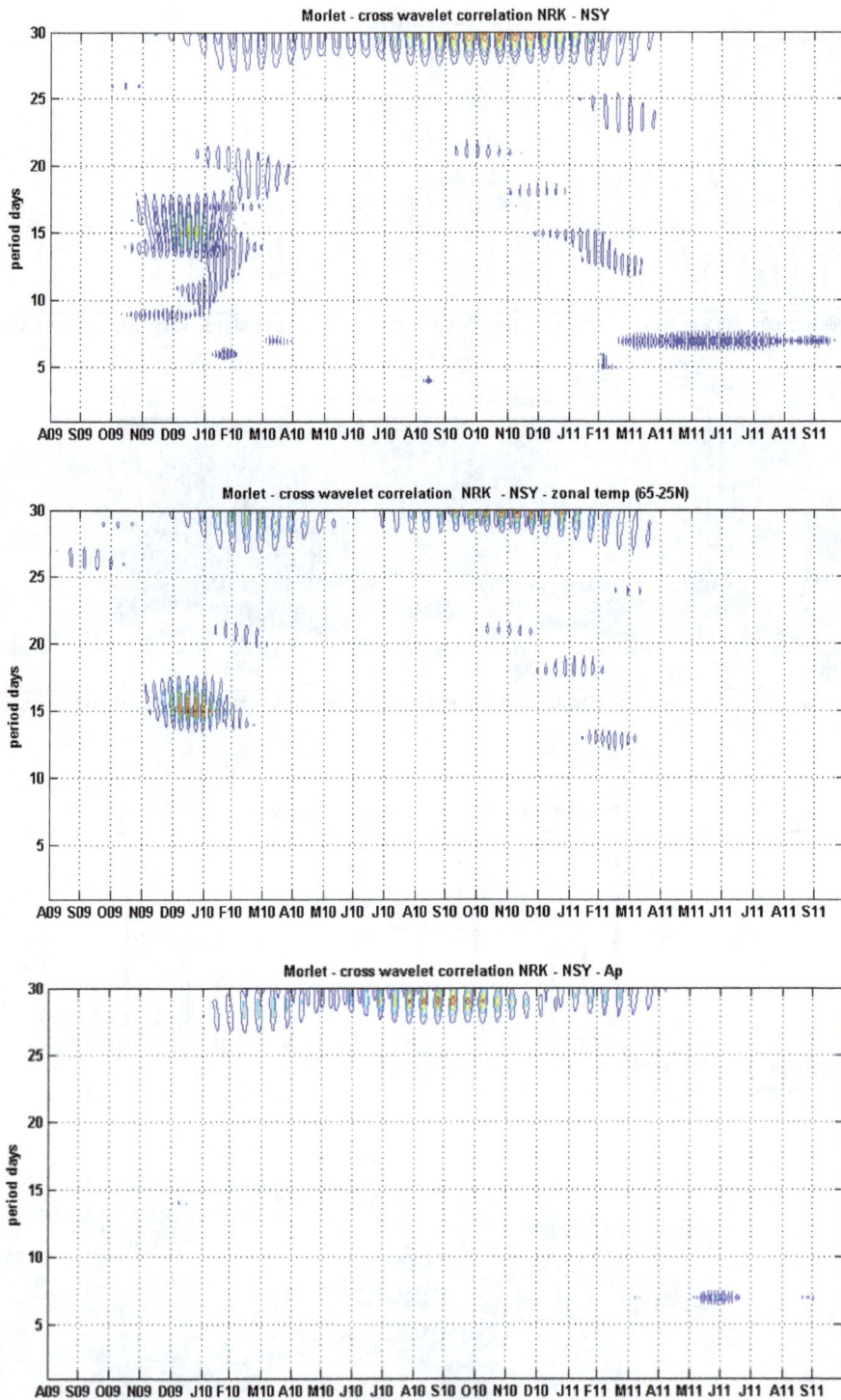

Fig. 4. Time range 1 August 2009 to 30 September 2011: top: cross wavelet analysis of the NRK and NSY transmitter signal day/night ratios. The 7 day period is due to weekly transmitter drop outs. Middle: NRK-NSY-zonal temperature cross correlation. Prominent are the near 16 day period correlations during winter 2009/2011 between the 2 propagation paths as well as with the temperature data. Less significant but nevertheless clear are the near 13 and 18 day period correlations during winter 2010/2011. Bottom: NRK-NSY-Ap cross correlation. There is no significant cross correlation between the Ap index and radio propagation data up to the 27 day range. However there is a clear correlation between all data in the 27–30 day range.

of the periodograms points to common time ranges of periodicity. The day/night amplitude ratio of 40 kHz low frequency MSK transmitters essentially is an indicator for day over absorption of radiation in the lower ionosphere in the height range 65–85 km. Some 2 yr of monitoring a northern and a southern propagation path to a midlatitude site clearly show periods of 12–17 days (quasi 16 day wave) in the wavelet analyzed data during winter time, not related to solar/geomagnetic activity, together with a strong correlation with seasonal stratosphere temperature changes over the year – more explicitly on the northern path. Traces of periods near 2, 5 and 10 days, presumably also normal planetary modes, can additionally be identified in the radio propagation data (also with the temperature data, however not explicitly correlating in time with the radio data). A modulation of ionospheric parameters, electron density and especially collision frequency via neutral density variations caused by planetary wave activity is endorsed by these results (Lastovicka et al., 1994a; Schmitter, 2011). Recurrent details like the nearly constant level in the amplitude ratios from August to September as well as an assessment of the winter time signal variability with respect to the fingerprint of stratospheric warmings need more data and explanation. Our observation time range (August 2009–September 2011) encompasses part of the long solar minimum and the start of activity within the new solar cycle 24. Future data of our ongoing continuous monitoring will show if and how the absorption behaviour, especially with regard to the day time absorption and its relation to generating planetary waves, will change with increasing solar activity and if other long term trends can be identified (Lastovicka, 2002 reports no sufficiently consistent observational pattern of trends in the lower ionosphere from his long term data). In any case data analysing low frequency MSK signal amplitude ratio measurements prove as an inexpensive additional method for atmosphere and climate research.

Acknowledgements. Topical Editor Matthias Förster thanks Dora Pancheva and Kirsti Kauristie for their help in evaluating this paper.

References

Emery, B. A., Richardson, I. G., Evans, D. S., Rich, F. J., and Wilson, G. R.: The Sun-Earth Connection Near Solar Minimum: Solar Rotational Periodicities and the Semiannual Variation in the Solar Wind, Radiation Belt, and Aurora, Sol. Phys., 274, 399425, 2011

Fraser, G. J.: The 5-day wave and ionospheric absorption, J. Atmos. Terr. Phys., 39, 121–124, 1977.

Goupillaud, P., Grossman, A., and Morlet, J.: Cycle-Octave and Related Transforms in Seismic Signal Analysis, Geoexploration, 23, 85–102, 1984.

Lauter, E. A., Taubenheim, J., and Von Cossart, G.: Monitoring middle atmosphere processes by means of ground-based lowfrequency radio wave sounding of the D-region, J. Atmos. Terr. Phys., 46, 775–780, 1984.

Lastovicka, J.: Trends in planetary wave activity in the upper middle atmosphere inferred from the nighttime LF radio wave absorption measurements, Stud. Geophys. Geod., 38, 206–212, 1994.

Lastovicka, J.: Long-term changes and trends in the lower ionosphere, Phys. Chem. Earth, Parts A/B/C 27, 497–507, 2002.

Lastovicka, J., Ebel, A., and Ondraskova, A.: On the transformation of planetary waves of tropospheric origin into waves in radio wave absorption in the lower ionosphere, Stud. Geophys. Geod., 38, 71–81, doi:10.1007/BF02296254, 1994a.

Lastovicka, J., Fiser V., and Pancheva, D.: Long-term trends in planetary wave activity (2–15 days) at 80–100 km inferred from radio wave absorption, J. Atmos. Terr. Phys., 56, 893–899, 1994b.

Lei, J., Thayer, J. P., Forbes, J. M., Sutton, E. K., and Nerem, R. S.: Rotating solar coronal holes and periodic modulation of the upper atmosphere, Geophys. Res. Lett., 35, L10109, doi:10.1029/2008GL033875, 2008.

Pogoreltsev, A. I., Kanukhinaa, A. Yu., Suvorova, E. V., and Savenkovaa, E. N.: Variability of planetary waves as a signature of possible climatic changes, J. Atmos. Sol.-Terr. Phy., 71, 1529–1539, 2009.

McIntosh, D. H.: On the annual variation of magnetic disturbance, Phil. Trans. Roy. Soc. Lond., 251, 525–552, 1959.

Mukhtarov, Pl., Andonov, B. ,Borries, C., Pancheva, D., and Jakowski, N.: Forcing of the ionosphere from above and below during the Arctic winter of 2005/2006, J. Atmos. Sol.-Terr. Phys., 72, 193–205, doi:10.1016/j.jastp.2009.11.008, 2010.

Schmitter, E. D.: Remote auroral activity detection and modeling using low frequency transmitter signal reception at a midlatitude site, Ann. Geophys., 28, 1807–1811, doi:10.5194/angeo-28-1807-2010, 2010.

Schmitter, E. D.: Remote sensing planetary waves in the midlatitude mesosphere using low frequency transmitter signals, Ann. Geophys., 29, 1287–1293, doi:10.5194/angeo-29-1287-2011, 2011.

Schwentek, H.: Wave-like structures in the variation of ionospheric absorption, J. Atmos. Terr. Phys., 36, 1173–1178, 1974.

Volland, H.: Atmospheric tidal and planetary waves, Kluwer, 1988.

30

Planetary wave characteristics of gravity wave modulation from 30–130 km

P. Hoffmann and Ch. Jacobi

Institute for Meteorology, University of Leipzig, Stephanstr. 3, 04103 Leipzig, Germany

Correspondence to: P. Hoffmann (phoffmann@uni-leipzig.de)

Abstract. Fast gravity waves (GW) have an important impact on the momentum transfer between the middle and upper atmosphere. Experiments with a circulation model indicate a penetration of high phase speed GW into the thermosphere as well as an indirect propagation of planetary waves by the modulation GW of momentum fluxes into the thermosphere. Planetary wave characteristics derived from middle atmosphere SABER temperatures, GW potential energy and ionospheric GPS-TEC data at midlatitudes reveal a possible correspondence of PW signatures in the middle atmosphere and ionosphere in winter around solar maximum (2002–2005). In the case of the westward propagating 16-day wave with zonal wavenumber 1 a possible connection could be found in data analysis (November–December 2003) and model simulation. Accordingly, GW with high phase speeds might play an essential role in the transfer of PW and other meteorological disturbances up to the ionospheric F-region.

1 Introduction

Ionospheric variations have a disturbing impact on GPS navigation systems and radio communication. Via the two carrier frequencies of the GPS satellites such electron density disturbances can be partly corrected and ensure the commercial application of this product. In turn ionospheric science community is interested in variability. New knowledge about characteristic of planetary wave (PW) like structures coming from the lower atmosphere are expected. For this purpose comparisons with other meteorological observations (reanalyses, satellites) will give more insight into a possible coupling (e.g. Borries et al., 2007; Borries and Hoffmann, 2010; Goncharenko et al., 2010; Hoffmann et al., 2011). It is supposed that fast small scale gravity waves (GW) excited in the troposphere are able to penetrate the thermosphere/ionosphere system and transfer momentum and also indirect properties of PW through GW modulation. This indirect signal of PW is usually identified as the PW signature (PWS). Planetary waves cannot exist in the ionospheric plasma. Dynamic features at periodicities of PW that are seen in the ionosphere therefore must be of indirect nature and forced by neutral atmosphere. Thus, they are termed as PW Type Oscillations (PWTO).

In this paper we present preliminary results from satellite data analyses of GW potential energy with respect to spectral PW characteristics and numerical model experiments by changing the spectrum of GW.

2 Planetary wave proxies

The long-term analysis of PW proxies (Borries and Hoffmann, 2010; Hoffmann et al., 2011) at midlatitudes ($\sim$50° N) from 2002 to 2008 using different data sets as stratosphere temperatures, lower thermosphere GW potential energy and ionospheric GPS-TEC maps (processed at DLR Neustrelitz, Jakowski et al., 2002), represents the seasonal behaviour of PW, PWS and PWTO (Fig. 1, left). The comparison of PW with PWTO (upper panel) and of PWS with PWTO (lower panel) indicate correspondence of wave activity between the middle and upper atmosphere. While the proxy of traveling PW activity in the stratosphere shows a less disturbed annual cycle with a maximum in winter, the PWTO variance in the ionosphere appears superposed by the solar activity at time-scales of several years (solar cycle) and of several weeks (solar rotation). However, both stratosphere and ionosphere show a maximum of wave activity during the winter season. If we look at the indirect PW signal GW potential energy, there are characteristic structures in PWS during winter 2003/04 and 2004/05 that fit quite well to the characteristics found in the ionosphere. This fact may have two causes. On the one hand, the solar rotation period may modulate GW in the lower thermosphere and, on the other hand, what we

(a)

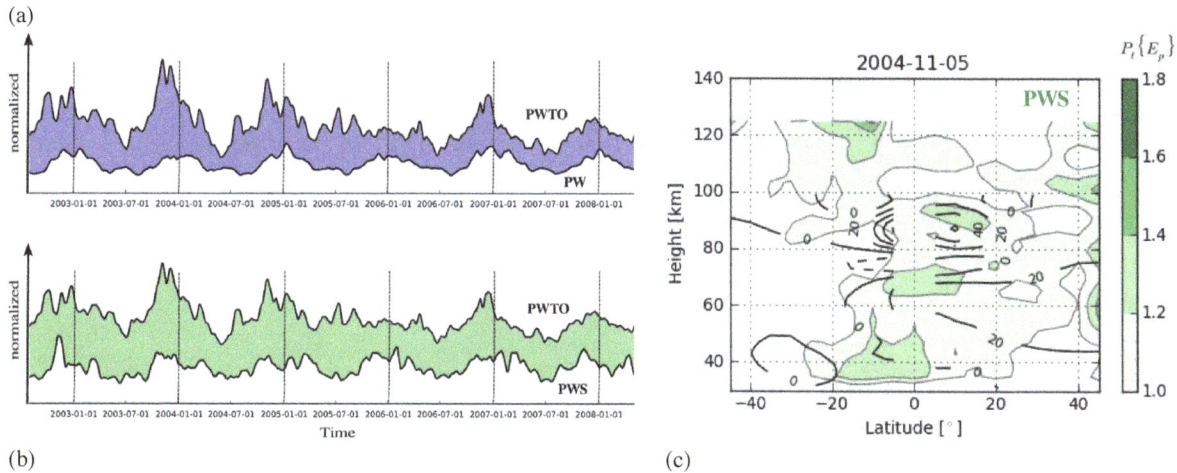

(b)

(c)

Fig. 1. Left: time series of PW proxies at midlatitudes from 2002 to 2008: stratospheric PW, ionospheric PWTO (**a**) and PWS and ionospheric PWTO (**b**). Right: height-latitude cross section of PWS derived from SABER temperatures on 2004-11-05 (**c**). The geostrophic zonal wind is given in black contours.

observe is the meteorological part on the total variability of the ionosphere.

The spatial structure of the GW modulation (Fig. 1, right) derived from Sounding of the Atmosphere using Broadband Emission Radiometry (SABER) data centered around 2004-11-05 indicates that PWS (green scaling) on the winter hemisphere (45° N) are able to penetrate the lower thermosphere. The vertical propagation of atmospheric GW is determined by the background wind and PW. Global wind wave disturbances lead to a periodic filtering of GW and a possible indirect transfer of PW into the ionospheric F-region. The geostrophic zonal wind component derived from SABER (black contours) indicates weak westerlies in early winter.

In order to determine the connection of PW properties in the stratosphere and ionosphere spectral components need to be considered. Furthermore, the spectrum, which in Fig. 1 is characterized by the vertical wavelength $\lambda_z < 6$ km, must be extended to longer ones ($\lambda_z = 5...10$ km). These GW are faster and more likely to penetrate the thermosphere. In the following we present a method to filter GW energy from satellite temperature profiles and to analyze PW characteristics spectrally in space and time (Sect. 3). The modeling part (Sect. 4) will demonstrate the role of GW parameters in the simulation of the middle and upper atmosphere variability.

3 Planetary wave characteristics

Unevenly spaced data from the SABER instrument on board the TIMED satellite (Mertens and et al., 2004) provide temperature profiles (30–130 km) with a global coverage (50° S–50° N) since 2002. In this section we apply a space-time method (Pancheva et al., 2009) to analyze PW from temperature fields and PWS from potential energy.

3.1 GW potential energy

For the calculation of the potential energy E_p we consider single profiles $T(z)$ and apply a Savitzky-Golay filter (Savitzky and Golay, 1964) to fit the background profile $T_B(z)$ (Fig. 2, left). An example of the residuals at 50.7° N, 267.1° E on 2003-12-20 are shown in the middle panel of Fig. 2. Other than on the summer hemisphere, the vertical wavelength spectrum is determined by PW, the semidiurnal tide and GW, which partly corresponds with each other. In principle, an additional horizontally filtering of global-scale waves from the background would be necessary in order to consider pure GW information (e.g. Preusse et al., 2006; Ern et al., 2011), but this is not applied here, because this method is not applicable to a running space-time analysis. Finally, the residuals are band-pass filtered and $T_F(z)$ includes fluctuations in the range of $\lambda_z = 5...10$ km. The Brunt-Väisälä frequency $N^2 = \frac{g}{T_B}\left(\frac{dT_B}{dz} - \Gamma\right)$ is determined, with Γ as the vertical temperature gradient and g the gravitational acceleration. The potential energy is then calculated as $E_p = \frac{1}{2}\left(\frac{g}{N}\right)^2 \left(\frac{T}{T_B}\right)^2$ and integrated over a sliding 15 km interval. From the dispersion relation of internal GW the horizontal phase speed $c_{ph} \simeq \frac{\lambda_z N}{2\pi}$ is proportional to the vertical wavelength. Consequently, fast GW have long vertical wavelength. For our configuration ($\lambda_z = 5...10$ km) the phase speed is in a range of $16...32$ m s^{-1} with $N = 0.02$ s^{-1}.

The atmospheric background near the Equator is mainly determined by the diurnal tide. At high latitudes of the summer hemisphere only the semidiurnal component dominates because PW cannot propagate upward during easterlies. In our further considerations we focus on midlatitudes on the winter hemisphere for investigating the characteristics of PWS.

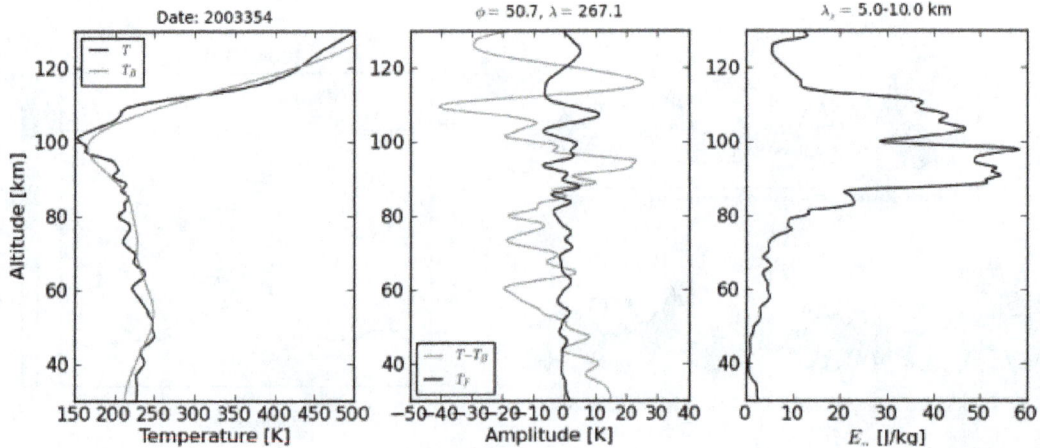

Fig. 2. Principle to derive GW potential energy E_p from SABER temperature profiles (50.7° N, 267.1° E) on the day of year 354 in 2003. Left: sounding of the temperature from 30–130 km (black line) and the fitted background (grey line); middle: residuals (grey line) and the band-pass filtered signal (black line); right: the potential energy profile for $\lambda_z = 5$–10 km.

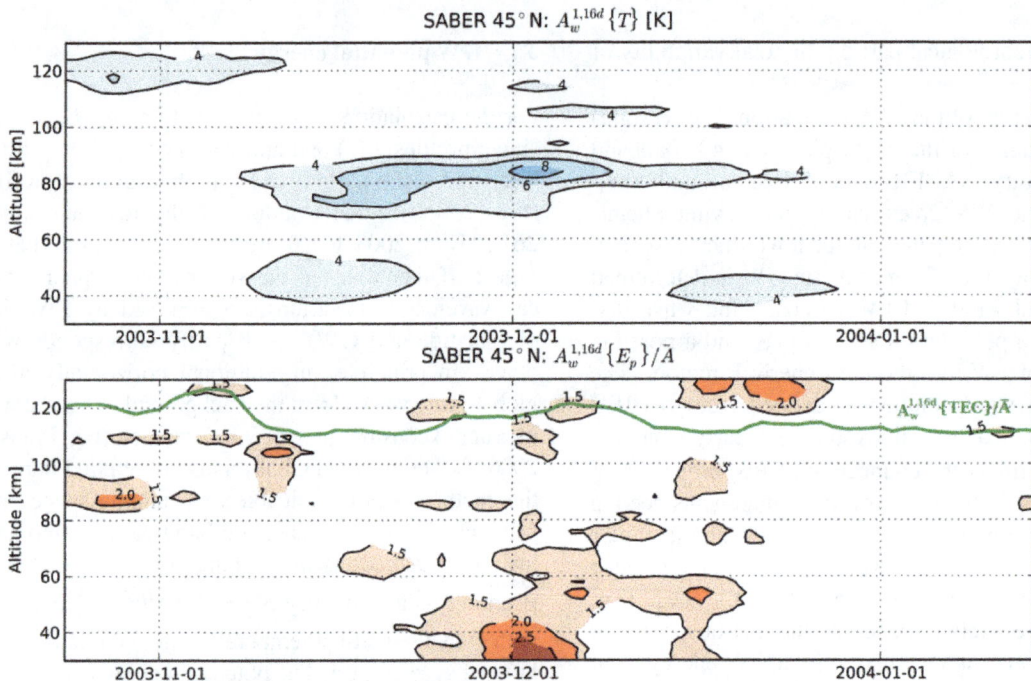

Fig. 3. Height vs. time cross section of the westward propagating 16-day wave ($k = 1$) amplitude at 45° N in temperature (top) and potential energy (bottom). The potential energy amplitudes are normalised. The course of this component derived from ionospheric GPS-TEC is shown as added (green curve).

3.2   Space-time spectral method

In this section we present results from space-time spectral analysis (Pancheva et al., 2009) of PW and PWS derived from SABER temperature and potential energy profiles in the latitude range centered around $\phi = 45°$ N $\pm 2.5°$ and averaged vertically over 2 km.

Because the variance analysis (Fig. 1) results in correspondences between PWS and PWTO around solar maximum, we consider the winter situation during November 2003 to February 2004. Figure 3 (top) shows the amplitude distribution of the westward propagating 16-day wave with zonal wavenumber ($k = 1$) in a height vs. time cross section. Two maxima in mid November and December at 40 km ($\sim$4 K) and a stronger one near 80 km ($\sim$8 K) exists.

The pattern of PW activity in the stratosphere and mesosphere changes with height. There are times when a lack of PW activity at 40 km (1 December) goes along with a strong signal at 80 km, but usually not at higher altitudes. Maybe the signals of the 16-day waves in early December above 80 km, which corresponds to the 16-day wave in the ionosphere, come from GW. These are included in the unfiltered temperature profile.

The indirect signatures of PW analyzed from GW modulation is shown in Fig. 3 (bottom). Here, the amplitudes are normalized with respect to the time average at each height. Additionally, the course of the 16-day wave in the ionospheric F-region is represented by the green line. A behaviour of PWS that is more variable and partly different from that of PW is seen. In general we observe characteristics of a long-periodic PW above 100 km in the modulation of GW energy that correspond with PWTO in the ionospheric F-region.

4 Numerical modeling

4.1 The Middle and Upper Atmosphere Model

The Middle and Upper Atmosphere Model (MUAM) is able to simulate the variability of the thermosphere (Pogoreltsev et al., 2007). Although the model does not includes the ionospheric component of the upper atmosphere an appropriate tuning of the parametrized GW will extend the application to processes that are responsible for the vertical coupling of the middle atmosphere with the thermosphere-ionosphere system. MUAM is usually forced at the lower boundary with climatological mean reanalysis fields for individual months of mean temperature cross sections (1000–30 hPa) and the first three stationary wave components in temperature as well as geopotential height at 1000 hPa. In order to study the response of the neutral upper atmosphere to changes in the GW parameters we run the GW scheme (Jacobi et al., 2006) in an offline mode.

Figure 4 depicts the simulated January mean meridional circulation (arrows) up to the F-region (~400 km). The mean zonal wind jets (contours) dominate the dynamics of the middle atmosphere (yellow filled). Above, at thermospheric heights the ionosphere is symbolized by the blue scaling. As shown by different model experiments (e.g. Miyoshi and Fujiwara, 2008; Yiğit and Medvedev, 2010) tropospherically forced fast GW have an important impact on the thermosphere circulation and this will be also examined with MUAM.

4.2 GW parametrization and offline simulations

Gravity waves parameterized in MUAM are described in Jacobi et al. (2006). In order to operate offline experiments, atmospheric background fields, GW parameters and constants must be transferred to the routine in order to

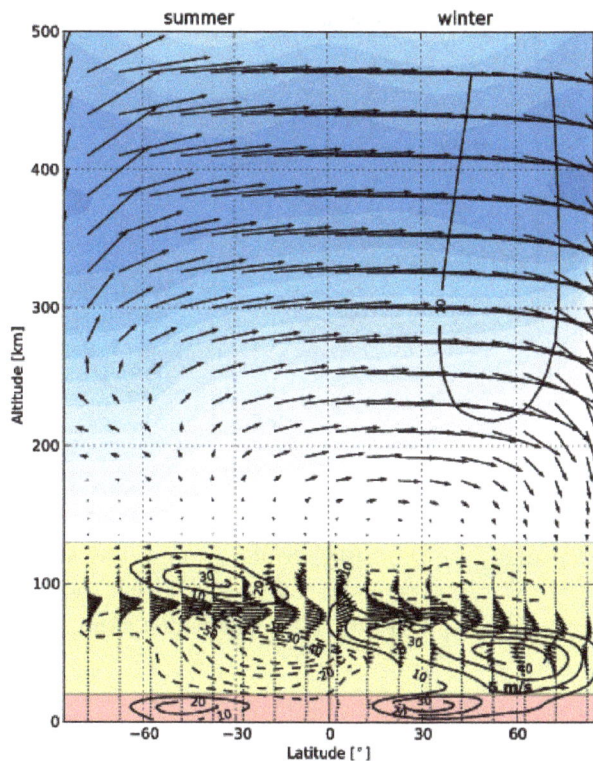

Fig. 4. Height vs. latitude cross section of the middle and upper atmosphere circulation up to the F-region height (~400 km) simulated with MUAM for January conditions: meridional circulation (arrows), mean zonal wind (contours).

obtain GW momentum fluxes and accelerations quasi instantaneously. Here we run the scheme with doubled phase speed of GW related to the reference one characterized by phase speed ($c_{ph} = [5, 10, 15, 20, 25, 30]\,\mathrm{m\,s^{-1}}$), horizontal wavelength ($\lambda_h = 300$ km) and phase angle ($\theta = [0, 45, 90, 135, 180, 225, 270, 315]$).

The prescribed spectrum consists of 48 different single GW with a constant horizontal wavelength. The current adjustments are tuned to the middle atmosphere in order to describe the background wind reversal in the lower thermosphere. Due to this relatively slow phase speeds of the spectrum ($\overline{c_{ph}} \approx 20\,\mathrm{m\,s^{-1}}$), all GW are filtered by the mean winds and break in the upper mesosphere. No momentum fluxes are transferred into the thermosphere. Cooling and heating due to GW are calculated based on Medvedev and Klaassen (2003).

This condition changes in the case of faster GW. Yiğit and Medvedev (2010) simulated the behaviour of GW in the thermosphere with changing solar activity using TIME-GCM. The spectrum that they used has phase velocities from 2 to 80 m s^{-1}. For our simulations we simply double the phase speeds that correspond to 10 to 60 m s^{-1}. This corresponds to periods between 1...8 h. Figure 5 (middle panel) shows

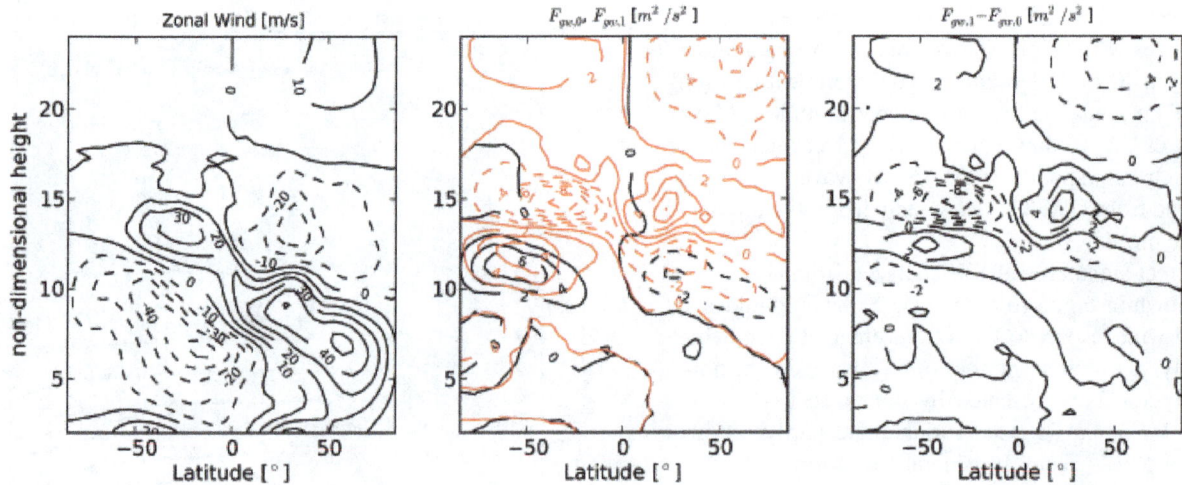

Fig. 5. Simulated atmospheric background zonal wind (left), and GW momentum fluxes (middle) for the standard phase speeds ($F_{GW,0}$, black) and the double phase speeds ($F_{GW,1}$, red). The differences pattern is given in the right panel. The non-dimensional height is defined as $x = -\ln\left(\frac{p}{p_0}\right)$ with $p_0 = 1000$ hPa ($x = 15$ about 100 gpkm).

the momentum fluxes. The comparison indicates a possible coupling between the middle and upper atmosphere by fast GW (red contours). Because small-scale GW are not only filtered by the mean wind (left panel) but also by the superposed global-scale disturbances also PW characteristics may be indirectly propagate upward in the thermosphere. This can be demonstrated by running the GW scheme offline with a modeled (non-stationary) background atmosphere.

4.3 Online simulations

A model simulation has been performed, in which the phase speeds are two times larger than in the standard configuration. Additionally, an externally forced 16-day wave, westward travelling with $k = 1$, was forced. The simulation represents January–February condition, when PW can directly propagate upward to the mesopause region (Fig. 6, colorscaling) on the Northern Hemisphere. The signatures of these PW characteristics in the modulation of GW fluxes are visible at greater altitudes. The applied analysis method is based on Pancheva et al. (2009). We show height vs. latitude cross sections of the mean (A_m), the stationary wave 1 (A_s^1) and the westward propagating 16-day wave ($A_w^{1,16d}$) amplitude in the temperature (T) and GW momentum flux in zonal direction (F_{GW}).

By comparing the pattern of the direct and indirect PW amplitude distribution we found an extension of the 16-day wave into the thermosphere through GW modulation (Fig. 6, right). The wave maxima are located on the Northern Hemisphere. The reason is that travelling PW in winds periodically filter momentum fluxes of fast GW. In the case of the SPW1 signature (Fig. 6, middle) the summer maximum is

connected with other processes that must be also considered, e.g. interaction with tides and in situ generations through instabilities.

5 Conclusions

The combination of satellite data analysis and model simulations were presented in order to physically explain a possible coupling mechanism through GW, which are able to carry PW characteristics into the thermosphere/ionosphere system. SABER temperature data (30–130 km) have been used to analyze PW and estimate GW energy at midlatitudes. From 2002 to 2008 proxies of PW indicate a possible connection between characteristics in the GW modulation and PW type oscillations (PWTO) in the total electron content (TEC) derived from GPS during the first three winters. A more detailed spectral decomposition when the proxy indicates a positive correspondence from November 2003 to February 2004 results in first insights into the spectral characteristics of GW modulation. Amplitude maxima of the westward travelling 16-day wave in the ionospheric F-region correspond with that one of the PW signature in the lower thermosphere. In order to draw more substantial conclusions more spectral components must be considered over a longer period of time.

However, starting model experiments with faster GW (double phase velocity) that is described in the parametrization of MUAM shows, in the case of an externally forced 16-day wave, signatures in the GW momentum flux modulation between 100 and 120 km. Such a mechanism could explain part of the variability in the thermosphere/ionosphere system to the pure meteorological origin. If model simulations can be operated with an observed distribution of GW

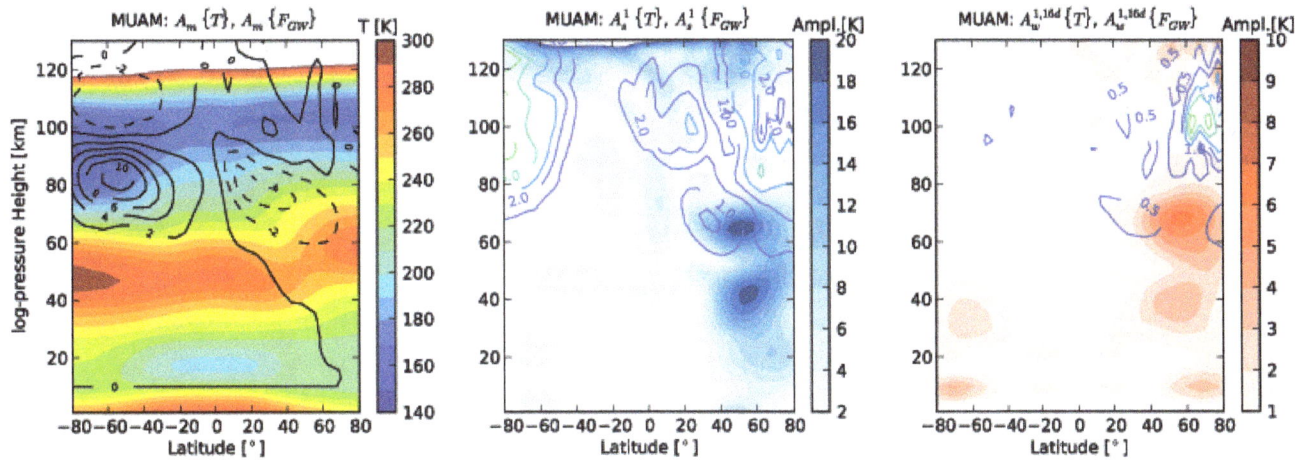

Fig. 6. Space-time spectra analysis of a MUAM simulation with double phase speeds. Shown are amplitudes in the height vs. latitude cross section. Left: mean A_m temperature (colorscaling) and GW flux (contours); middle: stationary wave 1 (A_s^1); right: westward propagating 16-day wave ($A_1^{1,16d}$).

derived from SABER using a ray-tracer model (Preusse et al., 2009) and assimilating stratospheric PW part of the resulting variability at F-region heights could be in phase with the observed ionospheric component.

Acknowledgements. We thank the SABER team for the data availability via Internet and the DLR-IKN Neustrelitz for providing the GPS-TEC data. Special thanks to the working group of A. Pogoreltsev for discussion about the potential application of MUAM. Topical Editor Matthias Förster thanks Ivan Karpov and an anonymous reviewer for their help in evaluating this paper.

References

Borries, C. and Hoffmann, P.: Characteristics of F2-layer planetary wave-type oscillations in northern middle and high latitudes during 2002 to 2008, J. Geophys. Res., 115, A00G10, doi: 10.1029/2010JA015456, 2010.

Borries, C., Jakowski, N., Jacobi, C., Hoffmann, P., and Pogoreltsev, A.: Spectral analysis of planetary waves seen in the ionospheric total electron content (TEC): First results using GPS differential TEC and stratospheric reanalyses, J. Atmos. Sol.-Terr. Phy., 69, 2442–2451, doi:10.1016/j.jastp.2007.02.004, 2007.

Ern, M., Preusse, P., Gille, J. C., Hepplewhite, C. L., Mlynczak, M. G., Russell III, J. M., and Riese, M.: Implications for atmospheric dynamics derived from global observations of gravity wave momentum flux in strato- and mesosphere, J. Geophys. Res., 116, D19107, doi:10.1029/2011JD015821, 2011.

Goncharenko, L. P., Chau, J. L., Liu, H.-L., and Coster, A. J.: Unexpected connections between the stratosphere and ionosphere, Geophys. Res. Lett., 37, L10101, doi:10.1029/2010GL043125, 2010.

Hoffmann, P., Jacobi, C., and Borries, C.: A possible planetary wave coupling between the stratosphere and ionosphere by gravity wave modulation, J. Atmos. Sol.-Terr. Phy., 75–76, 71–80, doi:10.1016/j.jastp.2011.07.008, 2011.

Jacobi, C., Fröhlich, K., and Pogoreltsev, A. I.: Quasi two-day-wave modulation of gravity wave flux and consequences for planetary wave propagation in a simple circulation model, J. Atmos. Sol.-Terr. Phy., 68, 283–292, 2006.

Jakowski, N., Heise, S., Wehrenpfennig, A., Schlüter, S., and Reimer, R.: GPS/GLONASS-based TEC measurements as a contributor for space weather forecast, J. Atmos. Sol.-Terr. Phy., 64, 729–735, 2002.

Medvedev, A. S. and Klaassen, G. P.: Thermal effects of saturating gravity waves in the atmosphere, J. Geophys. Res., 108, 4040, doi:10.1029/2002JD002504, 2003.

Mertens, C. J., Schmidlin, F. J., Goldberg, R. A., Remsberg, E. E., Pesnell, W. D., Russell III, J. M., Mlynczak, M. G., López-Puertas, M., Wintersteiner, P. P., Picard, R. H., Winick, J. R., and Gordley, L. L.: SABER observations of mesospheric temperatures and comparisons with falling sphere measurements taken during the 2002 summer MaCWAVE campaign, Geophys. Res. Lett., 31, L03105, doi:10.1029/2003GL018605, 2004.

Miyoshi, Y. and Fujiwara, H.: Gravity Waves in the Thermosphere Simulated by a General Circulation Model, J. Geophys. Res., 113, D01101, doi:10.1029/2007JD008874, 2008.

Pancheva, D., Mukhtarov, P., Andonov, B., Mitchell, N. J., and Forbes, J. M.: Planetary waves observed by TIMED/SABER in coupling the stratosphere-mesosphere-lower thermosphere during the winter of 2003/2004: Part 1 – Comparison with the UKMO temperature, J. Atmos. Sol.-Terr. Phy., 71, 61–74, doi:10.1016/j.jastp.2008.09.016, 2009.

Pogoreltsev, A. I., Vlasov, A. A., Fröhlich, K., and Jacobi, C.: Planetary Waves in coupling the lower and upper atmosphere, J. Atmos. Sol.-Terr. Phy., 69, 2083–2101, 2007.

Preusse, P., Ern, M., Eckermann, S. D., Warner, C. D., Picard, R. H., Knieling, P., Krebsbach, M., Russell, J. M., Mlynczak, M. G., Mertens, C. J., and Riese, M.: Tropopause to mesopause gravity waves in August: measurement and modelling, J. Atmos. Sol.-Terr. Phy., 68, 1730–1751, 2006.

Preusse, P., Eckermann, S. D., Ern, M., Oberheide, J., Picard, R. H., Roble, R., Riese, M., Russell III, J. M., and Mlynczak,

M.: Global ray tracing simulations of the SABER gravity wave climatology, J. Geophys. Res., 114, D08126, doi:10.1029/2008JD011214, 2009.

Savitzky, A. and Golay, M. J. E.: Smoothing and Differentiation of Data by Simplified Least Squares Procedures, Anal. Chem., 36, 1627–1639, 1964.

Yiğit, E. and Medvedev, A. S.: Internal gravity waves in the thermosphere during low and high solar activity: Simulation study, J. Geophys. Res., 115, A00G02, doi:10.1029/2009JA015106, 2010.

31

MAARSY – the new MST radar on Andøya: first results of spaced antenna and Doppler measurements of atmospheric winds in the troposphere and mesosphere using a partial array

G. Stober[1]**, R. Latteck**[1]**, M. Rapp**[1]**, W. Singer**[1]**, and M. Zecha**[1]

[1]Leibniz-Institute of Atmospheric Physics at the Rostock University, Schlossstr. 6, 18225 Kühlungsborn, Germany

Correspondence to: G. Stober (stober@iap-kborn.de)

Abstract. MST radars have been used to study the troposphere, stratosphere and mesosphere over decades. These radars have proven to be a valuable tool to investigate atmospheric dynamics. MAARSY, the new MST radar at the island of Andøya uses a phased array antenna and is able to perform spaced antenna and Doppler measurements at the same time with high temporal and spatial resolution. Here we present first wind observations using the initial expansion stage during summer 2010. The tropospheric spaced antenna and Doppler beam swinging experiments are compared to radiosonde measurements, which were launched at the nearby Andøya Rocket Range (ARR). The mesospheric wind observations are evaluated versus common volume meteor radar wind measurements. The beam steering capabilities of MAARSY are demonstrated by performing systematic scans of polar mesospheric summer echoes (PMSE) using 25 and 91 beam directions. These wind observations permit to evaluate the new radar against independent measurements from radiosondes and meteor radar measurements to demonstrate its capabilities to provide reliable wind data from the troposphere up to the mesosphere.

1 Introduction

Over the past decades mesosphere-stratosphere-troposphere (MST) radars have proven to be a powerful technique for the investigation of atmospheric dynamics from the troposphere up to the mesosphere (e.g. Hocking, 2011). On the island of Andøya first radar measurements with the mobile SOUSY radar were carried out on a campaign basis since 1983 (Czechowsky et al., 1988; Reid et al., 1988). The Middle Atmosphere Alomar Radar System (MAARSY) is a new scientific instrument on the island of Andøya and provides continued observations of tropospheric and mesospheric winds, which started with the Alomar-SOUSY radar in 1994–1997 (Singer et al., 1995) and the ALWIN radar from 1998–2009 (Latteck et al., 1999). In particular, the investigation of the polar mesosphere is a key objective of MAARSY. The system is designed to provide temporally and spatially highly resolved measurements of atmospheric winds and its variation by waves and turbulence.

The main focus of MAARSY is the investigation of small scale gravity waves and the underlying scattering processes at mesospheric altitudes. The polar mesosphere shows two characteristic radar signatures such as polar mesospheric summer and winter echoes (PMSE/PMWE) (e.g. Rapp and Lübken, 2004; Kirkwood, 2007). Besides being interesting phenomena in their own right, these radar returns provide an ideal tracer to gain information about the dynamics of the background atmosphere. MAARSY promises new insights into these phenomena due to its flexible beam steering capabilities and its higher power compared to the previous VHF radars operated at polar latitudes. Since May 2010 MAARSY has been in operation using an initial expansion stage (Latteck et al., 2010, 2012a,b). During summer 2010 first experiments were conducted to test the system's beam steering capabilities and to evaluate the wind measurements with radiosondes and co-located meteor radar observations.

Beside the mesospheric observations, the radar is also a helpful tool to study the polar troposphere and lower stratosphere and due to its location at the edge of the Scandinavian mountains it is possible to investigate the generation of mountain waves. However, before the radar can be used to investigate these phenomena the system has to be evaluated to ensure the quality of the wind observations.

Table 1. Experimental radar configurations for the presented data.

Experiment	meso	tropo1	tropo2
PRF	1250 Hz	5000 Hz	5000 Hz
coherent integrations	2	128	16
Pulse code	8 bit coco	mono	4 bit coco
number of Beams	25	1	8
duty cycle	2 %	1 %	4 %
data points	1024	1024	1024
range resolution	300 m	300 m	300 m
pulse length	2 μs	2 μs	2 μs

In this paper we focus on the analysis of prevailing winds using the spaced antenna (SA) (e.g. Röttger and Vincent, 1978; Röttger, 1981; Vincent and Röttger, 1980; Larsen and Röttger, 1982; Briggs, 1984; Holdsworth and Reid, 1997; Holdsworth, 1999) and the Doppler beam swinging (DBS) method (e.g. Woodman and Guillen, 1974; Briggs, 1980; Baelen et al., 1990; May, 1990) at tropospheric and mesospheric altitudes. In particular, MAARSY has the ability to perform wind measurements combining both techniques due to its multi-channel receiver system. This permits to analyze data from the same experiment with both techniques and to investigate eventual biases between the techniques. Such investigations were already performed with the SOUSY Radar (e.g. Röttger and Czechowsky, 1980; Röttger, 1983), but the new MAARSY permits to analyze differences between these techniques with a significantly increased temporal resolution by using the same data for the SA and DBS vector winds.

2   Experimental setup

MAARSY is a phased array operating at a frequency of 53.5 MHz. The array consists of 433 antennas each connected to its own transceiver module of 2 kW. The array is sub-structured in groups of 7 antennas called hexagons and groups of 49 antennas called anemones. The color code in Fig. 1 represents the initial expansion stage of the array. Each colored area marks a group of 49 antennas or one anemone. The smaller hexagon structure of 7 antennas is also visible. The term hexagon refers to the underlying geometric shape formed by the 7 antenna sub-array. More details are given in Latteck et al. (2010, 2012a,b).

The first experiments were conducted in summer 2010 and only used a partial array, consisting of 3 adjacent anemones and some additional hexagons (Latteck et al., 2010, 2012b). This expansion stage resulted in a transmitting power of approximately 294 kW and a beam width of 6°. This preliminary operational array had already its full beam steering capabilities for all off-zenith angles <30° and utilized a 16 channel receiver system. The radiation pattern for the vertical and for off-zenith pointed beam directions are given in Latteck et al. (2010, 2012b). As shown there the radar

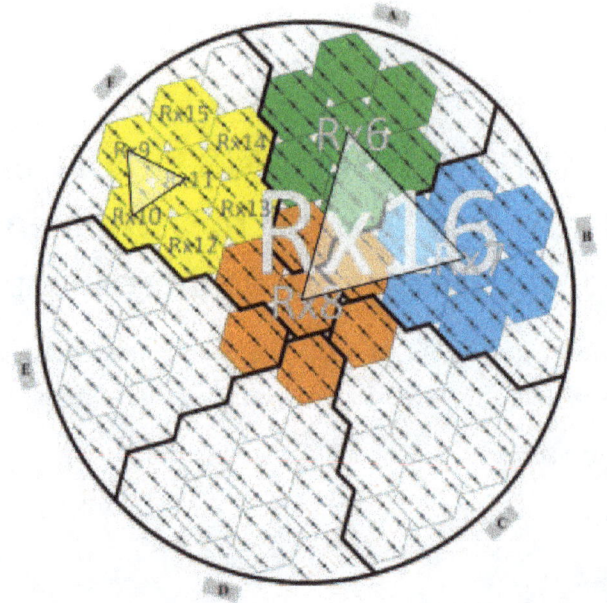

Fig. 1. Configuration of receiving channels for the experiment during summer 2010.

does not generate severe side or grading lobes for this off-zenith angular range. The radar parameters for the different experiments in the troposphere and mesosphere including PRF, range resolution, pulse coding, coherent integrations and pulse length are summarized in Table 1.

The configuration of the receiver channels is given in Fig. 1. One channel was used to sample the complete available array and 6 other channels were reserved for spaced antenna experiments. A full correlation analysis (FCA) of the spaced antenna experiment requires different baseline length in dependence of the correlation times one expects at the analyzed altitude. Therefore, a larger triangle is necessary to measure tropospheric winds and a smaller triangle is used for a mesospheric analysis, due to the much shorter auto- and cross-correlation times.

In this paper we present results from one tropospheric and two mesospheric experiments. The tropospheric experiment contained two sequences with different experiment parameters (see Table 1). One sequence contained 8 oblique beams at two different zenith angles (5° and 10°) and one sequence was used for a vertical beam only. The beam positions for the tropospheric experiment are shown in Fig. 2a with regard to the coastline of the island Andøya. The red circles indicate the beam width at an altitude of 10 km where it obtains a value of 1 km in diameter. The experiment was designed to compare simultaneous FCA spaced antenna winds and classical DBS derived winds using the multi-channel recording system of MAARSY.

The mesospheric observations consisted of a systematic scanning experiment of PMSE with the objective to observe the 3-D structure of the distribution of scatterers in such a target. The first experiment consists of 25 beams at four different zenith angles (0°, 5°, 10° and 15°). The beam positions are shown in Fig. 2. The red circle indicates the beam width at an altitude of 84 km, which is a typical altitude for the occurrence of PMSE. The experiment is subdivided into 4 sequences containing 7 beams each (Latteck et al., 2012a). Every experiment contains 6 oblique and a vertical beam, which results in a slightly better temporal resolution of the spaced antenna winds computed from the time series of the vertical beam than the DBS-winds. However, for the comparisons presented here the SA winds are interpolated to the same temporal resolution as the all sky fit (DBS) wind measurements.

The second mesospheric experiment sequence included 91 positions with 6 different zenith angles (0°, 5°, 10°, 15°, 20° and 25°). Figure 2c) indicates again the beam positions at an altitude of 84 km.

3   Radar wind observations

The multi-channel recording system of MAARSY allows to derive prevailing atmospheric winds by two different techniques. In the following we briefly summarize the basic differences among these techniques and provide a short overview of both methods.

The spaced antenna method relates to radars with multiple, but at least three, spatially separated receiving antennas, which each have to be connected to its own receiver. To derive horizontal winds from spaced antenna measurements several analysis methods have been developed, e.g. similar fades (Mitra, 1949), the spatial correlation analysis (SCA) (Briggs, 1968) or the full correlation analysis (FCA) (e.g. Briggs et al., 1950; Briggs, 1984; Phillips and Spencer, 1955; Fedor, 1967; Gregory et al., 1979). The MAARSY spaced antenna winds are determined by using the FCA method. The basic idea of such an analysis is given by the assumption that the motion of an anisometric diffraction pattern over the radar site is mainly governed by atmospheric winds. In contrast to other correlation methods, the FCA takes into account that spatial (cross correlation) and temporal (autocorrelation) variations of the diffraction pattern likely occur. However, the FCA method still assumes that these variations can be described by the same functional form (Baelen et al., 1990). A detailed investigation of the instrumental effects and biases of the FCA is given in Hocking et al. (1989); Holdsworth and Reid (1997); Holdsworth (1999).

The second commonly used method to derive atmospheric winds are Doppler beam swinging experiments, which require a narrow and steerable beam, but only one receiver channel. Such radars operate often as monostatic systems and use the complete array for transmission and reception. The transmitted electromagnetic wave is backscattered from variations of the refractive index. Below 100 km these variations are mainly caused by turbulence or atmospheric in-

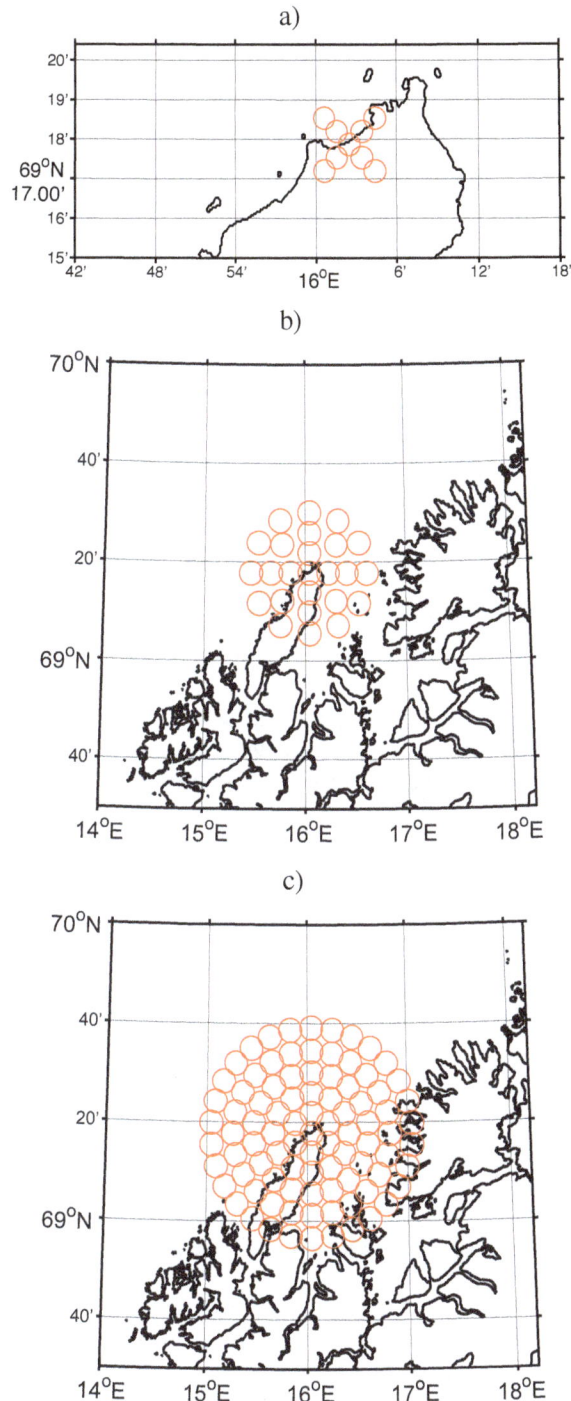

Fig. 2. (a) shows the beam positions at 10 km altitude for the tropospheric experiment. **(b)** and **(c)** show the beam positions for the sequential 25 and 91 beam experiments at an altitude of 84 km.

homogeneities traveling with a mean wind speed. Hence, the backscattered wave is Doppler shifted due to these atmospheric winds and can easily be converted into a line of sight velocity (radial velocity). Measuring the line of sight velocity in at least two different directions, which are ideally perpendicular to each other allows to derive a mean horizontal wind speed by solving Eq. (1).

$$v(\theta,\phi)_{\mathrm{rad}} = u\cos\phi\sin\theta + v\sin\phi\sin\theta + w\cos\theta, \qquad (1)$$

where u, v and w are the zonal, meridional and vertical wind speed. The angles theta and phi correspond to the zenith distance and azimuth angles with reference to east (mathematical convention). A more general form is called "all sky fit" and is often used to derive meteor radar winds (Hocking and Thayaparan, 1997; Hocking et al., 2001). The all sky fit minimizes the quantity under the assumption of a vanishing vertical mean wind ($w = 0$);

$$\Lambda = \sum_{i=1}^{N}(v_{\mathrm{rad}_i} - u\cos(\phi_i)\sin(\theta_i) - v\sin(\phi_i)\sin(\theta_i))^2 \qquad (2)$$

in a least squares sense and is ideally applied to derive a mean horizontal wind if a larger number of beam directions is available. The index i in Eq. (2) labels the number of the meteor or beam pointing direction. Hence, this generalized DBS seems to be suitable to determine prevailing winds from the multi-beam scanning experiments conducted with MAARSY. However, for large zenith angles ($>25°$) both DBS methods should only be used to derive a mean horizontal wind speed. Specular meteor radars often observe the majority of the echoes at zenith angles between 35–70° off-zenith. The huge horizontal extent of the measurement volume can exceed 500 km in diameter at 90 km altitude. For such large observation areas Hocking et al. (2001) suggested rather to assume $w = 0$ than to estimate a mean vertical velocity. In the mesosphere the horizontal wind speed can exceed the vertical velocity by up to 2 orders of magnitude, which leads for large off-zenith angles to an almost negligible contribution to the line of sight velocity. Small errors in the line of sight velocity translate into large errors in the vertical velocity. Estimating the vertical wind form such observations often tends to produce unrealistic values.

4   First results of tropospheric wind measurements

The first tropospheric observation campaign with MAARSY started the 4 August 2010. This tropospheric experiment focused on the evaluation of the new radar against wind observations with radiosondes launched at the nearby Andøya Rocket Range. This experiment contained 9 beam positions in the DBS sequence consisting of 8 oblique beams at two different zenith angles of 5° and 10° (see Table 1, tropo2) and one sequence containing the vertical beam (see Table 1, tropo1). The much higher number of coherent integrations for the vertical beam improved the signal-to-noise-ratio to perform a reliable spaced antenna wind analysis reaching a comparable altitude than for the Doppler observation. The temporal resolution of both sequences was approximately 2 min.

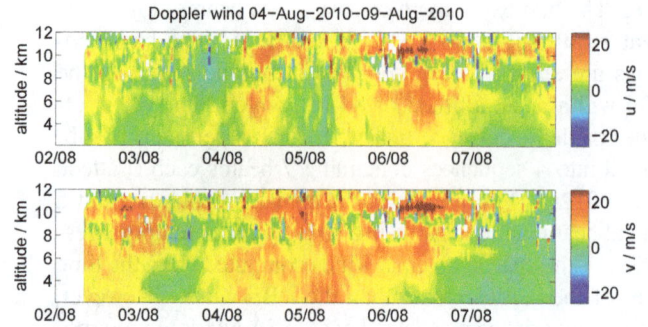

Fig. 3. Tropospheric wind situation at the beginning of August derived from a MAARSY Doppler experiment.

The tropospheric wind situation at the beginning of August 2010 is shown in Fig. 3. During this time the weather at Andenes was driven by several low-pressure systems over the Northern Atlantic and Northern Scandinavia. These low-pressure areas generated a prevailing wind from South/Southwest, which remained almost stable during the complete campaign period. The prevailing winds shown in Fig. 3 were computed using 15 min averages.

The first comparison focuses on the evaluation of the spaced antenna winds and the Doppler measurements. In Fig. 4 the FCA and DBS winds are compared using the complete campaign data and all available height gates. The correlation coefficients for the zonal wind ($r = 0.90$) and meridional wind ($r = 0.92$) confirm a reasonably good agreement with almost no systematic bias. However, the scattering in Fig. 4 around the red slope 1 line points out that there are some differences between the DBS and FCA winds, which are only partly explainable by the different observation volumes. There are systematic errors related to the FCA winds (Holdsworth and Reid, 1997) and the DBS determined winds. In particular, the DBS technique requires an accurate determination of the effective beam pointing direction, which can be altered by aspect sensitive scattering processes and the distribution of scatterers in the radar volume (Hocking et al., 1990; Worthington et al., 1999; Worthington, 1999; Worthington et al., 2000). However, Röttger and Czechowsky (1980) demonstrated that already at 7° off-zenith angle the aspect sensitivity shows a reduced impact. The tropospheric experiment used in this study uses up to 10° off-zenith angle, which should further decrease the impact of aspect sensitivity on the computed vector winds. In fact, the presented comparison is not sufficient to give a final answer on this question.

In parallel to the radar observations in the troposphere insitu wind measurements were conducted. A radiosonde campaign at the Andøya Rocket Range from 2 August–8 August 2010 provided an excellent opportunity for a detailed comparison of radiosonde and radar horizontal winds. During the campaign 18 radiosondes were launched. The trajectories (blue) as well as the MAARSY beam positions at 10 km altitude (red) are shown in Fig. 5.

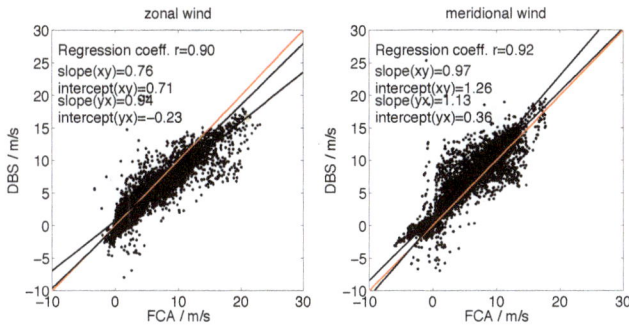

Fig. 4. Comparison of DBS and FCA zonal and meridional winds in the troposphere in the altitude range between 1.8–8 km.

The general problem in comparing radiosondes and radar measurements is the rapid motion of the balloon out of the observation volume of the radar. To avoid any bias in the comparison due to the rapidly increasing horizontal distance we defined spatial and temporal coincidence criterion. The spatial coincidence criteria is given by the green circle, which labels a horizontal distance of 8 km around the radar. For this area we assume that the balloon observes the same horizontal wind as the radar. As temporal coincidence we considered the time span where the radiosonde reached an altitude of 1.8 km (corresponding to the lowest radar range gate) and the time where the horizontal distance of the balloon exceeded 8 km. In addition, the radiosonde measurements were interpolated to the altitude resolution given by the radar. We computed an average wind for each 300 m altitude gate centered at the corresponding altitude gate of the radar.

In Fig. 6 the scatter plots of DBS wind versa the radiosonde observations are shown. The correlation coefficients are calculated for the wind speed ($r = 0.85$) and the wind direction ($r = 0.83$). In particular, the wind direction of both observations are in reasonable agreement, which is pointed out by the slopes for both fits being close to 1 and an intercept smaller than $1\,\mathrm{m\,s^{-1}}$. The black line indicate the slopes of both fits (changing xy-axis) to obtain an estimate of a possible bias. Further, it is surprising that the wind speed shows this good agreement, because we did not apply an aspect sensitivity correction. At least during this campaign the effective and nominal beam directions seemed to be in coincidence and were not significantly altered due to aspect sensitive backscatter.

The FCA determined wind speed shows a smaller correlation for the wind speed of $r = 0.84$ and $r = 0.84$ for the wind direction. Also the general agreement provides some evidence that the FCA winds tend to underestimate the radiosonde winds, which is in agreement with the study of Reid et al. (2005). In this work a data base of 3000 radiosonde ascents was used to determine the bias of the FCA winds to be in the order of 7 % for the altitude range between 2–6 km. However, the available data from the 18 radiosondes are not sufficient to analyze this in detail. The bias in the

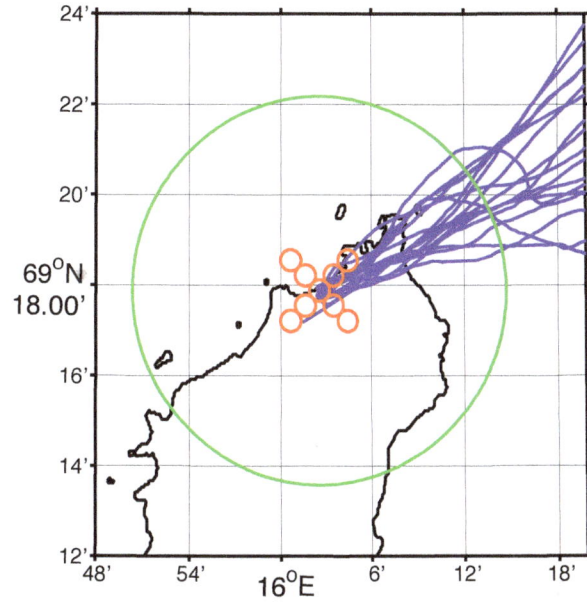

Fig. 5. Lambert projection of island of Andøya in Northern Norway. The position of the tropospheric beams at an altitude of 10 km is labeled by red circles. The blue lines are the trajectories of the radiosondes launched at the ARR.

wind direction is slightly larger compared to the DBS winds, but shows the same tendency.

5   First results of mesospheric winds during PMSE

Polar mesospheric summer echoes are a well-known radar tracer at Arctic and Antarctic latitudes. The backscatter from these echo structures permits to measure horizontal and vertical winds with high spatial and temporal resolution (e.g. Zecha et al., 2001; Rapp et al., 2011) at mesospheric altitudes. Over the past decades the investigation of the scattering processes and the dynamics within the PMSE have become an important field of research (see Rapp and Lübken, 2004, for a review).

The persistence and strength of these echoes provide a chance to gain information about an atmospheric region, which is otherwise rarely accessible by VHF radars due to a lack of scatterers. In particular, the height of occurrence of PMSE between 80–90 km altitude permits to investigate the atmospheric dynamics with a significantly improved temporal and spatial resolution compared to meteor radars and MF radars which are widely used in this height range.

Close to the MAARSY facility there is also a co-located meteor radar (MR). The Andenes MR operates at 32.55 MHz with a transmitting power of 20 kW. The receiving antennas are crossed Dipoles ensuring an almost perfect azimuthal radiation pattern. The system records between 8000 to 20 000 meteors per day. The typical observed volume used for the wind determination has a diameter of 500 km at

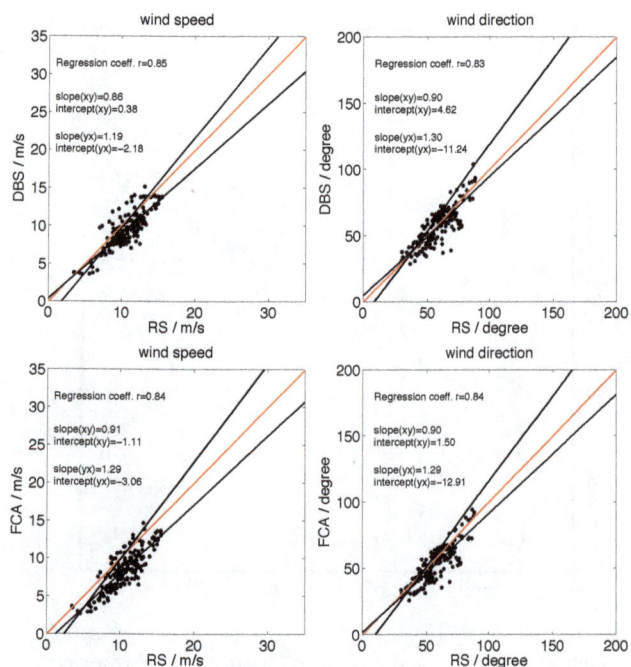

Fig. 6. Comparison of DBS and FCA prevailing winds with the horizontal wind measured by the radiosondes.

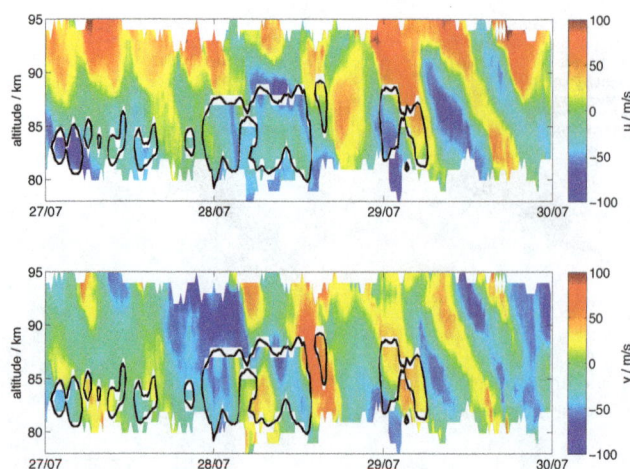

Fig. 7. Comparison of the prevailing zonal and meridional wind measured with the Andenes Meteor Radar. The black line represents a contour line of an existing PMSE. The wind field within the contour line is determined from a 25 beams MAARSY scanning experiment.

90 km altitude. A more detailed description of the meteor radar wind measurements and wave activities can be found in Stober et al. (2012).

At the end of July 2010 MAARSY recorded a rather strong PMSE employing a 25 beam scanning experiment. In Fig. 7 we compared the MAARSY and MR horizontal winds. The contour of the PMSE is given by the black line. Inside the contour line the wind is computed from the radial velocities measured by MAARSY using a Doppler analysis similar to those of the MR with a temporal resolution of 15 min. Outside the contour line the MR winds are shown. These winds were derived by using a running window of 1 h, which was shifted in steps of 15 min and centered at the reference time of the corresponding MAARSY wind analysis to get a similar temporal resolution than MAARSY. The color code in Fig. 7 visualizes a strong diurnal and semidiurnal tidal activity, which is considered to cause the minimum in the PMSE occurrence in the afternoon (e.g. Hoffmann et al., 1997, 1999; Barabash et al., 1998; Klostermeyer, 1999; Li and Rapp, 2011). However, taking further into account the MR winds the phase lines of the upward propagating tidal structure becomes even more obvious.

The reasonable agreement visualized in the contour plot is also confirmed by correlating the zonal and meridional wind of both radar measurements, which is shown in Fig. 8. The correlation coefficients for the zonal and meridional wind are $r = 0.78$ and $r = 0.79$, respectively. The intercepts of the fits indicate a reasonable good agreement. However, the slopes reveal some larger discrepancies between both radars, which

is explainable considering the spatial and temporal resolutions of both radars: MAARSY sampled at a range resolution of 300 m in the scanning experiment and the wind is computed on the basis of such 300 m height gates as well. In contrast, the MR wind measurement is interpolated to the same vertical resolution by using 3 km altitude bins shifted by 300 m centered at the same height as the appropriate MAARSY altitude gate.

In addition the observation volumes differ considerably. The meteor radar records meteors in a cone with almost 500 km diameter at PMSE altitudes, in contrast to the 80 km diameter of the PMSE scanning experiment. In principle it is possible to limit the observation volume of the MR to the same size as MAARSY considering only meteors with off-zenith angles smaller than or equal to the beam positions shown in Fig. 2. But such a small volume would dramatically decrease the number of meteors, which finally would lead to a worse temporal and vertical resolution of the wind measurement and is therefore not applicable to compare both observations.

Similar to the comparison shown in Zecha et al. (2001), we also compared the FCA and DBS mesospheric winds (data not shown). For this purpose we used a temporal resolution of 10 min and analyzed the same period as for the comparison with the MR. The zonal and meridional correlation coefficients are $r = 0.82$ and $r = 0.86$, respectively. In contrast to the tropospheric experiments the FCA winds show a slightly decreased correlation with the DBS winds at mesospheric altitudes. However, the scattering around a slope one still indicates some substantial differences, which may be related to the different observation volumes. As already mentioned above the DBS winds are computed on the basis of a

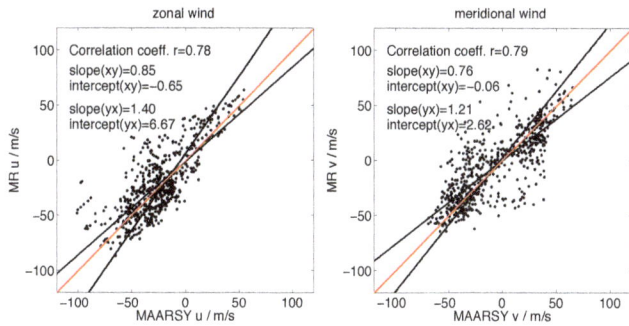

Fig. 8. Scatter plots of the common volume wind measurements at Andenes between the MR and MAARSY.

circle with 80 km in diameter whereas the characteristic volume used for the FCA wind is given by the beam width of 6° corresponding to a circle of 9 km diameter at 84 km altitude.

6 Conclusions

All shown results are based on the initial construction stage during summer 2010 employing a partial array. The detailed wind comparisons of tropospheric and mesospheric observations using MAARSY, radiosondes and the co-located meteor radar demonstrate the proper operation of the new radar. The capability of MAARSY to perform multichannel recording permits to combine spaced antenna and Doppler measurements in the same experiment to investigate atmospheric dynamics.

The radiosondes launched at the Andøya Rocket Range provided an ideal opportunity to evaluate our tropospheric wind measurements by an independent method. The excellent agreement of the Doppler winds (without aspect sensitivity correction) confirm the potential of MAARSY to investigate tropospheric dynamics at polar latitudes with high spatial and temporal resolution. In particular, the effect of aspect sensitive backscatter at tropospheric and mesospheric altitudes has to be investigated in much more detail to enhance the understanding of the different scattering process. The increased power and aperture using the next expansion stages will further enhance the accessible altitude range up into the lower stratosphere (Latteck et al., 2012a).

The conducted mesospheric experiments represent the first attempts of multi-beam operation with pulse to pulse steering to perform systematic scanning experiments revealing the 3-D structure of PMSE/PMWE (Latteck et al., 2012a; Rapp et al., 2011). The observed wind velocities have been evaluated by comparing the well established MR winds in the MLT region with the Doppler and spaced antenna wind measurements of MAARSY. Our results indicate an excellent agreement of the mean winds, but also show some evidence that on shorter time scales substantial differences can occur. These differences are more likely due to wave

activity within the observation volume than due to instrumental biases. The obtained results are similar to those derived in a previous wind comparison using the ALWIN system (Engler et al., 2008).

Acknowledgements. MAARSY was built under grant 01LP0802A by the German Bundesministerium für Bildung und Forschung. The support of the Andøya Rocket Range (ARR) is acknowledged. Further, we explicitly thank all the employees of the IAP who helped to build the antenna array and other parts of the infrastructure of MAARSY during summer 2009 at Andenes. Topical Editor Matthias Förster thanks Jürgen Röttger and an anonymous reviewer for their help in evaluating this paper.

References

Baelen, J. S. V., Tsuda, T., Richmond, A. D., Avery, S. K., Kato, S., Fukao, S., and Yamamoto, M.: Comparison of VHF Doppler beam swinging and spaced antenna observations with the MU radar: First results, Radio Sci., 25, 629–640, 1990.

Barabash, V., Chilson, P., Kirkwood, S., Réchou, A., and Stebel, K.: Investigations of the possible relationship between PMSE and tides using a VHF MST radar, J. Geophys. Res., 25, 3297–3300, 1998.

Briggs, B.: On the analysis of moving patterns in geophysics, J. Atmos. Terr. Phys., 30, 1777–1788, 1968.

Briggs, B.: Radar observations of atmospheric winds and turbulence: a comparison of techniques, J. Atmos. Solar-Terr. Phys., 42, 823–833, 1980.

Briggs, B.: The Analysis of spaced sensor records by correlation techniques, MAP Handbook, 13, 166–186, 1984.

Briggs, B., Phillips, G., and Shinn, D.: The analysis of observations on spaced receivers of the fading radio signals, Proc. Phys. Soc. London, Sect. B, 63, 106–121, 1950.

Czechowsky, P., Reid, I. M., and Rüster, R.: VHF radar measurements of the aspect sensitivity of the summer polar mesopause echoes over Andenes (69° N, 16° E), Norway, Geophys. Res. Lett., 15, 1259–1262, 1988.

Engler, N., Singer, W., Latteck, R., and Strelnikov, B.: Comparison of wind measurements in the troposphere and mesosphere by VHF/MF radars and in-situ techniques, Ann. Geophys., 26, 3693–3705, doi:10.5194/angeo-26-3693-2008, 2008.

Fedor, L.: A Statistical Approach to the Determination of Three-Dimensional Ionospheric Drifts, J. Geophys. Res., 72, 5401–5415, 1967.

Gregory, J., Meek, C., Manson, A., and Stephenson, D.: Developments in the Radiowave Drifts Technique for Measurement of High-Altitude Winds, J. Appl. Meteorol., 18, 682–691, 1979.

Hocking, W. and Thayaparan, T.: Simultaneous and co-located observation of winds and tides by MF and Meteor radars over London, Canada, (43° N, 81° W) during 1994–1996, Radio Sci., 32, 833–865, 1997.

Hocking, W., Fukao, S., Tsuda, T., Yamamoto, M., Sato, T., and Kato, S.: Aspect sensitivity of stratospheric VHF radio wave scatterers particularly above 15-km altitude, Radio Sci., 25, 613–627, 1990.

Hocking, W., Fuller, B., and Vandepeer, B.: Real-time determination of meteor-related parameters utilizing modern digital technology, J. Atmos. Solar-Terr. Phys., 63, 155–169, 2001.

Hocking, W. K.: A review of MesosphereStratosphereTroposphere (MST) radar developments and studies, circa 1997–2008, J. Atmos. Solar-Terr. Phys., 73, 848–882, 2011.

Hocking, W. K., May, P., and Röttger, J.: Interpretation, reliability and accuracies of parameters deduced by the spaced antenna method in middle atmosphere applications, Pure Appl. Geophys., 130, 571–604, 1989.

Hoffmann, P., Singer, W., Keuer, D., Bremer, J., and Rüster, R.: Mean diurnal variation of PMSE as measured with the ALOMAR SOUSY radar during summer, ESASP-397, 471–475, 1997.

Hoffmann, P., Singer, W., and Bremer, J.: Mean seasonal and diurnal variation of PMSE and winds from 4 years of radar observations at ALOMAR, Geophys. Res. Lett., 26, 1525–1528, 1999.

Holdsworth, D.: Influence of instrumental effects upon the full correlation analysis, Radio Sci., 34, 643–655, doi:10.1029/1999RS900001, 1999.

Holdsworth, D. and Reid, I.: An investigation of biases in the full correlation analysis technique, Adv. Space Res., 20, 1269–1272, 1997.

Kirkwood, S.: Polar mesosphere winter echoes – A review of recent results, Adv. Space Res., 40, 751–757, 2007.

Klostermeyer, J.: On the diurnal variation of polar mesosphere summer echoes, J. Geophys. Res., 26, 3301–3304, 1999.

Larsen, M. F. and Röttger, J.: VHF and UHF Doppler radars as tools for synoptic research, B. Am. Meteorol. Soc., 63, 996–1008, 1982.

Latteck, R., Singer, W., and Bardey, H.: The ALWIN MST radar – Technical design and performance, 14th ESA Symposium on European Rocket and Balloon Programmes and Related Research, Potsdam, Germany, 179–184, 1999.

Latteck, R., Singer, W., Rapp, M., and Renkwitz, T.: MAARSY the new MST radar on Andøya/Norway, Adv. Radio Sci., 8, 219–224, doi:10.5194/ars-8-219-2010, 2010.

Latteck, R., Singer, W., Rapp, M., Vandepeer, B., Renkwitz, T., Zecha, M., and Stober, G.: MAARSY – The new MST radar on Andoya: System description and first results, Radio Sci., 47, RS1006, doi:10.1029/2011RS004775, 2012a.

Latteck, R., Singer, W., Rapp, M., Renkwitz, T., and Stober, G.: Horizontally resolved structures of polar mesospheric echoes obtained with MAARSY, Adv. Radio Sci., 10, in press, 2012b.

Li, Q. and Rapp, M.: PMSE-observations with the EISCAT VHF and UHF-radars: Statistical properties, J. Atmos. Solar-Terr. Phys., 73, 944–956, 2011.

May, P.: Spaced antenna versus Doppler radars: A comparison of techniques revisited, Radio Sci., 25, 1111–1119, 1990.

Mitra, S.: A radio method of measuring winds in the ionosphere, Proc. Inst. Electr. Eng., 96, 441–446, 1949.

Phillips, G. and Spencer, M.: The Effects of Anisometric Amplitude Patterns in the Measurement of Ionospheric Drifts, Proc. Phys. Soc. London, Sect. B, 68, 481–492, 1955.

Rapp, M. and Lübken, F.-J.: Polar mesosphere summer echoes (PMSE): Review of observations and current understanding, Atmos. Chem. Phys., 4, 2601–2633, doi:10.5194/acp-4-2601-2004, 2004.

Rapp, M., Latteck, R., Stober, G., Hoffmann, P., Singer, W., and Zecha, M.: First 3-dimensional observations of polar mesosphere winter echoes: resolving space-time ambiguity, J. Geophys. Res., 116, A11307, doi:10.1029/2011JA016858, 2011.

Reid, I. M., R. Rüster, P. C., and Schmidt, G.: VHF radar measurements of momentum flux in the summer polar mesosphere over Andenes (69° N,16° E), Norway, Geophys. Res. Lett., 15, 1263–1266, 1988.

Reid, I. M., Holdsworth, D. A., Kovalam, S., and Vincent, R. A.: Mount Gambier (38° S, 141° E) prototype VHF wind profiler, Radio Sci., 40, RS5007, doi:10.1029/2004RS003055, 2005.

Röttger, J.: Investigations of lower and middle atmosphere dynamics with spaced antenna drifts radars, J. Atmos. Solar-Terr. Phys., 43, 277–292, doi:10.1016/0021-9169(81)90090-8, 1981.

Röttger, J. (Ed.): The correlation of winds measured with a spaced antenna VHF radar and radiosondes, 97–99, 21th AMS Rad. Met. Conf., 1983.

Röttger, J. and Czechowsky, P. (Eds.): Tropospheric and stratospheric wind measurements with the spaced antenna drifts technique and the Doppler beam swinging technique using VHF radar, 577–584, 19th AMS Rad. Met. Conf., 1980.

Röttger, J. and Vincent, R.: VHF radar studies of tropospheric velocities and irregularities using spaced antenna techniques, Geophys. Res. Lett., 5, 917–920, 1978.

Singer, W., Keuer, D., Hoffmann, P., Czechowsky, P., and Schmidt, G.: The ALOMAR SOUSY radar: Technical design and further developments, in: Proceedings of the 12th ESA Symposium on European Rocket and Balloon Programmes and Related Research, Lillehammer, Norway , ESA SP-370, 409–415, 1995.

Stober, G., Jacobi, C., Matthias, V., Hoffmann, P., and Gerding, M.: Neutral air density variations during strong planetary wave activity in the mesopause region derived from meteor radar observations, J. Atmos. Solar-Terr. Phys., 74, 55–63, 2012.

Vincent, R. and Röttger, J.: Spaced antenna VHF radar observations of tropospheric velocities and irregularities, Radio Sci., 15, 319–335, 1980.

Woodman, R. and Guillen, A.: Radar observations of winds and turbulence in the stratosphere and mesosphere, J. Atmos. Science, 31, 493–505, 1974.

Worthington, R. M.: Calculating the azimuth of mountain waves, using the effect of tilted fine-scale stable layers on VHF radar echoes, Ann. Geophysicae, 17, 257–272, 1999.

Worthington, R. M., Palmer, R. D., and Fukao, S.: Letter to the Editor: Complete maps of the aspect sensitivity of VHF atmospheric radar echoes, Ann. Geophys., 17, 1116–1119, doi:10.1007/s00585-999-1116-z, 1999.

Worthington, R. M., Palmer, R. D., Fukao, S., Yamamoto, M., and Astin, I.: Rapid variations in echo power maps of VHF radar backscatter from the lower atmosphere, J. Atmos. Solar-Terr. Phys., 62, 573–581, 2000.

Zecha, M., Röttger, J., Singer, W., Hoffmann, P., and Keuer, D.: Scattering properties of PMSE irregularities and refinement of velocity estimates, J. Atmos. Solar-Terr. Phys., 63, 201–214, 2001.

32

Comparison of ionospheric radio occultation CHAMP data with IRI 2001

N. Jakowski and K. Tsybulya

Deutsches Zentrum für Luft-und Raumfahrt (DLR)/Institut für Kommunikation und Navigation, Kalkhorstweg 53, D-17235 Neustrelitz, Germany

Abstract. GPS radio occultation measurements on board low Earth orbiting satellites can provide vertical electron density profiles of the ionosphere from satellite orbit heights down to the bottomside. Ionospheric radio occultation (IRO) measurements carried out onboard the German CHAMP satellite mission since 11 April 2001 were used to derive vertical electron density profiles (EDP's) on a routine basis. About 150 vertical electron density profiles may be retrieved per day thus providing a huge data basis for testing and developing ionospheric models. Although the validation of the EDP retrievals is not yet completed, the paper addresses a systematic comparison of about 78 000 electron density profiles derived from CHAMP IRO data with the International Reference Ionosphere (IRI 2001).

The results are discussed for quite different geophysical conditions, e.g. as a function of latitude, local time and geomagnetic activity.

The comparison of IRO data with corresponding IRI data indicates that IRI generally overestimates the upper part of the ionosphere whereas it underestimates the lower part of the ionosphere under high solar activity conditions. In a first order correction this systematic deviation could be compensated by introducing a height dependence correction factor in IRI profiling.

1   Introduction

Low Earth Orbiting satellites carrying a dual frequency GPS receiver offer a unique chance to monitor the actual state of the global ionosphere on a continuous basis by GPS radio occultation measurements. No other profiling technique unifies profiling through the entire F2 layer with global coverage.

The German CHAMP satellite has been launched successfully on 15 July 2000 (Reigber et al., 2000) and has provided first ionospheric radio occultation measurements on 11 April

2001. Preliminary IRO retrieval and validation results are discussed by Jakowski et al. (2002). The vertical sounding based F2 layer height and electron density estimations agree within 13% and 17% RMS deviation, respectively.

The International Reference Ionospere (IRI) is a global empirical model of the most important parameters of the ionospheric plasma such as the electron density, electron and ion temperature, and ion composition. The IRI model is developed as a joint URSI/COSPAR project since 1972. A first IRI code was published in 1978 by Rawer et al. (1978). Over the years testing and modification of IRI has permanently continued resulting in essential improvements made available to the international research community through several versions (Bilitza et al., 1996). Considering the electron density, the data sources are mainly vertical sounding, incoherent scatter and in situ measurements onboard satellites (Bilitza, 2001). Ionospheric radio occultation measurements represent a new data type that should be helpful to improve IRI in the future.

In this paper we discuss a comparison of about 78 000 electron density profiles derived from IRO measurements onboard CHAMP with corresponding model values computed from the International Reference Ionosphere.

2   Radio occultation data basis

The CHAMP science data are received in the DLR Remote Sensing Data Center Neustrelitz and than passed to the Geo-ForschungsZentrum (GFZ) Potsdam. A part of these data – the IRO GPS data – are automatically processed in the Institute of Communications and Navigation of DLR by an operational data processing system (Wehrenpfennig et al., 2001). The computed higher-level data products are made available to the international science community via the Information and Science Data Center (ISDC) of GFZ Potsdam.

Correspondence to: N. Jakowski
(norbert.jakowski@dlr.de)

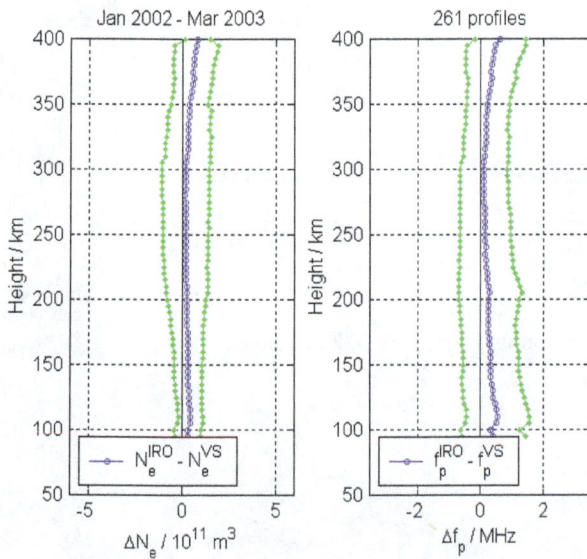

Fig. 1. Comparison of vertical electron density profiles derived from IRO data on CHAMP with vertical sounding data from the ionosonde station Juliusruh. Shown are the altitude dependent mean difference and the corresponding standard deviation. left: electron density, right: plasma frequency.

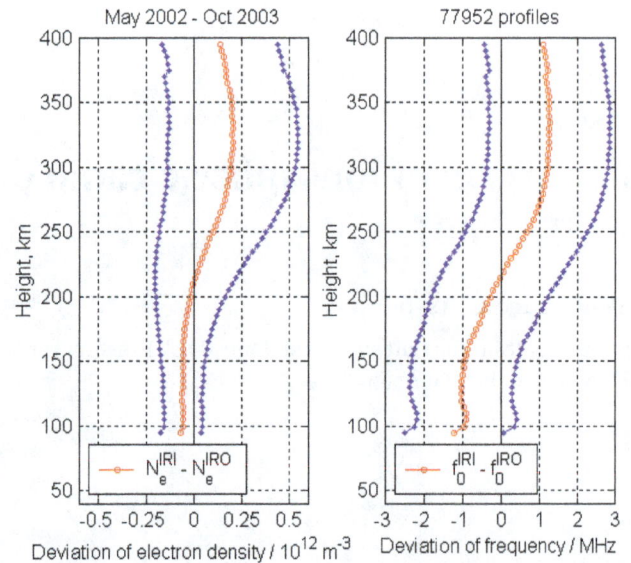

Fig. 2. Comparison of vertical electron density profiles derived from IRO data with corresponding profiles computed from IRI according to the difference EDP(IRI) – EDP(IRO) for all measurements obtained onboard CHAMP within the period May 2002–October 2003.

Due to the modular structure of the processing system high flexibility is achieved if retrieval modules shall be modified or replaced in the course of the CHAMP mission or if supplementary data shall be included.

To deduce vertical electron density profiles, a tomographic approach is established that uses spherically layered voxels with constant electron density (Jakowski, 1999; Jakowski et al., 2002).

Practically, the 1 Hz sampled dual frequency GPS measurements of TEC provide a system of equations for the electron density at the tangential points of the ray paths which is solved from above in a recursive way.

To overcome the difficulties arising due to the rather low orbit height of CHAMP (<450 km), the retrieval algorithm uses an adaptive model (Chapman layer superposed by a simple plasmasphere contribution) for estimating the topside ionosphere and plasmasphere above the CHAMP orbit.

From about 220 IRO measurements per day we retrieve about 150 vertical electron density profiles. For this comparative study with IRI 2001 we used about 78 000 profiles, which have been provided by the automatic processing system so far.

Before starting this analysis we present some validation results obtained after comparing the retrieved electron density profiles with independent ionosonde data. Former comparisons of f0F2 and hmF2 data using the SPIDR data base have revealed bias values of 0.18 MHz and 13.4 km, respectively.

To evaluate the accuracy of the entire profile from the CHAMP orbit height down to the E-layer, the IRO data were compared with vertical sounding data from the vertical sounding stations Juliusruh (53° N; 13° E). The results are

shown in Fig. 1 and indicate a systematic positive bias of the IRO data of less than 0.8 MHz (1×10^{11} m^{-3}) and a standard deviation σ from the mean of about 1 MHz (1.3×10^{11} m^{-3}) throughout the entire profile.

3   The IRI model

The IRI files were downloaded from NASA GSFC's National Space Science Center FTP server (ftp://nssdcftp.gsfc.nasa.gov/models/ionospheric/iri/iri2001/).

The source was translated with Fortran Power Station 4.0 compiler for PCs. To adapt the code to PC environment few insignificant changes in the code were made. None one of them changed the algorithms of the model.

To avoid low-level FORTRAN coding and facilitate generation of the graphs, MATLAB 6.0 was used for all high level programming. IRI 2001 was compiled to a dynamic link library, which had the main Ne calculating function exported and available for the MATLAB scripts. In all cases, ionospheric D-layer was not calculated (i.e. the faster and smaller IRI_SUB_NE function was used). In all cases, CCIR model was used for F peak, B0 option for the bottom side thickness and Standard option (no L condition) for estimating probability of F1 layer. Storm model was never used, as seriously impairing the processing speed. It's worth to notice that 12-month running mean sunspot number (Rz12) for year 2003, when the most of measurements were made, is based partially on measured Rz values for the past months and partially on predicted values for future months, so these predictions are implicitly included in the IRI calculations.

4   Results and discussion

Keeping in mind the validation results of IRO derived electron density profiles for mid-latitudes, the comparison with IRI data should provide valuable information on the quality of the IRI model if deviations exceed the measured bias and the 1σ dispersion range.

Figure 2 shows the bias and standard deviation between IRO and IRI data for more than one year of measurements from May 2002–October 2003. The comparison indicates a quite different behavior in the upper and lower part of the bottomside ionosphere. Whereas IRI generally exceeds the IRO measurements in the peak density range by about 1 MHz (2×10^{11} m^{-3}) IRI underestimates the IRO data by about 1 MHz (5×10^{11} m^{-3}) in the lower ionosphere (100–150 km). The standard deviation σ is generally less than 1.5 MHz (3.3×10^{11} m^{-3}).

To study the latitudinal and local time dependence of the difference, the data set has been divided into three parts representing high ($|\phi|>60°$), medium ($30°<|\phi|<60°$) and low latitude ($|\phi|<30°$) ranges. These data sets are subdivided in 4 groups corresponding to different local times (night: 21:00–03:00 LT, morning: 03:00–09:00 LT, day: 09:00–15:00 LT, evening: 15:00–21:00 LT) as it is shown in Fig. 3. The mean deviation of the plasma frequency fp is generally less than 1.6 MHz ($\sigma=1.4$ MHz) at high and less than 1.5 MHz ($\sigma=1.9$ MHz) at mid- latitudes under day-time conditions above 300 km height. Surprisingly the corresponding bias reaches only 1.1 MHz ($\sigma=2.1$ MHz) at low latitudes. It is difficult to comment this low latitude result because entire IRO derived electron density profiles have not yet been validated sufficiently for this latitude range. We are aware of the fact that the spherical symmetry assumption in IRO retrievals is violated in particular in the crest region due to strong horizontal gradients of the electron density distribution and therefore errors cannot be excluded. As expected, due to the enhanced ionization and variability at low latitudes in the evening hours, the bias and the dispersion reach high values in the order of 2 and 2.5 MHz, respectively. In the lower ionosphere (100–150 km height) bias and dispersion values are principally smaller than the corresponding F2 layer values. The smallest bias is found at low latitudes at day-time with 0.3 MHz ($\sigma=1.9$ MHz) whereas the biggest value is found also at low latitudes in the evening hours with 1.9 MHz ($\sigma=1.9$ MHz).

Considering the fact that the bias of the IRO measurements is slightly positive throughout the entire profile at least at mid-latitudes (Fig. 1), the IRI overestimation of the electron density in the F2 layer height range km is a clear finding at least at mid-latitudes. Due to the persistence of this feature at all latitude ranges and local times we assume that IRI generally overestimates the ionospheric ionization in the F2 layer height range under high solar activity conditions. Furthermore, it becomes evident that the IRI generally underestimates the electron density in the lower ionosphere in the 100–150 km height range.

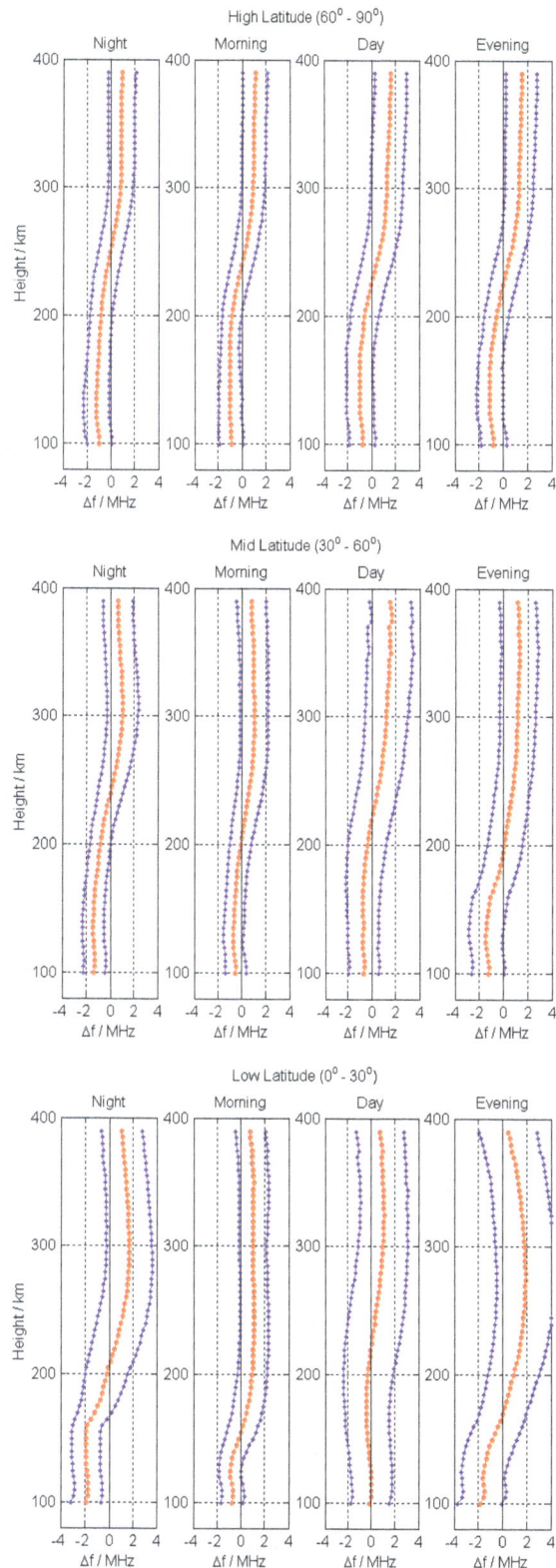

Fig. 3. Vertical profile of the difference fp(IRI) − fp(IRO) for IRO measurements at high (**a**), middle (**b**) and low (**c**) latitude sectors obtained onboard CHAMP within the period April 2002–March 2003.

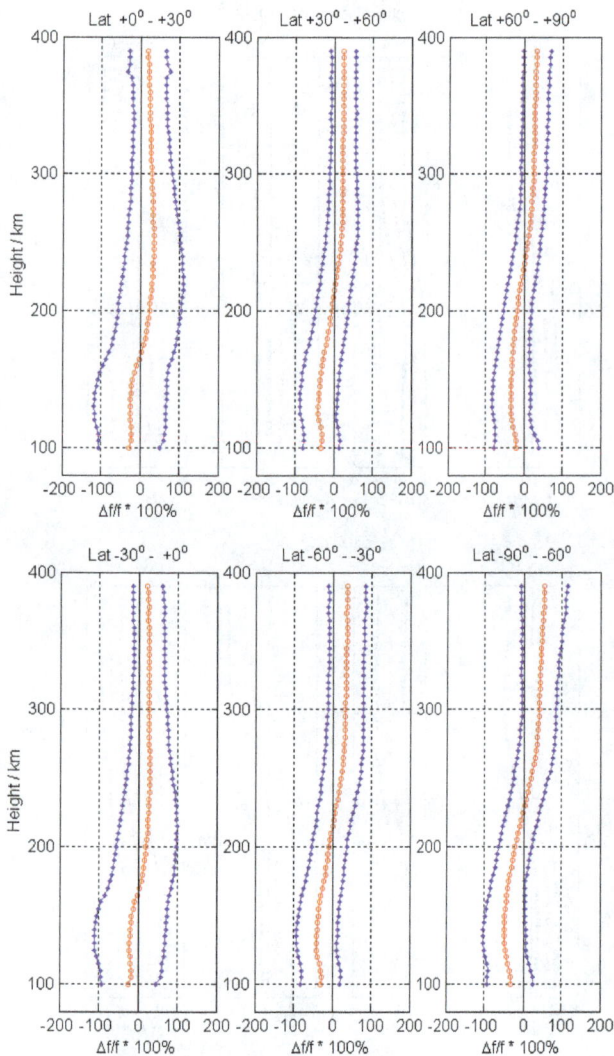

Fig. 4. Vertical profile of the percentage difference (fp(IRI) – fp(IRO))/fp(IRO) for IRO measurements at different latitude sectors at northern (**a**) and southern (**b**) hemispheres obtained onboard CHAMP within the period May 2002–October 2003.

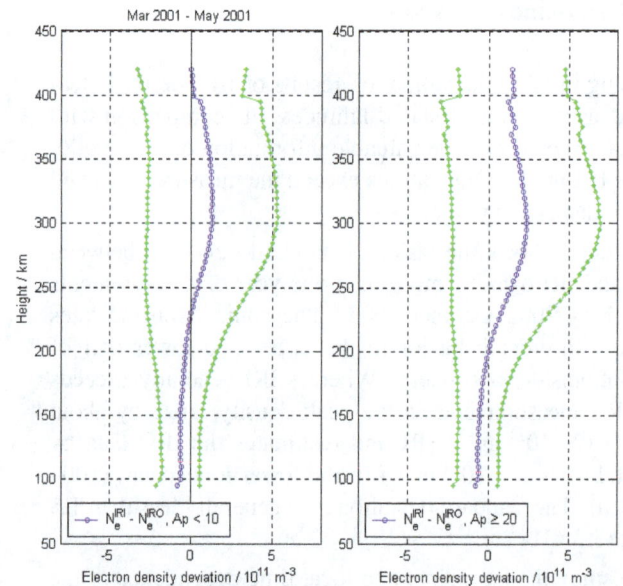

Fig. 5. Altitude profiles of the difference EDP(IRI) – EDP(IRO) between IRO derived electron densities and corresponding IRI values at different levels of geomagnetic activity (Ap<10 and Ap>20) within the period March–May 2001.

at the lower ionosphere) is persistent at all latitude ranges independent from the hemisphere. The neutral line crossing height grows from about 170 at low latitudes via 210 at mid- to about 235 km at high latitudes.

As Fig. 5 shows, it is interesting to note that the bias between IRO retrievals and IRI model values and the corresponding dispersion significantly grow up under conditions of enhanced geomagnetic activity. This result is understandable because the described use of this IRI version doesn't include ionospheric perturbations. More detailed studies of this type could help to implement/improve perturbation terms in the IRI model.

5   Summary and conclusions

Electron density profiles retrieved from ionospheric GPS radio occultations measured onboard the CHAMP satellite since 11 April 2001 were compared with corresponding IRI model values. Preliminary retrieval results obtained by applying a model assisted retrieval technique are promising. Because the radio occultation technique presents a new data type, a systematic validation of data products must be continued.

The comparison of IRO data with corresponding IRI data indicates a systematic overestimation of the IRI derived electron density above 250 km height compared with the IRO data, whereas IRI underestimates the IRO retrievals of GPS measurements onboard CHAMP in the lower ionosphere at about 100–150 km height. Although the IRO measurements are not yet fully validated, and therefore the presented results have a preliminary character, we believe that the indicated

In order to check the consistency of the results, computations have been made separately for the northern and southern hemispheres. Figure 4 shows the latitudinal dependence of the percentage difference between plasma frequency data sets obtained at both hemispheres separately. Comparing bias and dispersion values of corresponding latitude ranges, no substantial difference can be found.

As expected, the percentage deviation is smallest at the F2 layer height range. At the northern hemisphere the bias and the standard deviation at 300 km height are generally less than 25% and 50%, respectively.

The bias values are slightly enhanced in the southern hemisphere. At the lower ionosphere the bias is smallest with about 20% at both low latitude ionospheres accompanied by the highest dispersion value of about 90% found in the study.

It is important to mention that the main feature of the IRI – IRO difference (positive above 250 km height and negative general features of the IRI model compared with IRO retrievals might be helpful to improve IRI. So the height dependence of the IRI – IRO difference could simply be corrected by introducing an altitude dependent correction factor in IRI modeling.

Furthermore, the extraction of profile shape parameters such as fOF2, hmF2, and the bottomside slab thickness offer also unique chances for improving the IRI model.

It should be mentioned that the IRO derived vertical density profiles used in this study are directly taken from the automatically working IRO processing unit without further reviewing. A removal of unrealistic IRO outliers ($<0.5\%$) should principally lead to a better agreement between the IRI model data and the IRO measurements onboard CHAMP but would not essentially change the general finding.

Nevertheless, further improvements of the retrieval technique as e.g. the improvement of the topside ionosphere model are planned to enable more accurate observations.

Acknowledgements. The authors are very grateful to all the colleagues from the international CHAMP team keeping CHAMP in operation. We thank J. Mielich (IAP, Kühlungsborn) for providing vertical sounding data for comparison.

References

Bilitza, D.: International Reference Ionosphere 2000, Radio Sci., 36, 261–275, 2001.

Bilitza, D., Rawer K., Bossy L., and Gulyaeva T.: International Reference Ionosphere – Past, present and future, I, Electron density, Adv. Space Res., 13, 3, 23–32, 1996.

Jakowski, N.: Capabilities of radio occultation measurements onboard LEO satellites for ionospheric monitoring and research, Proc. 4th COST 251 Workshop "The Impact of the Upper Atmosphere on Terrestrial and Earth-Space Communications", edited by Vernon, A., 22–25 March, Funchal, Madeira, Portugal, 116–121, 1999.

Jakowski, N., Wehrenpfennig A., Heise S., Reigber Ch., Lühr H., Grunwaldt L., and Meehan T. K.: GPS radio occultation measurements of the ionosphere from CHAMP: Early results, Geophysical Res. Lett., 29, 10, 95(1)–95(4), 2002.

Reigber, Ch., Lühr H., and Schwintzer P.: CHAMP mission status and perspectives, Suppl. to EOS, Transactions, AGU, 81, 48, F307, 2000.

Rawer, K., Ramakrishnan, S., and Bilitza, D.: International Reference Ionosphere 1978, URSI, Brussels, 1978.

Wehrenpfennig, A., Jakowski N., and Wickert J.: A Dynamically Configurable System for Operational Processing of Space Weather Data, Phys. Chem. Earth, 26, 601–604, 2001.

33

Real-time ionospheric N(h) profile updating over Europe using IRI-2000 model

D. Buresova[1], Lj. R. Cander[2], A. Vernon[2], and B. Zolesi[3]

[1]Institute of Atmospheric Physics, Czech Republic
[2]Rutherford Appleton Laboratory, United Kingdom
[3]Istituto Nazionale di Geofisica e Vulcanologia, Italy

Abstract. In this paper a method for real-time updating of ionospheric electron density profile, N(h), over Europe using an ionospheric model and real-time measurements at ionosonde locations is presented. The N(h) profile update over European area has been simulated with the IRI-2000 ionospheric model and real-time N(h) profiles obtained from the EU COST271 Action Space Weather Database. Preliminary findings are shown for the geomagnetically quiet day on 4 May 2003 and disturbed day on 24 May 2002. Results are discussed in the context of real-time N(h) profile updating capabilities and effectiveness.

1 Introduction

The International Reference Ionosphere-IRI (Bilitza, 2001) is one of the most widely used empirical models. Among others, it describes the median values of electron density as a function of height for a given location, time and sunspot number. The IRI model is being refined following the annual IRI workshops and currently contains the foF2 storm model (Araujo-Pradere, et al., 2002a, b; Bilitza, 2003). This paper presents the comparative study of the electron density maps generated using a method for real-time ionospheric N(h) profiles updating and those obtained by the IRI-2000 model over Europe. The intent is to provide a possibility for a further IRI model improvement. Therefore, in this study the N(h) profile simulation has been performed by using online information available at the IRI web side http://nssdc.gsfc.nasa.gov/space/model/models/iri.html and at the COST271 Space Weather Database web side http://www.wdc.rl.ac.uk/cgi-bin/digisondes/cost_database.pl. (Zolesi and Cander, 2003). A few other stations operating in non real-time mode in Europe are used to provide additional IRI-2000 N(h) profiles required for mapping NmF2, hmF2 and electron density at

Correspondence to: D. Buresova
(buresd@ufa.cas.cz)

the F1-region heights over Europe. Ionospheric stations involved in this study are listed in Table 1.

Initial comparison of the measured and IRI generated N(h) profiles has been done for two stations: Pruhonice and Ebre. Some of the results obtained for the period of 14–21 February 1998 are given in Fig. 1. In this period the first three days were geomagnetically quiet, followed by the storm on 18 February and then a long lasting recovery phase. At Pruhonice station it can be seen that: (1) there is a good agreement between IRI model (red line) and NmF2 measured data (black line) for the whole period; (2) in general, there is no agreement between IRI model and electron density measured values at F1-region height. It is particularly true during the main phase of the geomagnetic storm and at the height of 180–190 km in the F1-region. The same is valid for the similar comparison made for Ebre station also shown in Fig. 1.

Examples at Fig. 1 demonstrate that it is possible to use IRI model for mapping ionospheric electron density at the certain heights over European region, as at hmF2 and at fixed F1-region heights, during quiet ionospheric conditions. During disturbed conditions IRI still cannot produce appropriate values mainly at F1 heights and therefore reasonably accurate N(h) profile as a whole. This is the main reason to introduce a method of the ionospheric 3D modelling by real-time N(h) profile updating with measured data.

2 Real-time N(h) updating results

Figure 2 shows the 4 May 2003 geomagnetically quiet study case at 11:00 UT. Left panel contains maps of electron density measured data at hmF2 and 190 km, while on the right panel maps are produced by IRI model using the same locations of five stations. It can be seen that NmF2 map over Europe obtained by measured values from selected stations was slightly different than IRI map itself. In the case of F1-region electron density maps it is clearly evident the significant discrepancy between maps of measured electron density data and IRI map. The IRI model underestimated electron

Table 1. List of contributing vertical incidence ionospheric stations.

VI station	Latitude (° N)	Longitude (° E)	Data available
Chilton	51.5	−1.3	Real-time mode
Athens	38.0	23.6	Real-time mode
El Arenosillo	37.1	−6.8	Non real-time mode
Ebre	40.8	0.5	Non real-time mode
Tromsø	69.9	19.0	Real-time mode
Roma	41.9	12.5	Real-time mode
Juliusruh	54.6	13.4	Real-time mode
Pruhonice	50.0	14.6	Non real-time mode
Warsaw	52.2	21.2	Non real-time mode
Dourbes	51.2	0.46	Non real-time mode
Uppsala	59.8	17.6	Non real-time mode

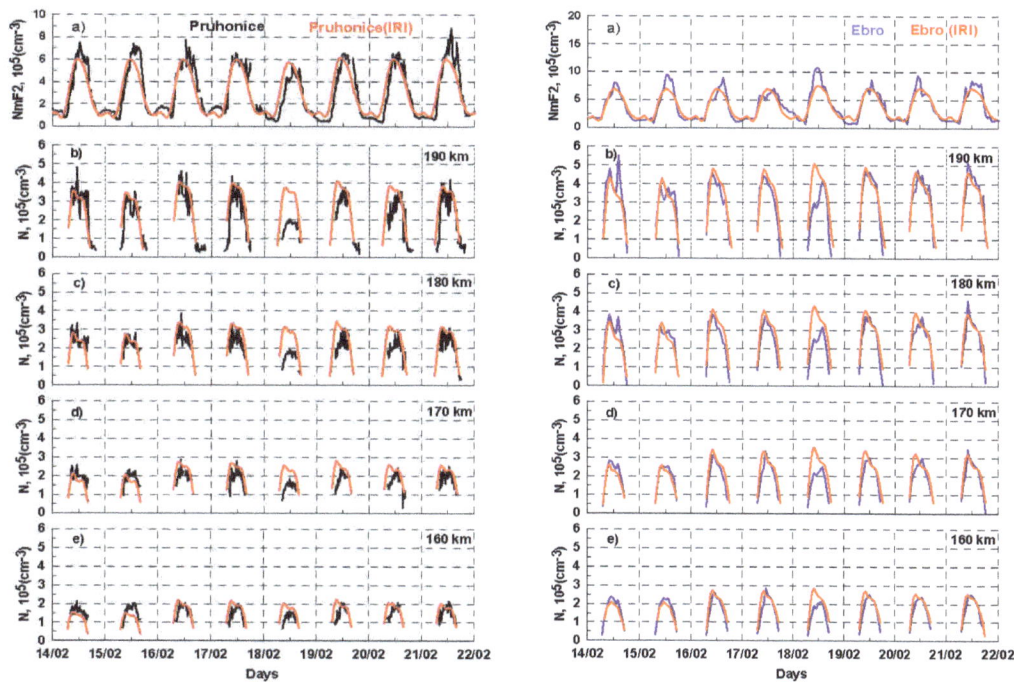

Fig. 1. Daily hourly NmF2 values and electron density values at specific heights obtained at Pruhonice (left panel) and at Ebre (right panel) stations and by IRI model during the period 14–21 February 1998.

density at 190 km height for all selected stations. To compare quiet and disturb periods in the same way as before, the disturbed day of 24 May 2002 was chosen based on the minimum value of the Dst index of −108 nT at late evening of the previous day. Figure 3 shows that at 11:00 UT on 24 May 2002 both NmF2 and the F1-region height maps over Europe obtained by measured values are significantly different than corresponding IRI maps. The advantage of using real-time data to update the IRI model values is obvious and most clearly seen at the F1-region height.

For the same days of 4 May 2003 and 24 May 2002 the simulation of an updating ionospheric model has been done by using 5 and 4 stations respectively, which offer the real-time measurements and the IRI N(h) profile values for other stations given in Table 1. Figures 4 and 5 show on the left side the simulated maps and on the right side maps produced

with IRI values only for both selected days. Updated NmF2 maps and IRI model itself show relatively good agreement. In contrast to NmF2, updated F1-region maps differ significantly from maps generated using IRI model data. In general, at European middle latitudes the F1-region response to spring-summer time geomagnetic storm is significantly smaller compared to other seasons (Buresova, et al., 2002; Mikhailov and Schlegel, 2003). Consequently, the updated IRI model maps for F1-region are quite similar for both selected quiet and disturbed days. The most important result seen at the Figs. 4 and 5 is that updating IRI model describes the actual ionospheric structure better then the IRI model itself. The quality of the maps will obviously depend on the number of ionospheric stations which can provide the real-time N(h) profiles.

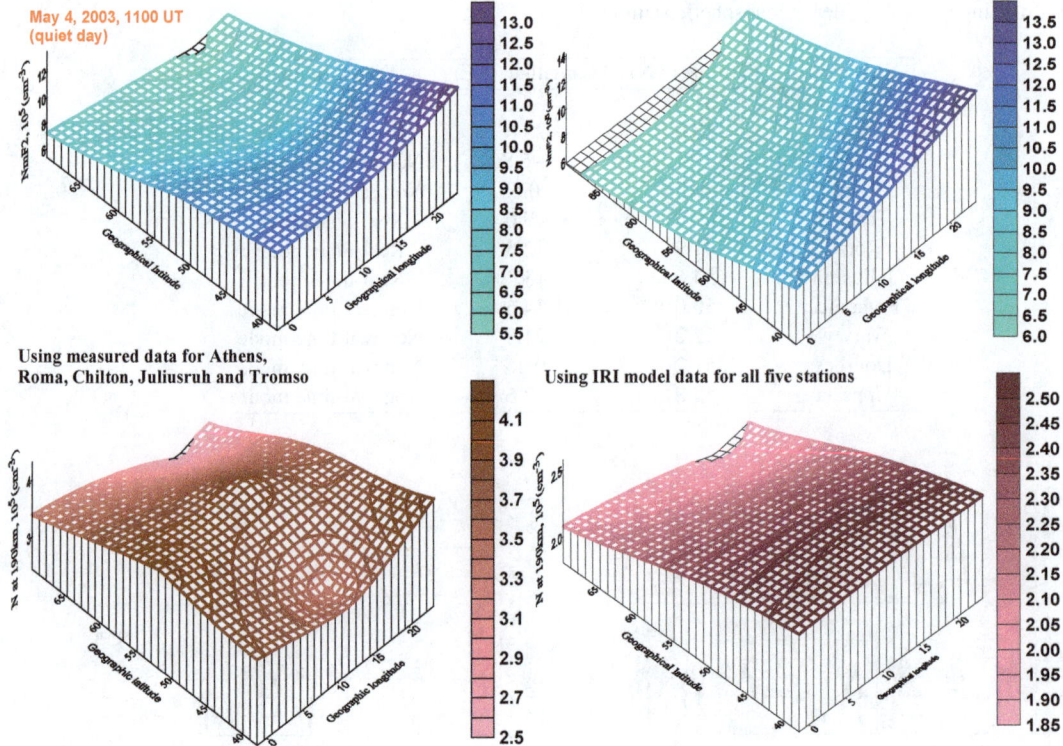

Fig. 2. Maps of the NmF2 and the electron density at 190 km values generated by measured data (left panel) and IRI data (right panel) on 4 May 2003 at 11:00 UT.

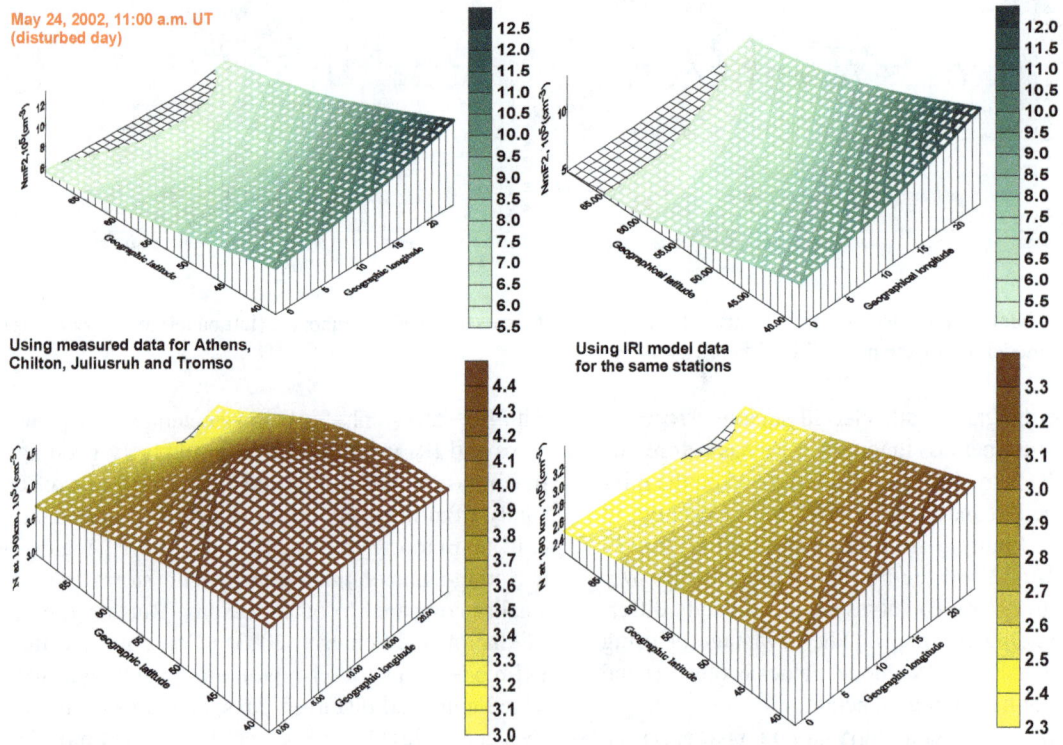

Fig. 3. Maps of the NmF2 and the electron density at 190 km values generated by measured data (left panel) and IRI data (right panel) on 24 May 2002 at 11:00 UT.

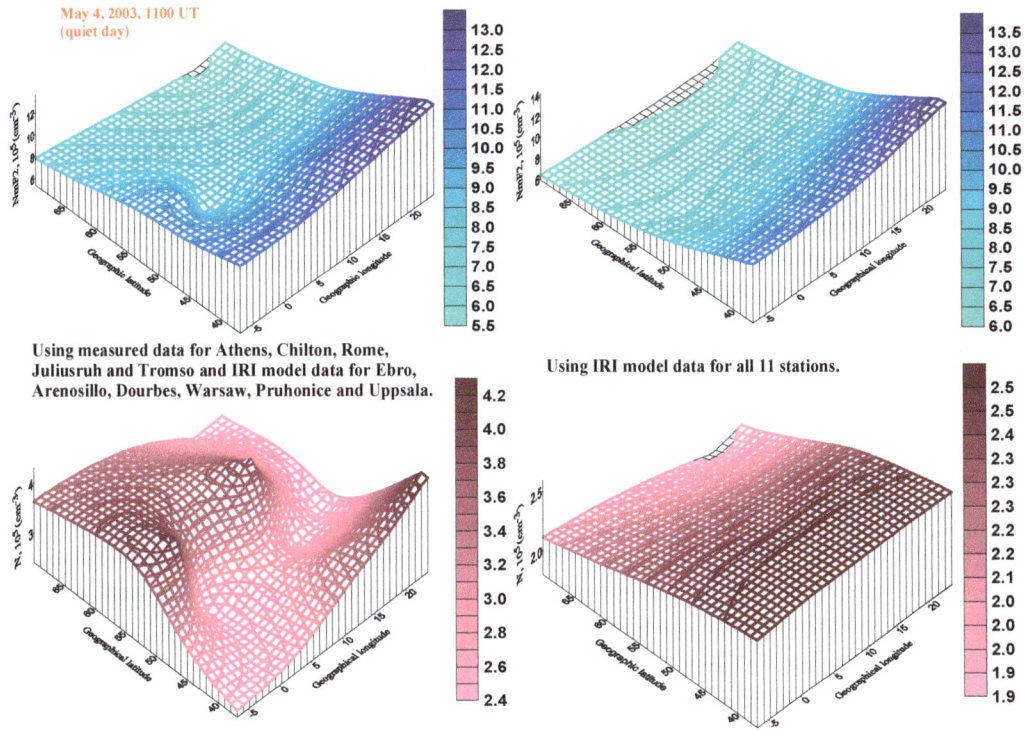

Fig. 4. Maps of the NmF2 and the electron density at 190 km values generated by measured data (left panel) for Athens, Roma, Chilton, Juliusruh and Tromsø and IRI data for Ebre, Arenosillo, Dourbes, Pruhonice, Warsaw and Uppsala on 4 May 2003 at 11:00 UT. Right panel represents the same ionospheric parameters calculated by IRI model for all selected stations.

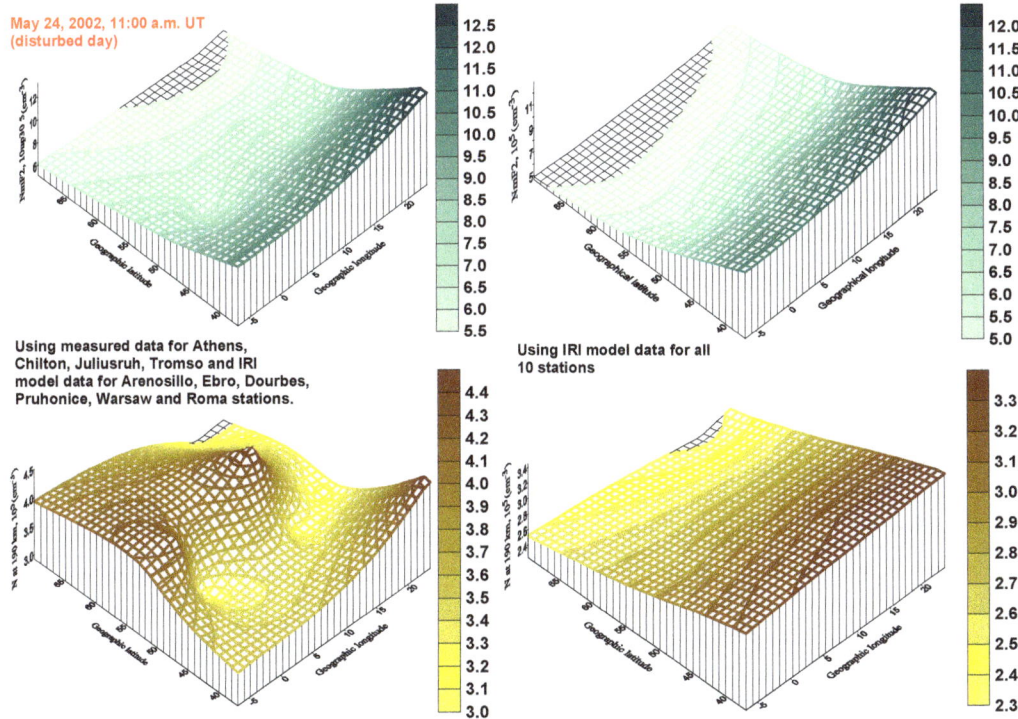

Fig. 5. Maps of the NmF2 and the electron density at 190 km values generated by measured data (left panel) for Athens, Chilton, Juliusruh and Tromso and IRI data for Ebre, Arenosillo, Dourbes, Pruhonice, Warsaw and Uppsala on 24 May 2002 at 11:00 UT. Right panel represents the same ionospheric parameters calculated by IRI model for all selected stations.

3   Conclusions

It is widely accepted that the IRI model needs improvement for better representation of the F1-region electron density distribution over Europe under the both geomagnetically quiet and disturbed conditions. Some of the results of our comparative analysis presented in this paper show that updated IRI-2000 model with real-time ionosonde data over Europe describes the actual ionospheric structure better than the IRI model itself. It is particularly valid during geomagnetically disturbed periods. The quality of the updated maps, which makes resulting N(h) profile more realistic, will however depend on the number of ionospheric stations that can provide the real-time N(h) profiles (Zolesi et al., 2004).

Acknowledgements. We are grateful to the Grant Academy of the Academy of Sciences of the Czech Republic and to the Ministry of Education, Youth and Sports, which supported this work. Grants No. A3042101 and OC.271.10, respectively.

References

Araujo-Pradere, E. A., Fuller-Rowell, T. J., and Codrescu, M. V.: Storm: An empirical storm-time ionospheric correction model – 1. Model description, Radio Science, 37, 5, Art. No. 070, 2002a.

Araujo-Pradere, E. A. and Fuller-Rowell, T. J.: Storm: An empirical storm-time ionospheric correction model – 2. Validation, Radio Sciences, 37, 5, Art. No. 1071, 2002b.

Bilitza, D.: International reference ionosphere 2002: Examples of improvements and new features, Adv. Space Res., 31, 3, 757–767, 2003.

Bilitza, D.: International Reference Ionosphere 2000, Radio Science, 36, 2, 261–275, 2001.

Buresova, D., Lastovicka, J., Altadil, D., and Miro, G.: Daytime electron density at the F1-region in Europe during geomagnetic storms, Ann. Geophysicae, 20, 1007–1021, 2002.

Mikhailov, A. V. and Schlegel, K.: Geomagnetic storm effects at F1-layer heights from incoherent scatter observations, Ann. Geophysicae, 21, 583–596, 2003.

Zolesi, B. and Cander, Lj. R.: Effects of the Upper Atmosphere on Terrestrial and Earth-Space Communications, EE Conference Proceedings No. 491, 2, 565–568, 2003.

Zolesi, B., Belehaki, A., Tsagouri, I., and Cander, Lj. R.: Real-time updating of the Simplified Ionospheric Regional Model for operational applications, Radio Science, in press, 2004.

Permissions

All chapters in this book were first published in ARS, by Copernicus Publications; hereby published with permission under the Creative Commons Attribution License or equivalent. Every chapter published in this book has been scrutinized by our experts. Their significance has been extensively debated. The topics covered herein carry significant findings which will fuel the growth of the discipline. They may even be implemented as practical applications or may be referred to as a beginning point for another development.

The contributors of this book come from diverse backgrounds, making this book a truly international effort. This book will bring forth new frontiers with its revolutionizing research information and detailed analysis of the nascent developments around the world.

We would like to thank all the contributing authors for lending their expertise to make the book truly unique. They have played a crucial role in the development of this book. Without their invaluable contributions this book wouldn't have been possible. They have made vital efforts to compile up to date information on the varied aspects of this subject to make this book a valuable addition to the collection of many professionals and students.

This book was conceptualized with the vision of imparting up-to-date information and advanced data in this field. To ensure the same, a matchless editorial board was set up. Every individual on the board went through rigorous rounds of assessment to prove their worth. After which they invested a large part of their time researching and compiling the most relevant data for our readers.

The editorial board has been involved in producing this book since its inception. They have spent rigorous hours researching and exploring the diverse topics which have resulted in the successful publishing of this book. They have passed on their knowledge of decades through this book. To expedite this challenging task, the publisher supported the team at every step. A small team of assistant editors was also appointed to further simplify the editing procedure and attain best results for the readers.

Apart from the editorial board, the designing team has also invested a significant amount of their time in understanding the subject and creating the most relevant covers. They scrutinized every image to scout for the most suitable representation of the subject and create an appropriate cover for the book.

The publishing team has been an ardent support to the editorial, designing and production team. Their endless efforts to recruit the best for this project, has resulted in the accomplishment of this book. They are a veteran in the field of academics and their pool of knowledge is as vast as their experience in printing. Their expertise and guidance has proved useful at every step. Their uncompromising quality standards have made this book an exceptional effort. Their encouragement from time to time has been an inspiration for everyone.

The publisher and the editorial board hope that this book will prove to be a valuable piece of knowledge for researchers, students, practitioners and scholars across the globe.

List of Contributors

K. Ruf
Max-Planck-Institut für Radioastronomie, Auf dem Hügel 69, 53111 Bonn

E. Fürst
Max-Planck-Institut für Radioastronomie, Auf dem Hügel 69, 53111 Bonn

K. Grypstra
Max-Planck-Institut für Radioastronomie, Auf dem Hügel 69, 53111 Bonn

J. Neidhöfer
Max-Planck-Institut für Radioastronomie, Auf dem Hügel 69, 53111 Bonn

M. Schumacher
Max-Planck-Institut für Radioastronomie, Auf dem Hügel 69, 53111 Bonn

K. Bibl
Center for Atmospheric Research, University of Massachusetts Lowell, 600 Suffolk St., Lowell, MA 01854-3625, USA

C. Jacobi
Institute for Meteorology, University of Leipzig, Stephanstr. 3, 04103 Leipzig, Germany

D. Kürschner
Institute of Geophysics and Geology, University of Leipzig, Collm Observatory, 04779 Wermsdorf, Germany

J. Bremer
Leibniz-Institut für Atmosphärenphysik, Schloss-Str.6, D-18225 Kühlungsborn, Germany

R. Beck
Max-Planck-Institut für Radioastronomie, Auf dem Hügel 69, 53121 Bonn, Germany

M. Förster
GFZ German Research Centre for Geosciences, Helmholtz Centre Potsdam, Germany

B. E. Prokhorov
GFZ German Research Centre for Geosciences, Helmholtz Centre Potsdam, Germany
University Potsdam, Institute for Applied Mathematics, Potsdam, Germany

A. A. Namgaladze
Murmansk State Technical University, Murmansk, Russia

M. Holschneider
University Potsdam, Institute for Applied Mathematics, Potsdam, Germany

O. A. Maltseva
Institute for Physics, Southern Federal Universiy, Russia

G. A. Zhbankov
Institute for Physics, Southern Federal Universiy, Russia

N. S. Mozhaeva
Institute for Physics, Southern Federal Universiy, Russia

M. V. Klimenko
Kaliningrad State Technical University, Kaliningrad, Russia

V. V. Klimenko
West Department of IZMIRAN, Kaliningrad, Russia

V. V. Bryukhanov
Kaliningrad State Technical University, Kaliningrad, Russia

W. Reich
Max-Planck-Institut für Radioastronomie, Auf dem Hügel 69, 53121 Bonn, Germany

C. Jacobi
Institute for Meteorology, University of Leipzig, Stephanstr. 3, 04103 Leipzig, Germany

N. Jakowski
DLR, Institute of Communications and Navigation, Kalkhorstweg 53, 17235 Neustrelitz, Germany

A. Pogoreltsev
Russian State Hydrometeorological University, 98 Maloohtinsky St., Petersburg 195196, Russia

K. Fröhlich
Institute for Meteorology, University of Leipzig, Stephanstr. 3, 04103 Leipzig, Germany

P. Hoffmann
Institute for Meteorology, University of Leipzig, Stephanstr. 3, 04103 Leipzig, Germany

C. Jacobi
Institute for Meteorology, University of Leipzig, Stephanstr. 3, 04103 Leipzig, Germany

D. Kürschner
Institute for Geophysics and Geology, University of Leipzig, Collm Observatory, 04779 Wermsdorf, Germany

C. Borries
Institute for Geophysics and Geology, University of Leipzig, Collm Observatory, 04779 Wermsdorf, Germany

Ch. Jacobi
Institute for Meteorology, University of Leipzig, Stephanstr. 3, 04103 Leipzig, Germany

T. Fytterer
Institute for Meteorology, University of Leipzig, Stephanstr. 3, 04103 Leipzig, Germany

C. Unglaub
University of Leipzig, Institute for Meteorology, Stephanstr. 3, 04103 Leipzig, Germany

Ch. Jacobi
University of Leipzig, Institute for Meteorology, Stephanstr. 3, 04103 Leipzig, Germany

G. Schmidtke
Fraunhofer IPM, Heidenhofstraße 8, 79110 Freiburg, Germany

B. Nikutowski
University of Leipzig, Institute for Meteorology, Stephanstr. 3, 04103 Leipzig, Germany
Fraunhofer IPM, Heidenhofstraße 8, 79110 Freiburg, Germany

R. Brunner
Fraunhofer IPM, Heidenhofstraße 8, 79110 Freiburg, Germany

T. Renkwitz
Leibniz Institute of Atmospheric Physics at the Rostock University, Schloss-Str. 6, 18225 Kühlungsborn, Germany

W. Singer
Leibniz Institute of Atmospheric Physics at the Rostock University, Schloss-Str. 6, 18225 Kühlungsborn, Germany

R. Latteck
Leibniz Institute of Atmospheric Physics at the Rostock University, Schloss-Str. 6, 18225 Kühlungsborn, Germany

M. Rapp
Leibniz Institute of Atmospheric Physics at the Rostock University, Schloss-Str. 6, 18225 Kühlungsborn, Germany

M. Y. Boudjada
Space Research Institute, Austrian Academy of Sciences, Graz, Austria

P. H. M. Galopeau
Université Versailles St-Quentin, CNRS/INSU, LATMOS-IPSL, Guyancourt, France

M. Maksimovic
LESIA – Observatoire de Paris-Meudon, Meudon, France

H. O. Rucker
Space Research Institute, Austrian Academy of Sciences, Graz, Austria

B. W. Reinisch
B.W. Reinisch, I. A. Galkin, G. Khmyrov, A. Kozlov, and D. F. Kitrosser

I. A. Galkin
B.W. Reinisch, I. A. Galkin, G. Khmyrov, A. Kozlov, and D. F. Kitrosser

G. Khmyrov
B.W. Reinisch, I. A. Galkin, G. Khmyrov, A. Kozlov, and D. F. Kitrosser

A. Kozlov
B.W. Reinisch, I. A. Galkin, G. Khmyrov, A. Kozlov, and D. F. Kitrosser

D. F. Kitrosser
B.W. Reinisch, I. A. Galkin, G. Khmyrov, A. Kozlov, and D. F. Kitrosser

J.-C. Jodogne
Institut Royal Météorologique, 3 Avenue Circulaire, B-1180 Bruxelles, Belgium

H. Nebdi
Institut Royal Météorologique, 3 Avenue Circulaire, B-1180 Bruxelles, Belgium

R. Warnant
Institut Royal Météorologique, 3 Avenue Circulaire, B-1180 Bruxelles, BelgiumRoyal Observatory of Belgium

B. W. Reinisch
Environmental, Earth and Atmospheric Sciences Department, Center for Atmospheric Research, University of Massachusetts Lowell, USA

X. Huang
Environmental, Earth and Atmospheric Sciences Department, Center for Atmospheric Research, University of Massachusetts Lowell, USA

A. Belehaki
Institute for Space Applications and Remote Sensing, National Observatory of Athens, Greece

R. Ilma
Jicamarca Radio Observatory, Lima, Peru

C. Jacobi
Institute for Meteorology, University of Leipzig, Stephanstr. 3, 04103 Leipzig, Germany

C. Arras
German Research Centre for Geosciences GFZ, Potsdam, Department Geodesy & Remote Sensing, Telegrafenberg, 14473 Potsdam, Germany

J. Wickert
German Research Centre for Geosciences GFZ, Potsdam, Department Geodesy & Remote Sensing, Telegrafenberg, 14473 Potsdam, Germany

D. Bilitza
Raytheon ITSS/SSDOO, GSFC, Code 632, Greenbelt, MD 20771, USA

C. Jacobi
Institute for Meteorology, University of Leipzig, Stephanstr. 3, 04103 Leipzig, Germany

C. Jacobi
Institute for Meteorology, University of Leipzig, Stephanstr. 3, 04103 Leipzig, Germany

E. G. Merzlyakov
Research and Production Association "Typhoon", Institute for Experimental Meteorology, 4, Pobeda Str 249038 Obninsk, Russia

R. Q. Liu
Institute for Meteorology, University of Leipzig, Stephanstr. 3, 04103 Leipzig, Germany

T. V. Solovjova
Research and Production Association "Typhoon", Institute for Experimental Meteorology, 4, Pobeda Str 249038 Obninsk, Russia

Y. I. Portnyagin
Research and Production Association "Typhoon", Institute for Experimental Meteorology, 4, Pobeda Str., 249038 Obninsk, Russia

S. Sommer
Leibniz Institute of Atmospheric Physics at the Rostock University, Schloss-Str. 6, 18225 Kühlungsborn, Germany

G. Stober
Leibniz Institute of Atmospheric Physics at the Rostock University, Schloss-Str. 6, 18225 Kühlungsborn, Germany

C. Schult
Leibniz Institute of Atmospheric Physics at the Rostock University, Schloss-Str. 6, 18225 Kühlungsborn, Germany

M. Zecha
Leibniz Institute of Atmospheric Physics at the Rostock University, Schloss-Str. 6, 18225 Kühlungsborn, Germany

R. Latteck
Leibniz Institute of Atmospheric Physics at the Rostock University, Schloss-Str. 6, 18225 Kühlungsborn, Germany

R. Latteck
Leibniz Institute of Atmospheric Physics at the Rostock University, Schloss-Str. 6, 18225 Kühlungsborn, Germany

J. Bremer
Leibniz Institute of Atmospheric Physics at the Rostock University, Schloss-Str. 6, 18225 Kühlungsborn, Germany

Ch. Jacobi
Institute for Meteorology, University of Leipzig, Stephanstr. 3, 04103 Leipzig, Germany

P. Hoffmann
Institute for Meteorology, University of Leipzig, Stephanstr. 3, 04103 Leipzig, Germany

M. Placke
Leibniz Institute of Atmospheric Physics at the Rostock University, Schlossstraße 6, 18225 Kühlungsborn, Germany

G. Stober
Leibniz Institute of Atmospheric Physics at the Rostock University, Schlossstraße 6, 18225 Kühlungsborn, Germany

G. Stober
University Leipzig, Institute for Meteorology, Stephanstr. 3, 04103 Leipzig, Germany

C. Jacobi
University Leipzig, Institute for Meteorology, Stephanstr. 3, 04103 Leipzig, Germany

D. Keuer
University of Rostock, Leibniz-Institute of Atmospheric Physics, Schlossstr. 6, 18225 Kühlungsborn, Germany

Ch. Jacobi
Institute for Meteorology, University of Leipzig, Stephanstr. 3, 04103 Leipzig, Germany

R. Latteck
Leibniz Institute of Atmospheric Physics at the Rostock University, Schloss-Str. 6, 18225 Kühlungsborn, Germany

W. Singer
Leibniz Institute of Atmospheric Physics at the Rostock University, Schloss-Str. 6, 18225 Kühlungsborn, Germany

M. Rapp
Leibniz Institute of Atmospheric Physics at the Rostock University, Schloss-Str. 6, 18225 Kühlungsborn, Germany

T. Renkwitz
Leibniz Institute of Atmospheric Physics at the Rostock University, Schloss-Str. 6, 18225 Kühlungsborn, Germany

G. Stober
Leibniz Institute of Atmospheric Physics at the Rostock University, Schloss-Str. 6, 18225 Kühlungsborn, Germany

E. D. Schmitter
University of Applied Sciences Osnabrueck, 49076 Osnabrueck, Germany

P. Hoffmann
Institute for Meteorology, University of Leipzig, Stephanstr. 3, 04103 Leipzig, Germany

Ch. Jacobi
Institute for Meteorology, University of Leipzig, Stephanstr. 3, 04103 Leipzig, Germany

G. Stober
Leibniz-Institute of Atmospheric Physics at the Rostock University, Schlossstr. 6, 18225 Kühlungsborn, Germany

R. Latteck
Leibniz-Institute of Atmospheric Physics at the Rostock University, Schlossstr. 6, 18225 Kühlungsborn, Germany

M. Rapp
Leibniz-Institute of Atmospheric Physics at the Rostock University, Schlossstr. 6, 18225 Kühlungsborn, Germany

W. Singer
Leibniz-Institute of Atmospheric Physics at the Rostock University, Schlossstr. 6, 18225 Kühlungsborn, Germany

M. Zecha
Leibniz-Institute of Atmospheric Physics at the Rostock University, Schlossstr. 6, 18225 Kühlungsborn, Germany

N. Jakowski
Deutsches Zentrum für Luft-und Raumfahrt (DLR)/ Institut für Kommunikation und Navigation, Kalkhorstweg 53, D-17235 Neustrelitz, Germany

K. Tsybulya
Deutsches Zentrum für Luft-und Raumfahrt (DLR)/ Institut für Kommunikation und Navigation, Kalkhorstweg 53, D-17235 Neustrelitz, Germany

D. Buresova
Institute of Atmospheric Physics, Czech Republic

Lj. R. Cander
Rutherford Appleton Laboratory, United Kingdom

A. Vernon
Rutherford Appleton Laboratory, United Kingdom

B. Zolesi
Istituto Nazionale di Geofisica e Vulcanologia, Italy